The Frontiers Collection

The books in this collection are devoted to challenging and open problems at the forefront of modern science and scholarship, including related philosophical debates. In contrast to typical research monographs, however, they strive to present their topics in a manner accessible also to scientifically literate non-specialists wishing to gain insight into the deeper implications and fascinating questions involved. Taken as a whole, the series reflects the need for a fundamental and interdisciplinary approach to modern science and research. Furthermore, it is intended to encourage active academics in all fields to ponder over important and perhaps controversial issues beyond their own speciality. Extending from quantum physics and relativity to entropy, consciousness, language and complex systems—the Frontiers Collection will inspire readers to push back the frontiers of their own knowledge.

Asmaa Eldesoukey

Len Pismen

Active Matter Within and Around Us

From Self-Propelled Particles to Flocks and Living Forms

Springer

Len Pismen
Department of Chemical Engineering
Technion Haifa
Haifa, Israel

ISSN 1612-3018 ISSN 2197-6619 (electronic)
The Frontiers Collection
https://doi.org/10.1007/978-3-030-68421-1

This Springer imprint is published by the registered company Springer Nature Switzerland AG
The registered company address is: Gewerbestrasse 11, 6330 Cham, Switzerland
(www.springer.com/mycopy)

To Yael, nothing without you

Preface

Active matter is within us, in our cells. It is around us, within living organisms and their gatherings, and even in sands and waters enlivened by winds, waves, and swimming plankton. The study of active matter is a new and rapidly expanding field on the frontier between physics and biology, also pertinent to fledgling biomorphic technologies. It is still young but already too extensive to be covered across-the-board.

The style of this book is less technical and less formal than in available reviews. I avoid mathematical equations (even though it is contrary to what I am used to in my own work) and include as many illustrations as possible, aspiring to reflect ideas with minimum technical details. Since the narrative is centered on modeling, I also avoid biochemical details, but always remember and at times remind the reader that molecular interactions, not yet understood in all relevant details, are the all-important players behind the scenes.

Studies of active matter, especially in their biological applications, are a part of mainstream twenty-first century science, which is an industry based on expensive precision experiments and computations, and dominated by large teams. Yet, the underlying ideas and methods of the theory of active matter are rooted in the nonlinear analysis that flourished late in the last century, and prominent figures in this field are schooled in statistical and nonlinear physics.

My own involvement in the field is marginal, but this makes it more interesting to write about it and leaves more freedom for unbiased judgement. I am awed by the inner workings of *life*, compared to which the field of active matter is a *game*. But it is a sophisticated and fascinating game, perhaps the most fascinating game played by the physicists of the early twenty-first century, between nine-eleven and the coronavirus pandemic. Its future stands open, as it penetrates ever deeper into the intricacies of *living* matter.

I appreciate the help of friends and colleagues for discussing details of their work and granting me permission to reproduce images in this book. My contacts with the group of Prof. Stanislav Shvartsman in New York and Princeton were particularly helpful in getting a better feel for the biological experiments.

Haifa, Israel, October 2021 — Len Pismen

Contents

Introduction

For the natural philosophers of the Enlightenment, the notion of *active matter* would be associated with the gnawing question of whether the newly discovered laws of mechanics could be extended to life itself. There were strong theological objections: the *élan vital* emanating from God had to be breathed into passive matter to bring it to life. Vitalism persisted throughout the rational 19th century; even Louis Pasteur supported his belief in vitalism by disproving the spontaneous generation of organisms from non-living matter. With time, matter became more and more familiar. The four classical elements, earth, water, air, and fire, could now be associated with the four states of matter, solid, liquid, gas, and plasma, governed on the macroscopic level by thermodynamic laws. Yet, it was not until the late 20th century that the wealth of behavior of far-from-equilibrium nonlinear systems was finally appreciated, and understood to be essential for life itself.

When does matter become active? The modern usage of this term is surprisingly novel: it first appeared in the paper by Ramaswamy and Simha (2006) as the appellation of the fledgling research field that has been rapidly expanding since then. It largely applies to interactions involving living organisms and their constituent parts, but also extends to biomorphic materials and designs. The studies of active matter could flourish only in this century: they are impossible without computer power and precision experiments, penetrating into the microscopic mechanisms of active motion.

Was there active matter before life? Perhaps, this is a matter of a definition. According to Chaté (2020), "*active matter physics is about systems in which energy is dissipated at some local level to produce work*". This definition, though part of a much more narrowly focused view, implies an affirmative answer, widening the notion of active matter to all non-equilibrium processes. On the other hand, Needleman and Dogic (2017) assert that "*active matter is different from traditionally studied non-equilibrium phenomena*", in which "*the entire system is driven away from equilibrium by energy provided through an external macroscopic boundary*". They cite the statement by Gottfried Wilhelm Leibniz in a letter to Damaris Masham: "*I define the Organism, or natural Machine, as a machine in which each part is a machine* [...], *whereas the parts of our artificial machines are not machines*".

Marchetti et al (2013), in an all-round review co-authored by seven physicists, define active matter as "*composed of self-driven units, active particles, each capable of converting stored or ambient free energy into systematic movement*", referring to Schweitzer (2003), who did not use this term, however.

Both points of view have their merit. Certainly, neither stars nor geologically active planets are driven externally. A star is also a structured "machine" driven by its own nuclear fusion energy, and governing its coterie of planets, some of which themselves are structured machines. Our Earth possesses, besides Sun's radiation, her own radioactive energy source that keeps her interior fluid, driving the magnetic dynamo and the continental plate tectonics. Her oceans are convective machines, with their network of mighty currents governing the climate, and occasional outbursts of hurricanes. And perhaps we should recall at this point Lovelock's (1979) *Gaia*, the living planet, of which we, as well as other living "machines", are constituent parts. Leibniz could not know anything like this, but if he were resurrected in our day, he couldn't fail to acquire a laptop and a smartphone and, with his natural curiosity, would agree that the parts of our artificial machines *are* machines. However, all this would bring us too far. As once remarked a fictitious Russian writer, one cannot embrace the unembraceable. The working definition of "active matter" may be confined to the problems investigated by scientists who are actually working in the field.

What does it include? First of all, indeed, self-driven "active" particles. What exactly constitutes a "particle" is interpreted very widely. It might just be a particle moving in a fluid under the action of a gradient of some field, e.g., the concentration of some species, or electric potential, or magnetic force. It might be a particle, not active in its own right, but entrained in a vibrating granular layer; part of a "dissipative structure" not unlike convective structures in the oceans, atmosphere, or laboratory studies. A granular layer was indeed the first medium in which a dissipative non-equilibrium structure was discovered, by Faraday (1831). It might be a microbe wandering in search of nutrients. It might be a bird or a fish, a part of a flock or a shoal. It might be a person in a crowd. Note that this list includes a rather loose notion of *self*-driving, but in any case, the "self" of the "particle" is always subjected to at least some degree to external forces and/or collective interactions.

There is a rather artificial distinction between "dry" and "wet" active matter, which originated more from the kind of modeling than from the nature of the "particles" themselves. Interactions in "wet" matter are mediated by the medium they are immersed in, while in the case of "dry" matter, the medium is ignored, sometimes rightly, when its influence is a minor factor, and sometimes just to make the problem tractable, e.g., ignoring fluid mechanics when modeling flocks.

The "particles" are often assigned certain intrinsic characteristics, most commonly, their *orientation*. The orientation may determine their preferred direction of motion, e.g., with respect to gradients of an external field, or the character of their interactions, e.g., the tendency to align with their neighbors. The basic types of orientation are *vector*, denoting the direction, or *nematic*, implying an alignment without a definite direction, as in a vector without an arrow. Both kinds of orientation may or may not be qualified by their strength.

But active matter is not necessarily dispersed in the form of particles. The notion of intrinsic orientation comes from the physics of condensed matter, which engaged with directed interactions in solids and fluids long before studies of active matter came on the scene. Moreover, most scientists active in these studies have been nurtured on problems in the physics of condensed matter. The orientation of magnetic or electric dipoles is a major factor determining the structure of crystalline solids – but solids, with their *solidity*, are not properly qualified to be active, unless dispersed. Liquids are more akin to active matter, and of all liquids, most kindred are partially ordered liquids, *liquid crystals*, combining various degrees of orientation with fluidity. This is the source of another research direction, active liquid crystals, which added the attribute of activity to the intrinsic properties of oriented liquids. The resulting models may involve assemblies of orientable particles but are largely "wet", and involve continuum rather than discrete description.

From here, the direct route for further expansion goes to *soft matter*, flexible solids and colloids endowed with some kind of activity. This area is most relevant for biology, as tissues, cells, and their components have a similar texture and consistency. Moreover, *activity* is their innate feature: they are machines consisting of machines, and themselves parts of greater machines, which are already beyond the realm of mere *matter*, even soft and active.

Living matter, driven by the complex interplay of genes and proteins, is far more complicated than models of active matter can afford. Specific molecular interactions are of most concern to biologists and biochemists, but it is gradually understood that factors of a general nature, common to active and living matter, play a substantial role in cellular and developmental processes. Finally, advanced biomorphic technology creates soft robots and artificial swimmers imitating, still on a basic level, soft natural machines. All of this will come out in our narrative.

Chapter 1
Polar Flocks

1.1 Birds of a Feather Fly Together

The most important feature of active "particles" (whatever their nature) is *collective motion*. It may arise surprisingly easily, without any leader, or external field, or geometrical constraints. The words "how birds fly together" appear in the title of a paper written by physicists (Toner and Tu, 1995), not by ornithologists. But what is a physicist's bird? It is a *vector* moving in the direction shown by its arrow. This is reasonable: animals move looking ahead, in the direction of their body axis. Animals are social, and capable of comprehending their surroundings and adjusting their behavior accordingly, thereby initiating collective motion.

The simplest example of the spontaneous emergence of directed motion based on these premises is the Vicsek model (Vicsek et al, 1995) sketched in Fig. 1.1. It lacks any specific physical mechanism, and just assumes that each particle (or "bird") adjusts its alignment to conform with the average alignment vector (shown by the arrows in the picture) of its neighbors within a certain range, and moves in the acquired direction. The adjustment is biased by random noise, and keeps repeating, as motion brings the particle to another neighborhood. The particles always move in the direction determined by their orientation with a constant speed; thus, polarity is

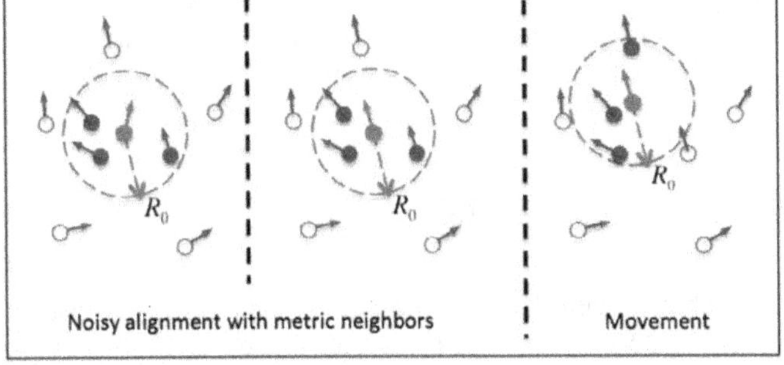

Fig. 1.1 Scheme of the Vicsek model. A test particle (*red*) aligns imperfectly with neighbors (*dark blue*) within a range shown by the *dashed circle*, and then moves in this direction (Ginelli, 2016)

L. Pismen, *Active Matter Within and Around Us*, The Frontiers Collection,
https://doi.org/10.1007/978-3-030-68421-1_1

Fig. 1.2 (**a**) The V-formation of Canada geese (Spedding, 2011). (**b**) A vortex trail shed by a fish (Weihs, 1973). (**c**) A flock of pigeons (Spedding, 2011)

not distinguished in this model from velocity. Vicsek's flock is a basic representative of what is called "dry" active matter abstracted from a surrounding medium.

The model seems to be oversimplified, but this was the key to its success (over 3600 citations at the time of writing). An advantage (or, depending on your point of view, a disadvantage) of models of this kind is that they may produce pictures bearing a superficial resemblance with observations even when they do not reflect the actual way the system in question operates. Alignment of migrating birds or fish is often motivated by hydrodynamics: in this way, they save propulsion effort. Canada geese migrate in a characteristic V-formation (Fig. 1.2a), just as plane squadrons fly, since such an orderly arrangement reduces drag. Fish also save energy when swimming in a shoal. A vortex trail shed by a fish or a bird induces immediately behind it a stream opposite to the swimming or flying direction (Fig. 1.2b), which would require the immediate follower to exert extra energy. However, if the follower's position is shifted laterally, it comes into the zone where the induced velocity is directed favorably. Weihs (1973) calculated that the best position is midway between two fish of the preceding row, obtaining a difference in relative speed of up to 30% between the best and worst lateral positions.

On the other hand, Usherwood et al (2011) have found that, for pigeons (Fig. 1.2c), flocking is energetically costly, so social factors apparently overrule hydrodynamics in this case. A simple universal model is the most reasonable choice when the alignment is of a social origin, arising from sensory inputs and information exchange, essential, e.g., when cohesion of the group is a necessary means of defense against attack by a predator.

1.2 Phenomenology of Vicsek's model

The Vicsek model is amenable to agent-based numerics. Depending on the average density and the level of noise, the particles may be perfectly aligned, or disordered, or ordered locally but disordered elsewhere (Fig. 1.3); the ordering can also be intermittent when particles align for a while and then succumb to noise. Even when the ordered state persists, the interaction network is permanently rearranged due to the active motion of individuals, continuously changing their neighbors.

Aligned particles tend to gather into dense self-propelling blobs, while particles in dilute surroundings move at random. Dense blobs, in their turn, tend to gather into ordered bands transverse to the alignment and, hence, propagation direction, whereupon they move more or less uniformly, but may dissipate if the local coherence is lost. Such a global ordering precedes slower velocity alignment. In the snapshot of the simulation of the Vicsek model (Katyal et al, 2020) shown in Fig. 1.4a, separation of dense and dilute domains is still very far from forming ordered bands. Nevertheless, the propagation directions are already well aligned in the two clusters, joined only by a narrow track of particles and far from forming dense ordered bands. After long transients, bands such as those shown in Fig. 1.4b periodically arrange in space, similar to the smectic phase in liquid crystals on a

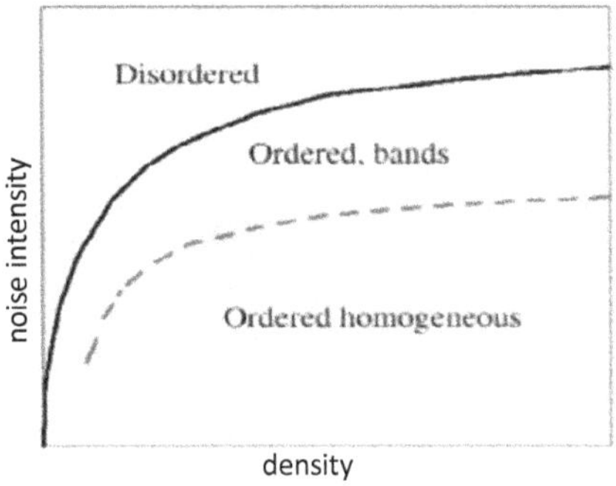

Fig. 1.3 Qualitative phase diagram showing the dependence of ordering on the flock density and the strength of noise (Ginelli, 2016)

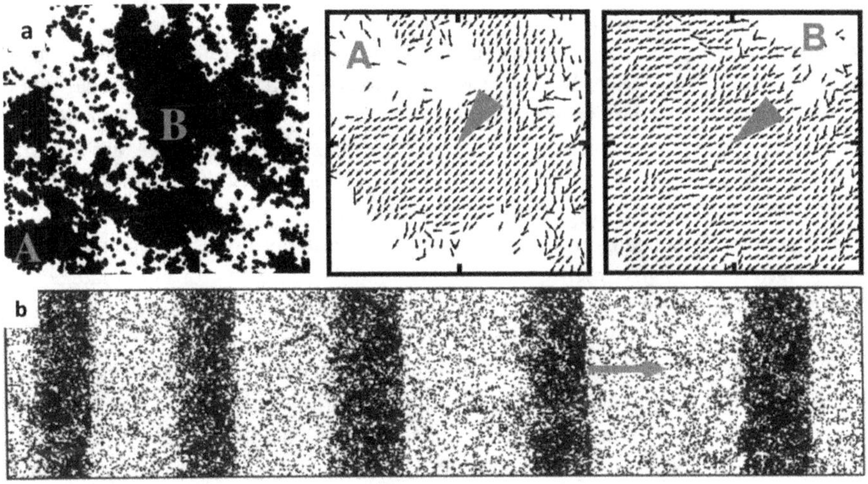

Fig. 1.4 (**a**) Snapshot of a part of the Vicsek flock, showing zooms of the density clusters A, B in the two panels on the right. The *red arrows* show the propagation direction (Katyal et al, 2020). (**b**) Propagating bands (Chaté et al, 2008)

macroscopic scale. The number of bands increases with increasing density at constant noise (Solon, Chaté, and Tailleur, 2015).

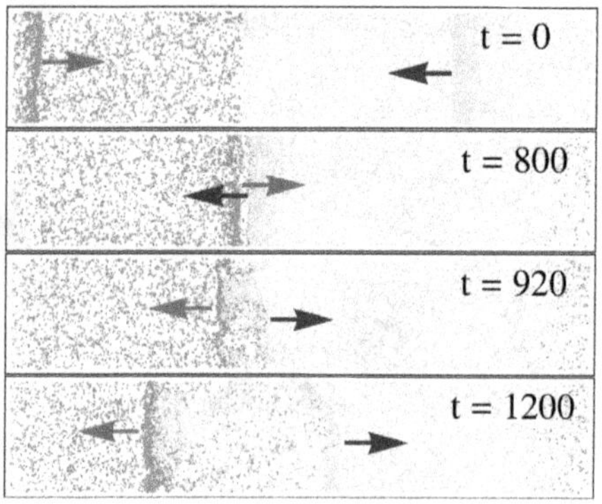

Fig. 1.5 Two counter-propagating bands passing each other in a soliton-like fashion (Ihle and Chou, 2014)

A rather unrealistic feature, arising in simulations by Ihle and Chou (2014) and demonstrated in Fig. 1.5, is a soliton-like behavior of counter-propagating bands, which pass each other with a minimal distortion following a collision. When the front of brown particles hits the front of turquoise particles, small groups of highly aligned brown particles tunnel through that front and continue going in the same direction. Once behind the turquoise front, they re-orient the oncoming turquoise particles, forming a new dense band. At the same time, brown particles that are left further behind their front, and hence less ordered and less dense, are also forced to return by groups of aligned turquoise particles. While returning, the freshly reoriented brown particles form a new dense front going in the opposite direction[1]. Real birds would hardly change their plans in such a way; anyway, two flocks, making use of the third dimension, would avoid each other.

Inhomogeneities (though not in the form of ordered bands) are prominent in actual bird flocks studied in the field. Field studies challenge the way the Vicsek model quantifies the interactions in animal aggregations. As a flock rearranges, sometimes even temporarily splitting, its density and structure are continuously changing but its coherence is never lost, as it would be in the Vicsek model when mutual distances exceed the interaction range. Ballerini et al (2008) confirmed, by quantifying their observations of large starling flocks, that interactions are actually based on *topological* rather than metric distance: each individual interacts with a fixed number of neighbors, commonly six to seven, irrespective of their spacing. This interaction mechanism allows the flock to maintain cohesion against strong perturbations. Of course, metric interactions should be relevant for inanimate active particles tied by physical forces, especially in "wet" active matter where interactions are carried by a surrounding medium.

Simulations of the "topological" version of the Vicsek model with metric-free interactions (Ginelli and Chaté, 2010) support the fact that they have a cohesive tendency. Unlike the metric model, the phase transition to collective motion with reduced noise is almost abrupt. There is no segregation into an ordered "liquid" and a disordered "gas" phase, because neighbors in dilute regions are never disconnected, and therefore low density does not necessarily induce disorder. The simulation with

[1] Thomas Ihle, private communication

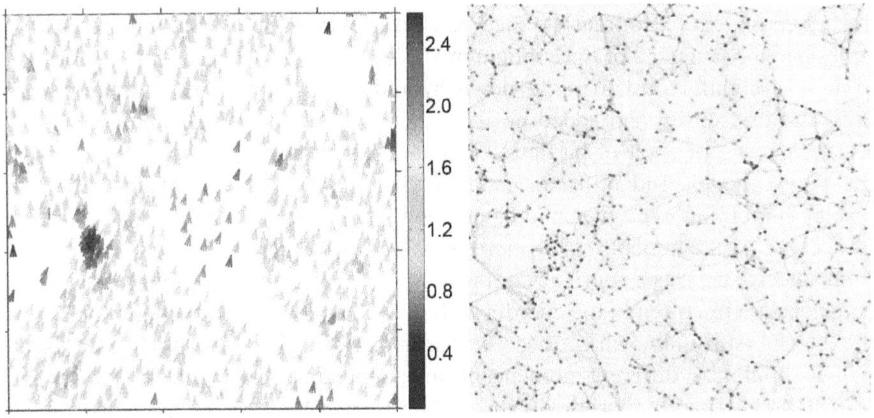

Fig. 1.6 Non-metric interactions with seven nearest neighbors. *Left*: A snapshot of particle positions; the colors show the interaction range. *Right*: An instantaneous interaction network (Komareji and Bouffanais, 2013)

non-metric interactions among seven nearest neighbors illustrated by Fig. 1.6 reveals many different metric interaction ranges correlated with density fluctuations.

Of course, reality is more complicated than anything that a basic model can predict. Observations showed waves of "information transfer" within a group (Attanasi et al, 2014), influenced by the rank of individual birds (Nagy et al, 2010). Leadership was also shown to be important in cooperative transport of large cargo by ants that necessarily requires alignment of their pulling force (Feinerman et al, 2018). We may recall the tendency of humans to obey their leaders. Experiments with groups of volunteers (Dyer et al, 2009) showed that a small informed minority (5%) could effectively guide a large uninformed group to a target by making decisions by consensus. We know from political experience that humans often follow even incompetent leaders, with or without coercion. Animal herds and flocks manage to sustain group integrity during long migrations with no verbal communication and no enforcement.

1.3 The Flock as a Continuum

Models defining motion and interaction rules are very handy for carrying out computations; cheap computer power has even encouraged the study of classical kinetic and hydrodynamic problems by racing millions of simulated molecules in the bowels of a computer (even though any number even the most powerful computer can handle is diminutive compared with the actual number of molecules in a water droplet we can see with the naked eye, and molecular interaction potentials are far more sophisticated than those commonly used in computer models). Nevertheless, there is something that cannot be obtained in this way: the analytical insight gained by the great physicists of the past, which necessitates a continuum description. Toner and

Tu (1995) undertook to translate the Vicsek model (still fresh from the press at the time) into the language of the continuum.

It is straightforward to write down hydrodynamic equations of motion with anisotropic viscosity and added noise in such a way as to define the local velocity and density of a set of particles. The task is simplified by a special feature of the Vicsek model: it identifies orientation with velocity. This eliminates a difficult task of combining hydrodynamic equations with "elastic" equations that take care of keeping the alignment intact (more on this in Sect. 2.1). The continuum model immediately suggests a heuristic argument for the stabilizing effect of motion. If the equations are rewritten in the coordinate frame moving with the average velocity of the "flock" (still unknown), it becomes clear that neighbors of a particular "bird" will be different at different moments of time, depending on inhomogeneities in the velocity field. Therefore originally distant "birds" may interact at a later time, which effectively extends the interaction range and stabilizes the ordered phase.

Yet, there are inconvenient facts, which make the entire undertaking questionable. Active fluids have no equation of state, and *pressure* cannot be defined in a constructive way, except within a narrow class of models (Solon, Fily, et al, 2015). One can compress an active fluid, increasing its average density. In a common fluid, the required work is unequivocally determined by pressure, but in an active medium it depends on the way particles interact with the confining walls. Different forces and hence different amounts of work are needed to reach the same final density when compressing with a hard wall or with a soft enclosure, into which particles bump gently. This can be demonstrated quantitatively by separating two parts of a container by a mobile wall with asymmetric interaction potentials on its two sides. The partition moves to equalize the two wall-dependent pressures, resulting in a steady state with unequal densities in the two chambers (Fig. 1.7). In equilibrium fluids, even oriented ones like liquid crystals, the normal force per unit area on any part of the boundary is independent of its orientation. This is not so in active media, as long as the propulsion speed is anisotropic, even if the particles are oriented isotropically.

Moreover, realignment of Vicsek's particles *does not conserve momentum*, while the hydrodynamic equations of Toner and Tu, like those of standard hydrodynamics, are based on momentum conservation, which is presumed to be valid in some average sense, and include the gradient of this ill-defined variable, pressure. The Vicsek model does not even possess the Galilean invariance inherent in classical

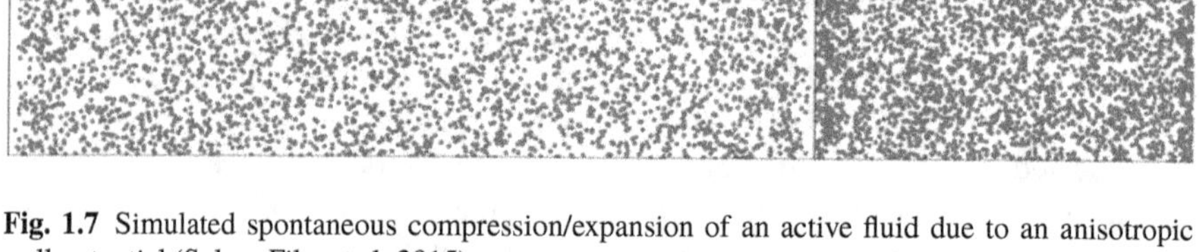

Fig. 1.7 Simulated spontaneous compression/expansion of an active fluid due to an anisotropic wall potential (Solon, Fily, et al, 2015).

mechanics, since the particles move with a constant velocity defined in a specific ("laboratory") frame of reference. An easy way to forget these objections is to tell oneself that the Vicsek model is after all a model rather than a law of nature, and the continuous equations of Toner and Tu can be viewed as just another model, related in some respects but different in others.

Once a system of differential equations, justified or not, is in place, it can be studied in a standard way. One can test the linear stability of its "trivial" homogeneous solutions propagating with a certain speed, and find their symmetry-breaking bifurcations leading to a family of propagating bands. Solon, Caussin, et al (2015) worked in this way with a modified system, simplified in some respects compared with that of Toner and Tu but including a nonlinear term that tends to keeps the absolute value of the velocity constant, as assumed in the Vicsek model. They obtained what Chaté (2020) calls an *embarrassingly* large family of linearly stable solutions: periodic patterns, solitary bands, phase-separated domains. All this variety disappears when a noise term, missing in the simplified system, is reintroduced, causing a unique solution to be selected. This is consistent with the role of noise in equilibrium systems where noise facilitates transition from metastable states to a state with the minimal energy, but energy is neither well defined nor conserved in active matter. Chaté (2020) lists this among current riddles, asking: "How do we understand the selection of a unique solution observed at microscopic and fluctuating hydrodynamic levels but which is not present at the deterministic hydrodynamic level?"

Clearly, noise is an essential component, and formulating phenomenological hydrodynamic equations (even if fully justified) is only the beginning of the road. Theories of phase transitions (Landau et al, 1980) have to account not just for fluctuations but for their *correlations* in time and space. This cannot be done in a straightforward way, since pair correlation functions depend on triple correlation functions, and so on, necessitating a cut-off under some assumptions. A powerful method in the theory of critical phenomena is the renormalization group, based on invariance to scale transformations (Goldenfeld, 1992). The theory is precise in 4D, and applications to physical dimensions are commonly based on the $4-\epsilon$ expansion, where ϵ is assumed to be small, even though $4-3=1$ is not what would normally be treated as a small parameter. Toner and Tu (1998) boldly applied the renormalization group further down, in 2D, even though admitting that some parameters of their theory are not scale-invariant, and in spite of all above mentioned shortcomings, claimed that their predictions (or retrodictions?) are in agreement with the computations by Vicsek et al (1995).

1.4 Towards Statistical Description

The central theoretical question is understanding both similarities to and distinctions from the behavior of passive matter obeying thermodynamic laws. The phenomenology of the Viscek model, as reflected in Fig. 1.3, looks at first sight to be not so very different from that of ordinary fluids: the liquid and gas phase differ by density, and the phase transition happens with a changing temperature that determines the level of random noise. Liquid and gas can coexist, as dense (ordered) and dilute (disordered) domains in Fig. 1.4. Even giant intermittent fluctuations are possible in common fluids near a critical point.

Another passive system similar in some respects to the Vicsek model is the magnetic XY model describing interactions of spins on a lattice (Kosterlitz and Thouless, 1973). Its microscopic "particles" are vectors of a fixed length, like Vicsek's "birds", and likewise the orientation of these vectors tends to adjust to their immediate environment, but there are no density inhomogeneities and, of course, no active motion. It is known that fluctuations at *any* non-zero temperature prevent formation of a phase with the long-range order in this system. A typical simulation of the XY model, like the one in Fig. 1.8, shows a disordered state with a number of topological defects – vortices with orientation rotating by 2π around their cores. Although activity often brings about disorder, it turns out in this case that it is *motion* which is responsible for the long-range order in the Vicsek model, and topological defects do not appear in simulated flocks.

Contrary to all distinctions between active and equilibrium systems, the notion of *entropy* appears to play a special role in collective motion. Bialek et al (2012) posed the question of how to derive the overall distribution of velocities, given the matrix of correlations between velocities of individuals measured in an actual flock. Infinitely many distributions are consistent with the measured correlations, but the successful choice was the one describing a system that is as random as it can be, i.e., having the maximal entropy, while still matching the data. The maximum entropy model correctly predicted, with no free parameters, the propagation of order throughout a flock of starlings based on pairwise interactions between birds. It also confirmed the conclusion by Ballerini et al (2008) that interactions are ruled by topological rather than metric distance.

Yet, the absence of meaningful definitions of such basic thermodynamic variables as energy and pressure is a clear sign that thermodynamics is actually a misnomer when it comes to active matter. Thermodynamics of equilibrium processes is based on statistical mechanics, but activity violates basic principles like equipartition of energy among various degrees of freedom and detailed balance, and noise in assemblies of

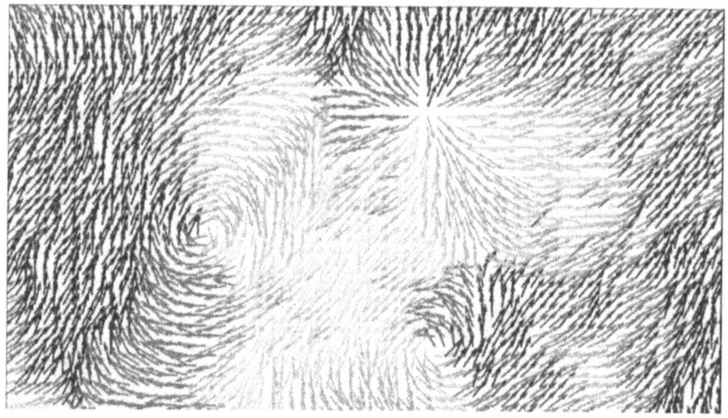

Fig. 1.8 Monte Carlo simulation of the XY model, showing a disordered state with a number of topological defects (vortices). The coloring shows the direction of vectors (by ChrisJLygouras - Own work, CC)

active particles is likely to be more structured than thermal noise. It makes it even more challenging to derive macroscopic dynamics from underlying microscopic interaction rules. This has prompted sophisticated statistical theories aspiring to explain on a deeper level the dynamics of the utterly simple and not too realistic model central to this chapter. The motivation was that the results might be applicable to a wider class of phenomena in "dry" active matter.

Bertin et al (2006) set the aim of building a statistical description of a modified version of the Vicsek model, wherein particles may change their propagation direction (but not the magnitude of their velocity) either by a random "kick" or as a result of a binary collision that aligns the velocities of the two particles to their average direction shifted by random noise. Recall that the original model allows interactions between many particles within a set distance, but is less realistic in another respect, as each particle adjusts its direction independently, while in the model by Bertin et al, binary collisions conserve the momentum before being biased by thermal noise. Restricting to binary interactions is common in molecular dynamics, and momentum conservation, with the random factors accounted for by effective viscosity, leads to hydrodynamic equations far better justified than those of Toner and Tu. All coefficients, including inertial effects, linear and nonlinear viscosities, and even a cubic term setting the magnitude of the velocity, are computed from the microscopic parameters of the model.

Yet, the theory is problematic in one important aspect: it is based on Boltzmann's hypothesis of "molecular chaos", which assumes that colliding particles are *uncorrelated.* This is justified when the mean free path is large compared to the radius of interactions, something which may be true in the "gas" phase but is not true at realistic densities when order is established. Indeed, Boltzmann's hypothesis fails even in equilibrium liquids, and the theory of the liquid state has to involve sophisticated approximations accounting for molecular correlations. Ihle (2011) pointed out this drawback and put forward an alternative theory based on the weak gradient expansion, mirroring the theory of weakly inhomogeneous media (Chapman and Cowling, 1970). His derivations led to more complicated nonlinear hydrodynamic equations than those of Bertin et al, but with coefficients also expressed through the microscopic parameters of the model.

Unlike Bertin et al, Ihle took as the basis of his theory the original Vicsek model evolving, as in standard agent-based computations, by discrete steps – indeed, by large steps, as he relied on the assumption that the free path between collisions was much larger than the interaction range. Such a highly discrete character of the motion is a questionable point in Ihle's theory, as time steps of agent-based computations are relatively small and, of course, real flocks interact continuously. The continuum statistical theory of flocks (or, more generally, of "dry" active matter) remains unsettled, as reflected, in particular, by the two contending papers in a discussion issue on active matter (Peshkov et al, 2014; Ihle, 2014), which contain both justification of these efforts and convincing mutual criticisms of the rival theories.

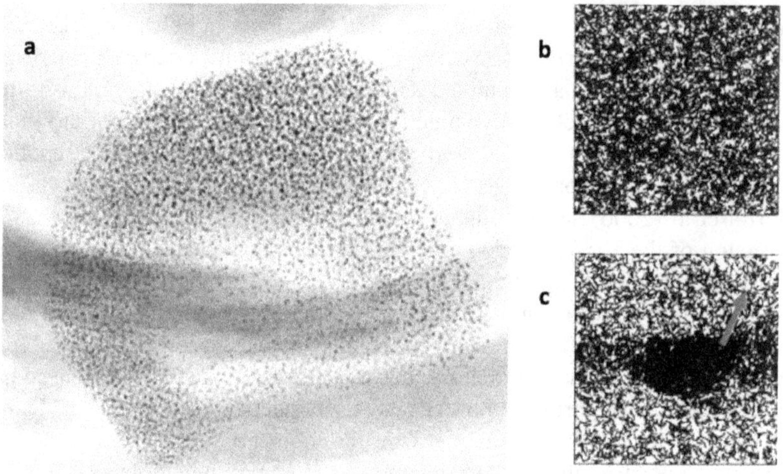

Fig. 1.9 (**a**) Snapshot of a 3D flock, with short-range attractive interactions; sky and cloud colors are added to make it look like a real bird flock. (**b**) A homogeneous 2D flock with medium strength long-range cohesive interactions. (**c**) Consolidated flock with strong cohesion. The *red arrow* indicates the mean direction of motion (Chaté et al, 2008)

1.5 Variations on Vicsek's model

There have been a number of attempts to make the basic Vicsek model model more realistic and versatile. Collective motion is possible only at finite density: if the Vicsek model evolved in an infinite domain, the "birds" would eventually fly apart. Grégoire et al (2003) added pairwise short-range attractive interactions to the Vicsek model. Simulations of this model produced compact flocks superficially similar to a dense flock of birds, especially when images of clouds and blue sky are added, as in Fig. 1.9a. An alternative attempt to consolidate a flock was to add long-range cohesive interactions of hydrodynamic origin, introduced in a rather vague way through advection by a fluid flow generated by the particles themselves and applicable to bacteria rather than to birds (Chaté et al, 2008). In the simulations of

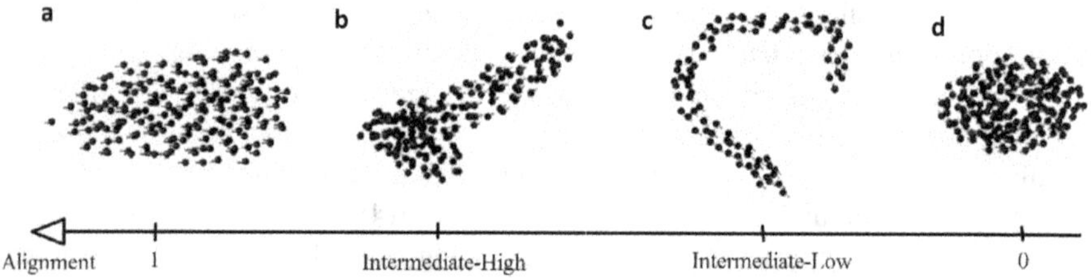

Fig. 1.10 Changes in the shape of a cohesive flock with changing alignment strength (Strömbom et al, 2015)

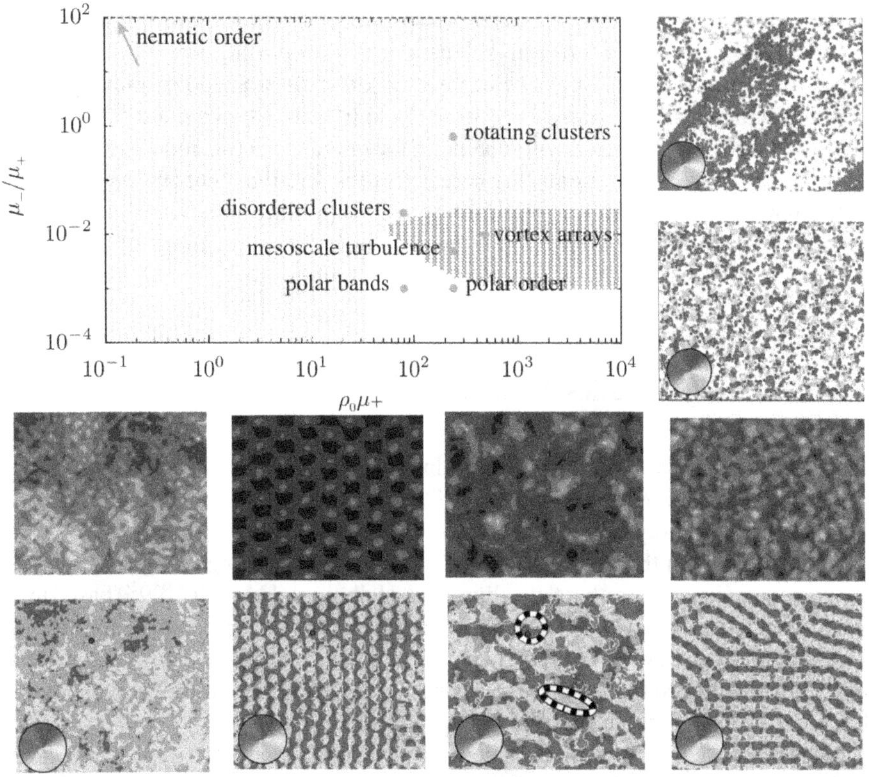

Fig. 1.11 Patterns obtained in the model with competing alignment and anti-alignment interactions. *Upper left*: Estimated domains of different patterns in the parametric plane spanned by the average density ρ_0 multiplied by the strength of the aligning interaction μ_+ and the ratio μ_-/μ_+ of anti-aligning to aligning interactions. *Blue points* indicate the parameter values for which simulation results are shown. *Upper right*: Alignment fields in the regimes of polar bands (*top*) and disordered clusters (*bottom*). Density (*middle row*) and alignment (*lower row*) maps in other regimes, from *left to right*: large-scale polar order; vortex lattice; mesoscale turbulence, with the black-white ellipse and ring indicating a jet and a transient vortex, respectively; and dense rotating clusters. In the density color maps, density grows from dark to light shading; the orientation field is color-coded as shown in the insets (Großmann et al, 2015)

this model, both the polar order and the band structure were destroyed with growing interaction strength (Fig. 1.9b), but for still greater cohesion, order was restored and the flock consolidated into a compact group (Fig. 1.9c).

Combining attraction to the center of a local group, defining the alignment with local repulsive interactions, and grading the alignment strength produces different shapes of simulated flocks (Strömbom et al, 2015). In the snapshots shown in Fig. 1.10, the shape changes from a compact rigid propagating flock at high alignment (a) to flocks of a more irregular shape with lively internal dynamics (b and c) as the strength of alignment decreases, and to a stationary but fluctuating group (d) when the alignment is switched off.

Großmann et al (2015) modified the Vicsek model still more radically, combining velocity alignment with nearest neighbors and *anti-alignment* with particles which are further away, such as might be caused by some unspecified complex hydrodynamic interactions. The inspiration for this model was the symmetry-breaking in reaction–diffusion systems due to competing short-range activation and long-range inhibition (Turing, 1952). Replacing here activation for alignment and inhibition for anti-alignment gives a ready recipe for complex patterning, and this is indeed what this model provides. A variety of patterns coming up in simulations of this model are shown in Fig. 1.11.

1.6 Crowding Impedes Motion

Physicists are not particularly interested in animal social relations, even when they are trying to amend the basic model to fit field observations – but human crowds behave in a special way, being driven by psychology rather than by mechanical forces. Though vectorial in their orientation and motion, freely moving humans are largely motivated by avoidance rather than alignment. Helbing and Molnár (1995) based their model of pedestrian motion on the *social force*, including three principal components: acceleration to the desired velocity of motion, avoiding other pedestrians and borders, and attraction toward the target. The paper gained over 2000 citations, and the model has been tested in both normal and emergency situations.

When moving in a constrained space like a corridor, people self-organize in lanes with a uniform walking direction (Fig. 1.12a). However, the lanes are not necessarily well ordered, and as shown in Fig. 1.12b, dynamically disordered multiple lanes, marked by green crosses, prevail at higher densities. Over some critical density,

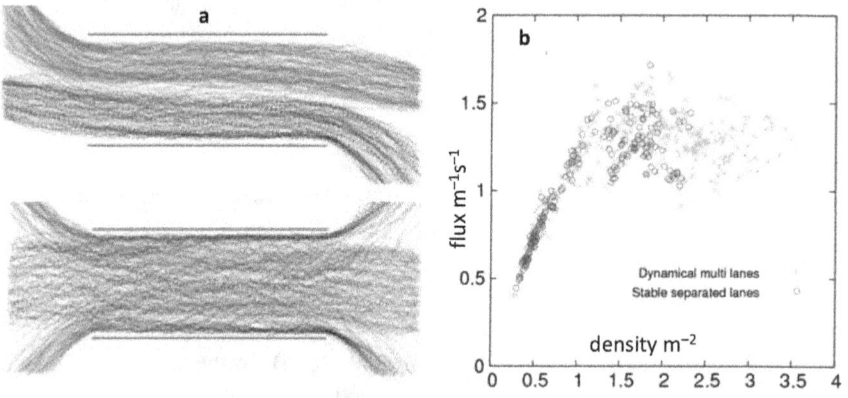

Fig. 1.12 (**a**) Recorded trajectories of pedestrians in a corridor, color-coded according to their direction, with stable separated lanes (*top*) and dynamical multiple lanes (*bottom*). (**b**) Density–flux diagrams for the two types of bidirectional flow (Zhang et al, 2014)

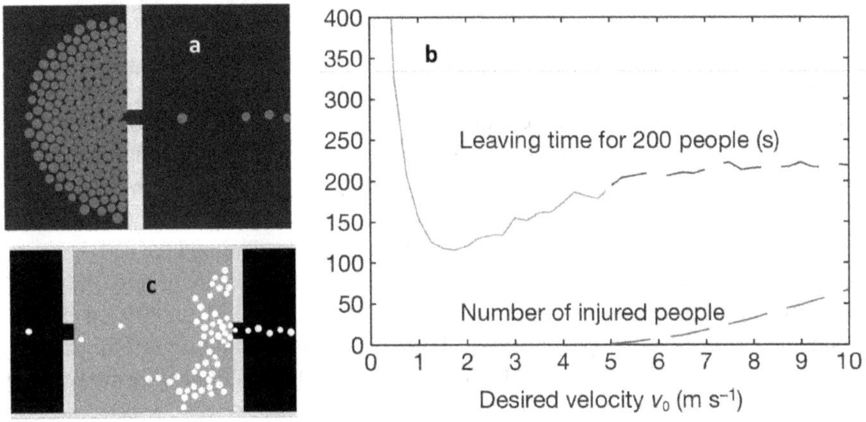

Fig. 1.13 (**a**) Snapshot of a simulation of pedestrians escaping towards the 1 m wide exit from a 15 m^2 room. (**b**) The dependence of the escape time and the number of injured people on the desired velocity. (**c**) Pedestrians trying to escape a smoky room through two invisible doors (Helbing et al, 2000)

the total flux flattens out, which means that the average velocity falls off in inverse proportion to the density. It is not helped at all by increasing disorder, which is a sign of frustration rather than foresight. The gap between two adjacent lanes is clearer in the stable flow. I recall from my own subway experience that, thanks to this gap, the velocity is maximal at the location of its highest gradient – quite a contrast to hydrodynamics, which was once naïvely viewed as a model for human flows. I used this observation to move forward more quickly – this kind of egotistic behavior may be responsible for the transition to disordered lanes.

The rational principles underlying the behavior of conscious and purposeful humans built into the social force model become counterproductive in crowded and constrained settings. Helbing et al (2000) simulated emergency escape through a narrow exit. Quite naturally, the density grows near the opening (Fig. 1.13a), but this can only be detrimental. The plot in Fig. 1.13b shows that the leaving time *increases* with the desired escape velocity, alongside the increasing number of injured people. The results of this simulation are both trustworthy and sad: the more dangerous the emergency, the harder it is to escape from it due to the ensuing panic.

The simulation in Fig. 1.13c demonstrates a mixture of individualistic and herding behavior in the situation when the escape route is invisible. Initially, each person randomly selects a desired walking direction, but one's direction subsequently becomes influenced by the way one's neighbors go, so the majority will be crowded near a single exit. The strength of this herding effect grows with increasing panic.

The assumption of a constant magnitude of the velocity is the most unrealistic feature of the Vicsek model: this is what allows the dense bands in Fig. 1.4b to march along like soldiers on a parade. Increasing density inhibits motion; this is what happens when panic freezes a crowd. It does not depend on human idiosyncrasies: active particles generically accumulate where they move more slowly, hindered by steric obstacles or repulsive interaction. A positive feedback between slowdown-induced accumulation and accumulation-induced slowdown leads to *motility-induced phase separation* (MIPS)[2], just as particles stuck in jams create dense regions, while others move more or less freely in any remaining voids (Cates and Tailleur, 2015).

Phase-separated domains tend to coarsen, just like liquid and gas phases in equilibrium systems, but this process may not lead to complete phase separation. Equilibrium systems are symmetric under time reversal, and the key to the unparalleled behavior of active particles is in abandoning this principle. Tjhung et al (2018) modified the standard statistical model of phase separation (Hohenberg and Halperin, 1977) by adding lowest-order terms that break time-reversal symmetry. In the sequence shown in the upper panels of Fig. 1.14, obtained in simulations of this model, the system is at a steady state, but a dense macroscopic "liquid" cluster formed during coarsening contains mesoscopic vapor bubbles, which are continuously created in its bulk, coarsen, and are ejected into the exterior "vapor" phase. The lower panels of the same figure taken from the same source show a picture of arrested coarsening, coming to a steady state in the two right-hand panels.

Attempts by theorists nurtured on statistical physics to connect the passive and active worlds are frustrated due to fundamental differences of this kind. As Cates and Tailleur admit in the review cited above, "we do not have a framework to combine MIPS-type effective attractions with standard (i.e., passive) colloidal interactions". Gompper et al (2020) assert that "a major theoretical undertaking is to develop a better understanding for the statistical foundations of active matter, ideally comparable to equilibrium statistical mechanics." This aim still remains a vision (or mirage?), but a

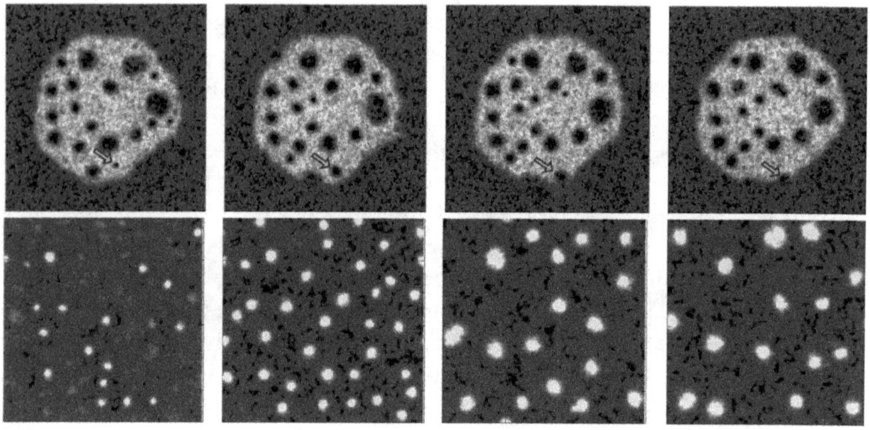

Fig. 1.14 *Top*: Bubbly phase coexistence of the "boiling liquid" (*yellow*) with the vapor phase (*blue*). The *green arrow* tracks a bubble created inside the liquid bulk, growing, moving toward the interface, and getting expelled. *Bottom*: Arrested phase separation: the average cluster size saturates at a finite steady-state level (Tjhung et al, 2018)

[2] This lush term reminds me of Molière's Jourdain being told that he is speaking prose.

phenomenological picture is nevertheless emerging from numerous simulations and carefully designed experiments.

1.7 Active Dipolar Rollers

The crowding phenomenon has also been observed on a microscopic scale. Bricard et al (2013) experimented with inanimate polar walkers – rollers invented by Quincke (1896), moving as explained in the caption of Fig. 1.15, and observed emergence of macroscopic directed motion with almost constant magnitudes of the velocity (Fig. 1.16a), similar to Vicsek's flocks, in populations of microscopic rollers. Both hydrodynamic and electrostatic interactions promote alignment of roller velocities, as do hydrodynamic and social interactions among fish (Sect. 1.1), in spite of all the discrepancies between the mechanisms and sizes.

Geyer et al (2019) demonstrated density effects in later experiments by the same group, running three million colloidal rollers of the radius $a = 2.4$ μm on the microfluidic racetrack shown in Fig. 1.16b. At low density, there is an isotropic "gas" with no net particle current, since in the absence of interactions each roller chooses its direction of motion independently. As the average packing fraction ϕ grows above 0.02, the rollers undergo a flocking transition, forming polar liquid bands propagating at a constant speed through the homogeneous isotropic gas, and as ϕ increases further, the liquid phase fills the entire system and flows steadily and uniformly.

However, above $\phi = 0.55$, which is still much smaller than the random close-packing density, the liquid solidifies, and collective motion is suppressed locally, as rollers stop their collective motion and jam, like panicking people in a crowd. The jammed bands are amorphous solids where particles are largely at rest, but they melt continuously at one end while growing at the other end, thereby propagating the flow of the liquid phase upstream, while preserving their approximate shape and length. Multiple jams nucleate on the track, propagating at almost the same speed, and are

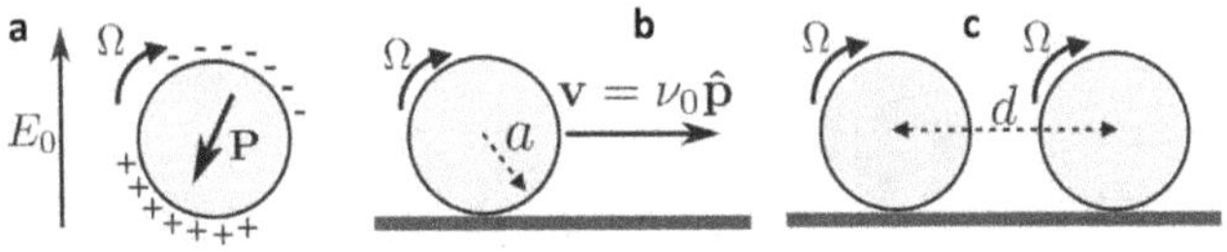

Fig. 1.15 Quincke rollers. When applying an electric field E_0 to an insulating sphere with a radius a immersed in a conducting fluid, an electric dipole forms at its surface (**a**). Above a critical voltage, the dipole inclines at a finite angle to the electric field, causing steady rotation of the sphere with the angular speed Ω (**b**). When the sphere settles on an electrode, the rotation is converted into translation with the speed $\mathbf{v} = v_0\mathbf{p}$ directed along some 2D vector $\mathbf{p}$. When isolated, it rolls without sliding at a constant speed $v_0 = a\Omega$. When two colloids rolling in the same direction are close to each other (**c**), the lubrication torque acting on the two spheres separated by a distance d hinders their rolling motion (Geyer et al, 2019, CC)

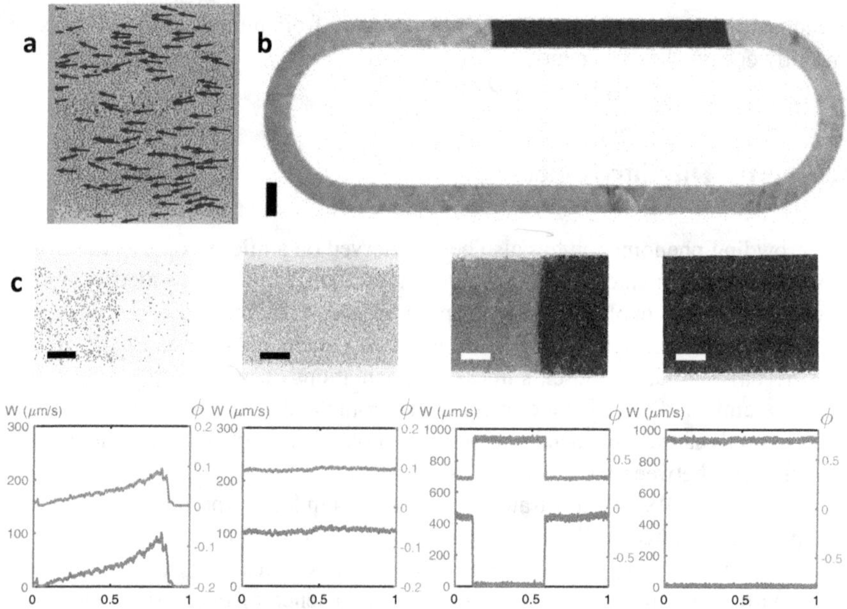

Fig. 1.16 (**a**) Collective motion of Quincke rollers. *Arrows* show the roller's displacement between two subsequent video frames (Bricard et al, 2013). (**b**) An active solid (*black*) propagates through a polar liquid (*gray*) on a microfluidic racetrack; scale bar 2 mm. (**c**) *Top*: Snapshots at different densities; scale bars 250 μm. *Bottom*: The dependence of the average particle flux W (*blue*) and density ϕ (*red*) on time (normalized by the time taken by an active solid to circle around the race track). *Left to right*: Coexistence between the active gas and a dense polar band at $\phi = 0.033$; propagating polar liquid phase at $\phi = 0.096$; coexistence between the polar liquid and amorphous active solid at $\phi = 0.49$; the homogeneous active solid phase at $\phi = 0.70$ (Geyer et al, 2019)

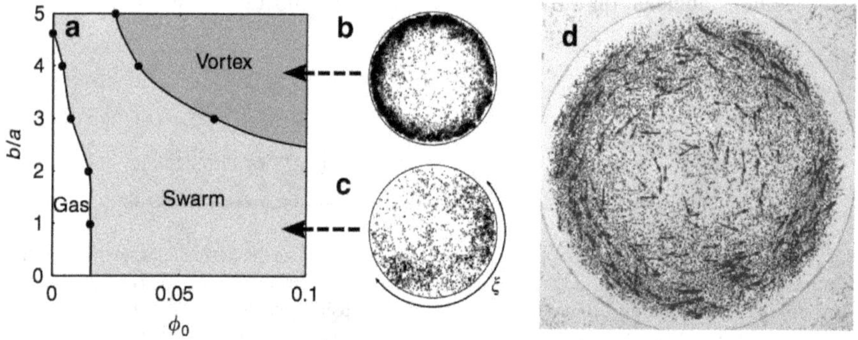

Fig. 1.17 (**a**) Computed phase diagram dependent on the average density ϕ_0 and the ratio b/a of the exclusion range to the roller's radius. (**b**) Snapshot of a computed vortex pattern. (**c**) A snapshot of a computed swarm of length ξ. (**d**) A snapshot of the experimental vortex pattern. *Arrows* show the roller's displacement between two subsequent video frames (Bricard et al, 2015)

therefore unable to catch up and coalesce. Finally, at a still higher packing, the entire system solidifies. Snapshots (upper row) and plots of the average roller velocities and densities (lower row) in different regimes are shown in Fig. 1.16c.

When running the same rollers in a circular corral (Bricard et al, 2015), another pattern was observed, namely, a vortex occupying the entire enclosure with a higher density at the periphery (Fig. 1.17d). The same pattern came about in simulations of collective dynamics (Fig. 1.17b) at higher densities and when the values of the exclusion range b (see Fig. 1.15) were substantially larger than the particle radius a. At lower ratios b/a, a swarm cruising around the circular corral (Fig. 1.17c) is seen instead of a vortex, and as usual, there is a disordered gas at low densities (Fig. 1.17a).

In another experiment, carried out by Kaiser et al (2017), magnetic rather than electric dipoles were induced to excite various patterns of rolling motion. A few hundred 6 μm-sized ferromagnetic colloidal spheres were energized by a vertical alternating magnetic field (Fig. 1.18a). The container had the form of a concave circular lens which served to prevent the particles from escaping or accumulating at the walls as they do in Fig. 1.17d.

The system is similar to the Quincke roller system, but the difference is that the rotation speed of magnetic particles depends, not on the magnitude of the driving field, but only on its frequency. Unlike Quincke rollers, even individual ferromagnetic

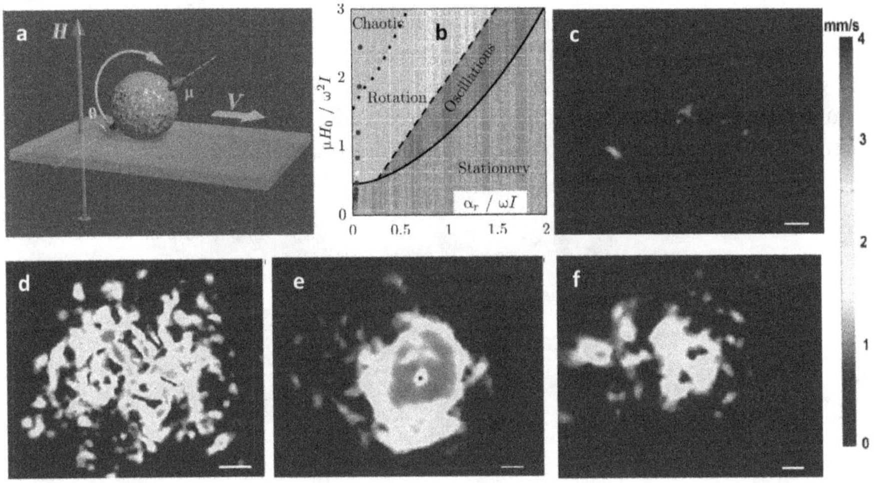

Fig. 1.18 (**a**) The scheme of the experiment, showing the directions of the applied magnetic field H, the induced dipole μ, the rotation angle θ, and the velocity V of a roller. (**b**) Dynamic states of individual particles. The *solid line* shows the stability limit of the locked state with the particle's magnetic momentum aligned with the field direction. The *dashed line* is the existence boundary of the rotating state. Chaotic regimes, characterized by spontaneous reversal of the rotation direction, exist above the *black diamond line*. Symbols on the left show the conditions realized in the experiments. (**c**)–(**f**) Coarse-grained velocity magnitude in the regimes of disordered gas at the frequency 20 Hz, flocking at 30 Hz, coherent vortical motion at 40 Hz, and reentrant flocking at 50 Hz. The color scale is shown on the right (Kaiser et al, 2011)

rollers exhibit complex dynamics characterized by spontaneous direction reversals. Different modes of their behavior are shown in Fig. 1.18b. The diagram is not easy to read. The field intensity H_0 increases along the ordinate, but the frequency $\omega = 2\pi f$ enters both dimensionless numbers, which also contain the magnetic moment of the particle μ, its moment of inertia I proportional to its mass and squared radius, and the rotational drag coefficient α_r, proportional to its volume and the viscosity of the surrounding fluid. Therefore the line H_0 = const at changing frequency, such as the sequence of dots corresponding to the collective regimes in Fig. 1.18c–f, is parabolic.

As the frequency of the magnetic field increases, going down the sequence of dots on the left in Fig. 1.18b, the observed dynamics evolves from a disordered gas to flocking to coherent vortical motion and back to reentrant flocking (Fig. 1.18c–f). Interactions of dipolar particles depend on their mutual orientation, and the onset of large-scale collective behavior is caused by increasing coherence through synchronization of particle orientations by the applied magnetic field. In the gas phase, the particles move randomly. Their orientations with respect to the field are uncorrelated, and the angle distribution is almost uniform. With the onset of flocking, all particle orientation angles converge to two well-separated narrow bands, with spontaneous reversals between the two directions. The directions become stationary and fully correlated in the vortex phase. High-frequency flocking is a noise-activated process when individual particles remain oriented along the field direction.

It should be noted that experiments with both Quincke and magnetic rollers have been carried out within a fluid medium, but the prevailing interactions are determined here by the driving field rather than hydrodynamics, so these systems can be still classified as "dry". Yet, the observed dynamics, especially in the magnetic system, is more variegated than anything encountered earlier in this chapter, and gives a foretaste of the genuine "wet" physics of colloidal systems in Sect. 3.

Chapter 2
Active Nematics

2.1 Liquid Crystals

Anisotropic liquids display varieties of order different from polar alignment, while retaining their fluidity. The most common kind of order is *nematic*, arising when molecules are elongated and stiff. The uniaxial nematic alignment is characterized by a *director*, which differs from a vector by lacking an arrow, and therefore is invariant under rotation by π. Further kinds of ordering, sketched in Fig. 2.1, are more restrictive. In the cholesteric phase, the prevailing orientation rotates around some axis. In smectics, molecules are layered; within each layer, they are oriented normally in the A phase, at a certain angle in the C phase, and at an angle rotated between one layer and the next in the C* phase; within each layer, their positions are disordered (de Gennes and Prost, 1995).

On the face of it, nematic, sometimes called *apolar*, order is less organized than the polar (vectorial, or magnetic) order of flocks, but its description is actually more involved. The director is not a well-defined geometric object. When it is presented as a vector **n**, its paired combinations have to appear in all expressions; otherwise they would be invariant under rotation by 2π, as vectors are, but not under rotations by π.

Fig. 2.1 Liquid-crystalline phases. *Left to right*: Nematic, cholesteric, smectic A, smectic C, and smectic C* (by Kebes – Own work, CC)

L. Pismen, *Active Matter Within and Around Us*, The Frontiers Collection,
https://doi.org/10.1007/978-3-030-68421-1_2

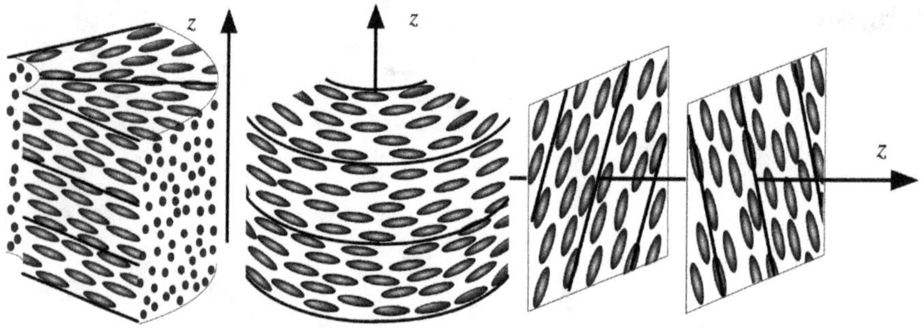

Fig. 2.2 *Left to right*: Splay, bend, and twist distortion of nematic alignment (Kleman and Lavrentovich, 2003)

Thus, the standard expression for the elastic energy density of a uniaxial nematic is

$$F = \frac{1}{2}\left[K_1(\operatorname{div}\mathbf{n})^2 + K_2(\mathbf{n}\cdot\operatorname{curl}\mathbf{n})^2 + K_3(\mathbf{n}\times\operatorname{curl}\mathbf{n})^2\right]. \tag{2.1}$$

The three terms in this expression correspond to the energies of splay, bend, and twist distortion of nematic alignment, sketched in Fig. 2.2, which depend on the invariant quadratic combinations of the director with its divergence and curl; the coefficients are the respective elasticities. In 2D, only splay and bend distortions remain, and the respective energies are expressed through the partial derivatives of the components of the 2D director n_x, n_y as $(\operatorname{div}\mathbf{n})^2 = (\partial_x n_x + \partial_y n_y)^2$ and $(\operatorname{curl}\mathbf{n})^2 = (\partial_y n_x - \partial_x n_y)^2$.

Notably, in this formulation there is no difference between distortions of the polar or nematic order. Yet, what this definition of the elastic energy is lacking is that, because **n** is presumed to be a *unit* vector, it cannot be extended to situations where the liquid is not perfectly oriented locally. This is a serious flaw, since distorted textures commonly include *defects*, where alignment is completely lost, and defects in polar and nematic order are quite different (see Sect. 2.4).

What fully characterizes the nematic state is not a vector but a symmetric tensor Q with zero trace. In 2D, it is expressed as

$$\mathbf{Q} = \frac{\rho}{\sqrt{2}}\begin{pmatrix}\cos 2\theta & \sin 2\theta \\ \sin 2\theta & -\cos 2\theta\end{pmatrix}, \tag{2.2}$$

where θ is the inclination angle and ρ measures the deviation from isotropy, whence $\rho = 1$ corresponds to the perfectly aligned nematic with the director having vector components $n_x = \cos\theta$, $n_y = \sin\theta$. The tensorial description can also be extended in 3D to the case where the alignment is not axially symmetric, e.g., where microscopic elements are shaped as ellipsoids with three unequal axes rather than like the spindles in Fig. 2.1. In this formalism, elastic interactions, taking account also of the energy of defects, are characterized by a tensor of the *fourth* rank, which retains far more terms than those present in (2.1), even when this number has been reduced to the minimum using the available symmetries.

Since the fluid is anisotropic, so are viscosities, which, similar to elasticities, have to be defined by a fourth-order symmetric tensor rather than a single scalar coefficient as for isotropic fluids. The hydrodynamic equations of motion of nematic liquid crystals are well established in the Ericksen–Leslie theory (Kleman and Lavrentovich, 2003), but are hardly ever used in their full form due to their complexity and lack of data on their numerous parameters. Activity brings about additional complications, and ingenuity in model building is required to arrive at realistic results without getting bogged down in technicalities.

2.2 Rod-like Particles

Liquid-crystalline order can be reproduced on a larger scale in assemblies of active particles. Nematic alignment arises when particles are stiff and elongated. These might be macroscopic rods in a vibrating granular layer, or rod-like bacteria or viruses, or stiff filaments in the cellular cytoskeleton, or migrating cells, what Gruler et al (1995) called "living liquid crystals". In this chapter, we restrict to two approaches to active nematics, parallel to the approaches to "dry" polar active media in Chap. 1: dynamics of discrete rod-like particles and of active nematic continua. "Wet" active media, comprising particles interacting with their surroundings, will be discussed in later chapters dedicated to the motion of self-driven colloids and living matter.

The most straightforward way to model "flocks" of self-propelled rod-like particles is similar to Vicsek's approach, in which velocity is identified with alignment and the magnitude of the velocity is constant. However, the interaction rules have to be modified, since the connection between velocity, which is a vector, and nematic alignment is not as straightforward as between two vectors in a polar medium. Mobile rod-like particles colliding at an obtuse angle tend to anti-align, and Peruani et

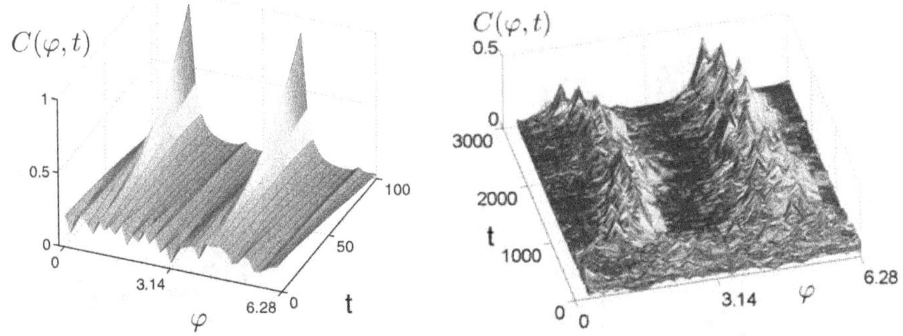

Fig. 2.3 Evolution of the distribution $C(\varphi, t)$ of the alignment angles φ with time t in a flock of particles with nematic symmetry in continuous (*left*) and agent-based (*right*) computations (Romanczuk et al, 2012, after Peruani et al, 2008)

al (2006) adopted this interaction rule in their simulations. Rather than averaging the orientation angle of a particle at each interaction step over the propagation directions of the particles within a metric neighborhood, as in a polar flock, they inverted these directions when they were at an obtuse angle to the direction of motion of the test particle.

This rule allows the particles to reverse their propagation direction, while always moving at constant speed. The conservation of momentum, dubious for Vicsek's vector particles, makes no sense whatsoever in this case. Therefore, the continuous model of Peruani et al (2008) imitating the particle-based computations was restricted to a sole equation for the orientation distribution function. Its solution implied evolution to two peaks of alignment angles, as can be seen in the left panel of Fig. 2.3, in contrast to the single peak for vectorial particles. Agent-based computations arrive at a qualitatively similar distribution, as shown in the right panel of Fig. 2.3, but with the smooth alignment peaks made rugged by noise.

Simulations of self-propelled rod-like particles have to take into account steric interactions that prevent particle self-intersections. This makes observed patterns dependent on the particles' aspect ratio, in addition to their density, propulsion velocity, and the level of noise. Rod-like particles, similar to polar ones (Sect. 1.2), tend to cluster. Clusters are more compact and dense at low levels of noise (Fig. 2.4a and b). At high densities, orientation ordering sets in before clustering, as for polar particles in Fig. 1.4, but at low particle densities, the onset of orientation ordering and clustering occurs at the same value of the noise. In some 3D simulations (Chaté et al, 2008), the ordered phase has been observed to take the form of a single high-density elongated domain (Fig. 2.4c) with ill-defined fluctuating interfaces. The picture is reminiscent of the large particle number fluctuations in a vibrated layer of elongated grains observed by Narayan et al (2007). The cause of these fluctuations lies in velocity reversals: if they are not too frequent, and the system is not too large, a single steady band can be observed (Chaté, 2020) – but velocity reversal is the characteristic feature of the nematic state that ensures evolution to a double-peak orientation distribution.

Abkenar et al (2013) relaxed the non-intersection rule, discretizing rod-like particles by treating them as chains of beads with repulsive interactions that discouraged but did not prevent their intersection. They were motivated by experiments with

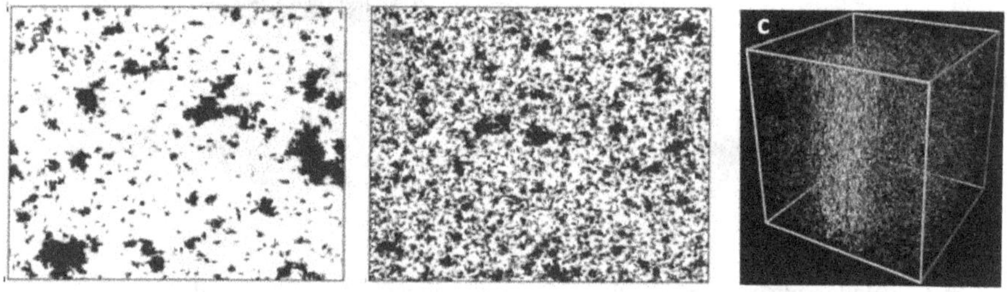

Fig. 2.4 Snapshots of clustering at low (**a**) and high (**b**) levels of noise (Peruani et al, 2010). (**c**) Separation of the ordered and disordered phases in 3D (Chaté et al, 2008)

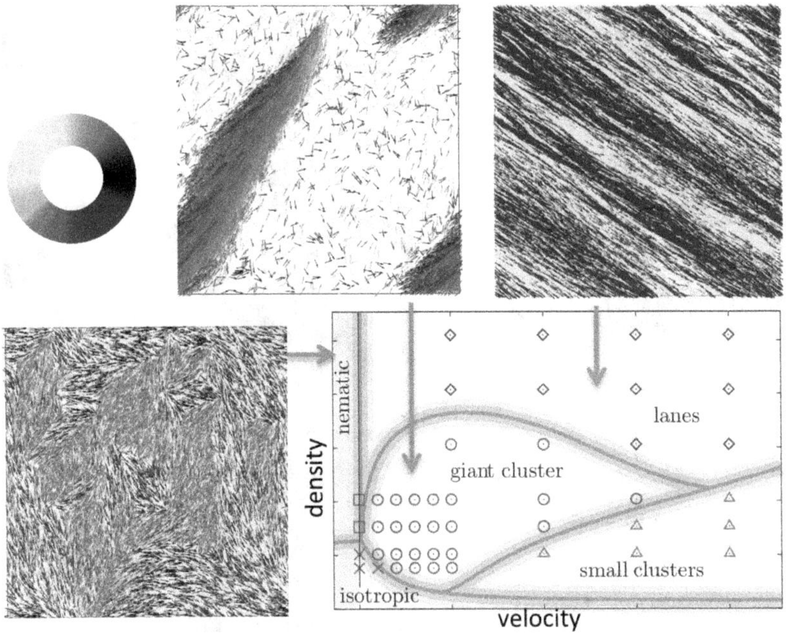

Fig. 2.5 Phase diagram of the regimes of self-propelled rod-like particles corresponding to the snapshots shown in the insets. The particle orientations are color-coded as shown in the *upper left corner* (Abkenar et al, 2013)

filaments in motility assays (Sect. 5.4) where intersections were not excluded. The detailed phase diagram of regimes dependent on the scaled density and velocity (or Péclet number) is shown in Fig. 2.5.

A statistical theory of "dry" active nematics has been developed (Bertin at al, 2013) along the same lines as the corresponding theory of polar particles, based on Boltzmann's hypothesis of "molecular chaos" (Bertin at al, 2006) and leading to hydrodynamic equations with coefficients derivable from the microscopic model. It was amended by Shi et al (2014) to account for particle diffusion. Pure nematic symmetry can be ensured in this model by a high direction-reversal rate. The limitations mentioned in Sect. 1.4 remain in place, but the advantage of having continuous equations lies in the ease in carrying out linear stability analysis. In this way, Shi et al computed both the critical density, below which the disordered solution is stable, and the stability limit of the ordered solution. The latter, naturally, stabilizes as the density increases, but the lower limit depends on the relative magnitudes of the effective longitudinal and transverse diffusivities $D_0 = (D_\parallel - D_\perp)/(D_\parallel + D_\perp)$ and turns out to be different when derived from hydrodynamic or kinetic equations, as shown in the master panel of Fig. 2.6.

The patterns obtained in computations by Shi et al (2014), based on the kinetic model, show nematic bands forming when a slightly perturbed isotropic homogeneous state breaks into pieces at a later time. The same is observed in very long

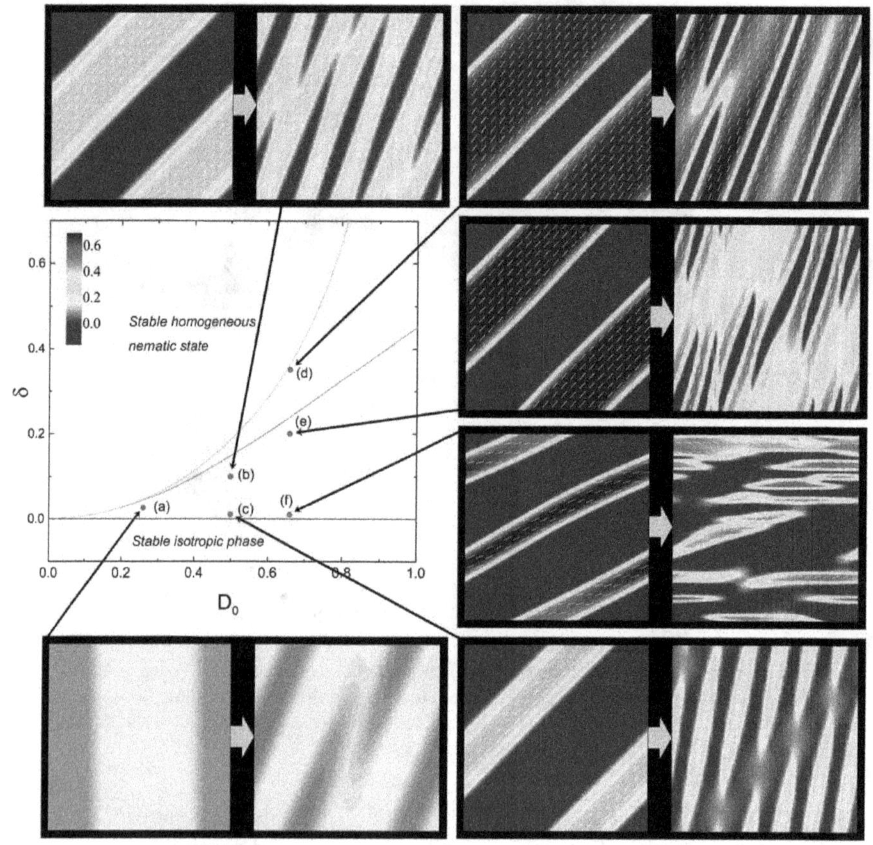

Fig. 2.6 Phase diagram of the kinetic equation dependent on the ratio of the effective diffusivities D_0 and the scaled deviation from the critical density δ, shown by the color scale in the side panels. The homogeneous ordered solution of the hydrodynamic equations is stable above the *orange line*. The *green curve* marks the same limit derived from the kinetic equations. Each *side panel* shows two consecutive snapshots for the corresponding parameter values: one for a transient regular nematic band forming from a slightly perturbed isotropic homogeneous state, and the other forming at a later time when this transient band has broken into several pieces (Shi et al, 2014)

agent-based simulations with about 10^6 particles evolving over 10^7 to 10^8 time steps, revealing a spatio-temporally disordered regime with slow, never-ending bending, splitting, and merging of bands on very long characteristic time and length scales (Chaté, 2020). Unfortunately, there are no phase diagrams drawn for an identical set of parameters that would allow one to compare results obtained by the different methods.

2.3 Vibrated Granular Layers

Experiments with rod-like particles, definitely dry ones, have been carried out in a vertically vibrated granular layer. This is a venerated device in which the first non-equilibrium patterns were observed by Faraday (1831). The activity of inanimate matter is facilitated in this setting by external forcing. There is no directed motion here, but when grains have an elongated shape with nematic symmetry, they spontaneously align while forming patterns in the plane of the vibrating layer.

However, unlike thin layers of passive nematic liquids, the alignment of elongated particles in a single dense layer is three- rather than two-dimensional. Blair et al (2003) observed, alongside a variety of patterns (Fig. 2.7a), spontaneously formed moving ordered domains of slightly inclined nearly vertical rods on the background of immobile horizontal rods. The ordered domains coarsened with time to form large vortices slowly rotating in the direction of the tilt (Fig. 2.7b). Since no motion was observed in a horizontally vibrated layer, the experimentalists tentatively attributed this remarkable breaking of chiral symmetry to the interactions of inclined rods with the bottom of the vertically vibrated container.

The model by Aranson and Tsimring (2003) fitting these observations included the momentum conservation equation with dissipation caused by friction rather than viscosity, and the particle density conservation equation with the velocity directed along the local tilt of the rods. If they had stopped at these two equations, it would have been a model similar to that of Toner and Tu. However, they were not satisfied with the simplistic assumption identifying the alignment with velocity, but closed the description by adding the equation for the evolution of the tilt. The form of this

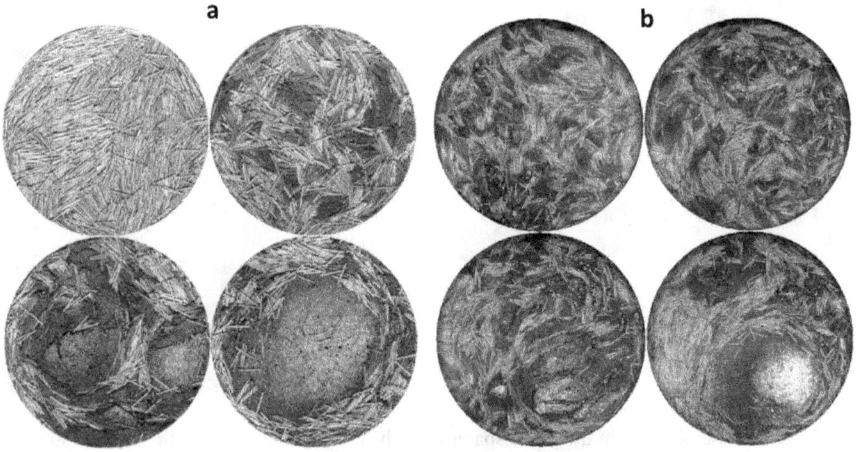

Fig. 2.7 (**a**) Examples of patterns in a vibrated layer of granular rods. (**b**) Snapshots of the development of a vortex from an initial random state, showing the formation and coarsening of domains of vertical rods (Blair et al, 2003)

equation reflected the unusual combination of 2D geometry with 3D orientations of the rods. It included the cubic term $-\mathbf{n}|\mathbf{n}|^2$ limiting the absolute value of the 2D projection $\mathbf{n}$ of the 3D director, with $|\mathbf{n}|$ equal to zero for vertical and unity for horizontal rods. The 2D orientation is now polar rather than nematic, and the influence of splay and bend distortions is accounted for by the Laplacian of $\mathbf{n}$ and the gradient of its divergence. The alignment is also coupled to the gradient of density in this model. This realistic model is too complicated for analysis, but computations faithfully reproduce the qualitative features of experimental observations.

The motion is inhibited when 3D alignment is suppressed by restricting the layer of rods by a lid that leaves the particles only a little vertical room to move (Narayan et al, 2006). The particles arranged themselves in a variety of almost static patterns: not only nematic but smectic or tetradic, depending on details of their shape; layers of cylindrical particles turned out to be more disordered than more easily rearranging layers of rice grains. The authors attributed a tendency to slow global rotation to some "stray chirality" rather than to the spontaneous breaking of chiral symmetry that naturally occurs in a layer of rods free to align in the vertical direction. Only small tapered rods, probably less restricted by the lid, showed a strongly distorted dynamic swirling pattern. Experiments by Aranson et al (2007) have also shown great sensitivity of swirling to the shape of the particles, but swirling motion of horizontally aligned particles was shown to be caused by an unintended and uncontrolled horizontal vibration component. Sensitivity to details is natural in real-life problems but unwelcome in studies of no practical significance.

Das et al (2017) compared the character of the order–disorder transition in active and passive nematics in the more artificial setting of a lattice model. Particles with 2D nematic alignment were allowed to move with equal probability to any of the four neighboring sites if "passive", but only to one of the two in the direction of their alignment if "active". This is similar to associating alignment with mobility,

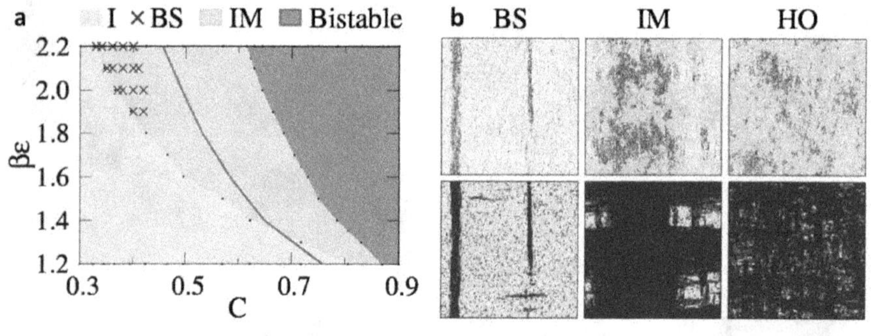

Fig. 2.8 (**a**) Phase diagram in the plane spanned by the concentration C and the relative strength of interactions $\beta\varepsilon$. The *red solid line* marks the locus of the isotropic/nematic phase transition for passive particles. The domains of isotropic (I), ordered bands (BS), inhomogeneous mixed (IM), and homogeneous globally ordered (HO) states of active particles are color-coded as indicated. (**b**) *Upper row*: Map of alignment in the BS, IM, and HO regimes, color-coded in fractions of π. *Lower row*: Concentration map in these regimes, color-coded as shown on the right (Das et al, 2017)

as in the above-mentioned computations by Peruani et al (2008). Both active and passive particles interacted with their nearest neighbors by minimizing the energy dependent on their alignment.

The resulting phase diagram in the plane spanned by the concentration C and the relative strength of interactions, measured by the product of the effective interaction energy ε and the inverse temperature β, is shown in Fig. 2.8a. The equilibrium system is isotropic at low density and high temperature and nematic at high density and low temperature; the red solid line marks the locus of the phase transition. Of course, this is unlike the isotropic/nematic phase transition in liquids, where the density does not change. In vibrated layers, on the other hand, as in the lattice model, increasing density enhances the alignment, but there is no direct analogue to temperature, since the vibration strength is the source, not only of disorder, but of activity itself.

The behavior of the active system is more complicated. As density grows and temperature decreases, it first transforms from the disordered isotropic (I) state to the locally ordered inhomogeneous mixed (IM) state. At low temperatures and densities, this transition takes place through the appearance of dense ordered bands (BS). At higher densities, the active nematic shows bistability between the mixed (IM) and the homogeneous globally ordered (HO) states. Typical patterns in these regimes are shown in Fig. 2.8b; evidently, the HO state is not really highly ordered.

2.4 Topological Defects

Order is hardly ever perfect even in passive nematic liquids, let alone active ones. The nematic texture may be influenced by boundary conditions that impose a certain alignment at the confining walls. In the two schemes in Fig. 2.9a, the orientation of a 2D nematic at the elliptical boundary is either normal or parallel. In both cases, the alignment direction makes a full revolution by 2π around the confining line. The only way to resolve it is through the formation of *defects*, which are said to be *topological* since they are determined exclusively by geometric properties preserved under continuous deformations, independently of any physical interactions. However, the number of defects does depend on their energy, which is scaled in the common one-constant approximation as the square of their *topological charge*, equal to the circulation along a surrounding contour. Therefore the charges should be minimal, and more highly charged defects are unstable to splitting[1].

For a nematic, the minimal charge is 1/2, which corresponds to circulation by π, and the circulation around the boundary in Fig. 2.9a is therefore compensated by the formation of two defects with charge $+1/2$. If the alignment was polar, a single defect of unit charge would form, as in the XY model (Fig. 1.8) where the 2D alignment is vectorial, or as in the experiments by Blair et al (Sect. 2.2) where the 2D alignment of rods became vectorial due to their freedom to rise vertically and therefore a single vortex was formed. In the region with two holes with normal alignment on

[1] Only in the limit of an infinite ratio of bend to splay energies does splitting a unit charge leave the energy invariant (Shin et al, 2008).

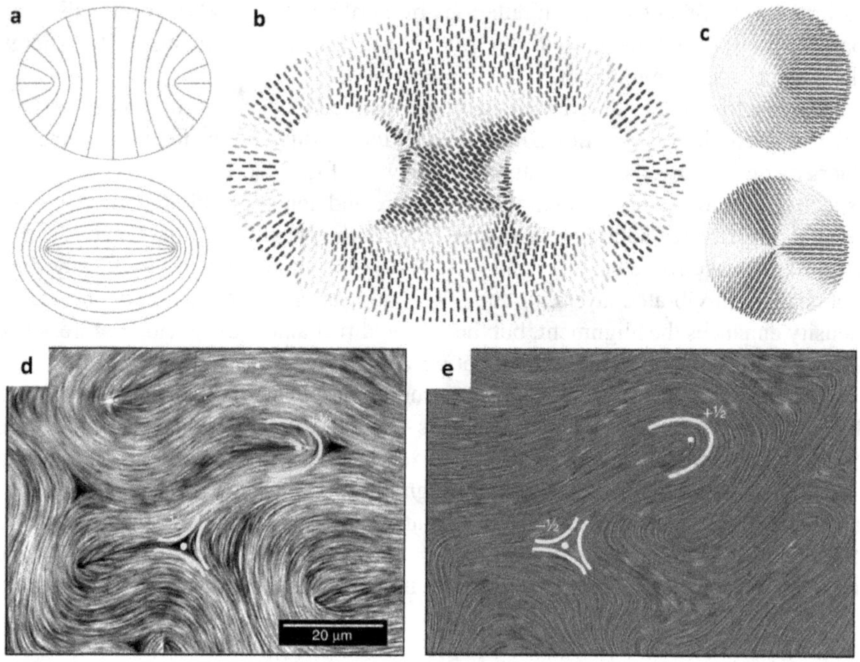

Fig. 2.9 (**a**) Two defects with charge $+1/2$ in an elliptic domain with normal (*top*) and parallel (*bottom*) nematic alignment on the boundary. (**b**) Two defects with charge $-1/2$ in a domain with two holes and normal nematic alignment on all boundaries. (**c**) Nematic texture near a positive (*top*) and negative (*bottom*) half-charged defects (Zakharov and Pismen, 2015). (**d**), (**e**) Half-charged defects of opposite signs in an experimental picture and a continuum simulation, respectively (Doostmohammadi et al, 2018)

all boundaries shown in Fig. 2.9b, the circulation around the holes, as viewed from the interior, is *negative*, so that the total circulation around all boundaries is -2π, compensated by the formation of two defects with charge $-1/2$.

The location of defects in passive nematics minimizes the overall energy. The configuration shown in Fig. 2.9a is solvable analytically, and the defects are located at the two foci of the ellipse. In Fig. 2.9b, the texture is computed numerically. The defects happen here to be located asymmetrically near the two holes, but the location changes with the size of the holes even if their positions remain fixed. In 3D, half-charged point defects turn into lines. In addition, there are point *hedgehog* defects, structured as this appellation suggests.

Half-charged defects are commonly observed in both experiments and simulations with both passive and active nematics. While positive and negative defects in polar media have the same symmetry, half-charged defects, accentuated in Fig. 2.9d and e, are not alike at all. The nematic alignment near the defect cores is seen under magnification in Fig. 2.9c. Positive defects are asymmetric, with a stripe of nearly parallel alignment, called a *comet tail*, on one side. Contrariwise, negative defects

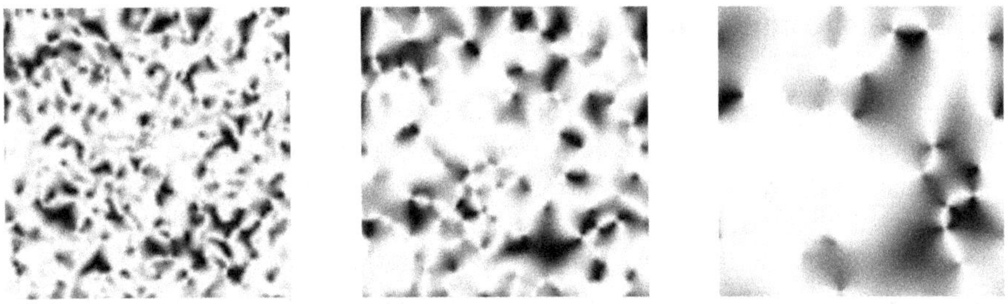

Fig. 2.10 A coarsening sequence (*left to right*) of the texture in a nematic layer following a rapid quench. The pictures show *schlieren* patterns observed in polarized light passing through a cross-polarizer. The transmitted intensity is maximal when the director is oriented at 45° to the polarizers and minimal when it is aligned with either (Fukuda, 1998)

have a threefold symmetry. This bears on their dynamics, as we shall see in the following.

In common nematic fluids, many defects appear when they are rapidly quenched into the nematic state, so that incompatible local alignments emerge in spatially separated domains. Subsequently, the texture coarsens, as defects of the opposite sign attract each other and annihilate to reduce the overall energy. A typical coarsening sequence is shown in Fig. 2.10. In active nematics, the situation is different, as the energy is not minimized, and defect pairs may emerge spontaneously. Defects do not appear in models of the Vicsek type with nematic as well as polar symmetry, since they are destroyed by the imposed motion, but they are prominent in simulations in some way imitating the dynamics of granular layers, where sterically constrained rod-like particles are intermittently driven in opposite directions along their long axes (Shi and Ma, 2013). Spontaneous emergence of a defect pair and the motion of interacting defects in this model are illustrated in Fig. 2.11. We shall see presently that they play a crucial role in the dynamics of active nematic fluids.

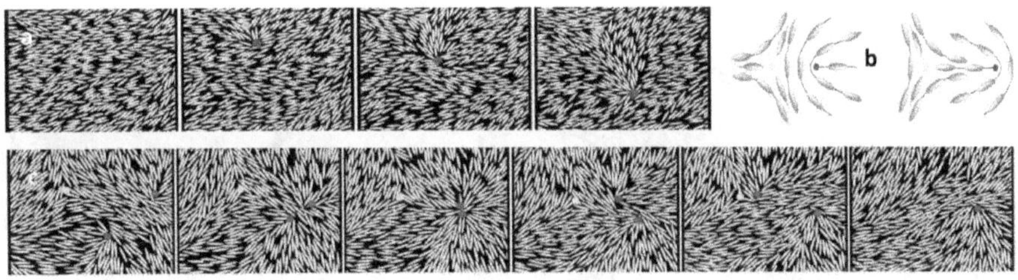

Fig. 2.11 (**a**) Emergence and separation of a defect pair. (**b**) Configurations of $\pm 1/2$ defects before annihilation (*left*) and following the creation of a pair (*right*). (**c**) Motion of interacting defects terminating in annihilation of a pair (Shi and Ma, 2013)

2.5 Extensile and Contractile Activity

Activity may be expressed not in autonomous motion but in forces applied by "agents" on the medium they are immersed in. This is characteristic of "wet" active matter dominated by interactions carried by a surrounding medium. Most complex active assemblies, from colloidal suspensions and emulsions to living cells, colonies, and tissues, belong to this class. On a basic level of description, both the medium and the "agents" can be lumped into a continuous active fluid, where the "agents" are characterized only by their symmetry and the matching action. A particle with nematic symmetry can exert a force along its axis, which can be either *tensile* or *contractile*.

Converting this action into a continuous description does not require a sophisticated (and not quite reliable) kinetic theory. The flow of active nematics, whatever the origin of activity (which may come from active colloidal particles, bacterial suspensions, filaments driven by molecular motors, or living cells), can be modeled by viewing them as homogeneous fluids and solving hydrodynamic equations of motion amended by adding the force exerted by microscopic particles uniformly distributed in the fluid. In such a "wet" medium, active agents stir the fluid they are immersed in and are advected by the collectively generated flow. The continuous approach is justified by the assumption that, similar to passive liquid crystals, the flow and alignment patterns develop on a scale far exceeding the size of individual anisotropic active entities, even though the latter now far exceed the molecular scale. Commonly, the motion is slow, so inertia is neglected, and the model is based on the Stokes equation of viscous motion supplemented by the effects of nematic elasticity and activity.

In addition to the pressure and mechanical stress tensor σ_m that defines the flow pattern of isotropic fluids, nematic fluids are also subject to the elastic stress σ_n

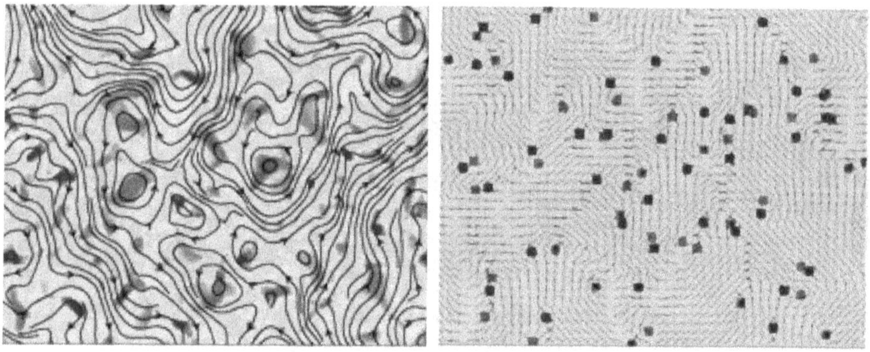

Fig. 2.12 Snapshot of the flow (*left*) and alignment (*right*) patterns in a nematic fluid with high tensile activity. In the left panel, *lines with arrows* indicate streamlines; the coloring is graded from red to blue, which correspond to high positive and negative vorticies, respectively, thereby revealing the direction of rotation. *Small bars* in the right panel show the director field; positive half-charged defects are marked by *red*, and negative, by *blue dots* (Thumpi et al, 2014)

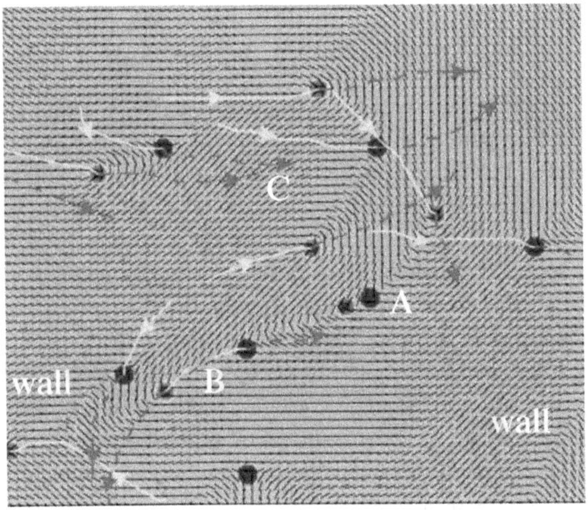

Fig. 2.13 Snapshots of defect dynamics in a tensile system. Positive half-charged defects are marked by *red*, and negative, by *blue dots*. The past and future trajectories of the defects are shown by *continuous yellow* and *dashed magenta lines*, respectively. The labels mark: A – a newly created defect pair; B – a pair of defects moving from their creation site along a wall; C – a future annihilation site (Thumpi et al, 2014)

caused by nematic misalignment. The mechanical stress driving passive flows is of external origin; internally, it includes only viscous dissipation. In autonomously motile active fluids, only the dissipative viscous stress is left in σ_{m}, while the cause of motion is the *active stress* $\sigma_{\mathrm{a}} = \zeta \mathbf{Q}$, proportional to the nematic state tensor $\mathbf{Q}$ defined by (2.2). The coefficient ζ is positive when activity is *tensile* and negative when it is *contractile*. The motion induced by activity is counterbalanced by the elastic stress and dissipated by the bulk and wall friction.

Dynamic equations of the nematic field are obtained by varying the Landau–de Gennes energy functional that includes distortion energy, with added advection. Of course, there are no data on anisotropic viscous and elastic coefficients in active nematics (they are lacking even in passive nematics), and the common reasonable choice, which does not infringe on a qualitative picture, is a basic hydrodynamic model with isotropic viscosity and the nematic energy defined in a "one-constant" approximation, wherein splay and bend elasticities are assumed equal. The most important thing is to base the simulation on the full representation of the nematic tensor $\mathbf{Q}$, which allows for an imperfect local alignment near defects.

A snapshot of the simulation of the alignment and flow in a 2D nematic fluid with high tensile activity (Thumpi et al, 2014) is shown in Fig. 2.12. High activity brings about a highly distorted turbulent alignment and flow patterns with the characteristic scale of inhomogeneities determined by the nonlinear dynamics of the director field. The picture for a contractile activity is qualitatively similar, but it turns out to contain a lower density of defects for equal magnitude of the activity. In the blowup picture of this simulation shown in Fig. 2.13, trajectories of defects are traced by yellow arrows from their creation sites, where defect pairs are born like electron–positron pairs from the quantum field vacuum. Some future trajectories traced by magenta arrows terminate in annihilation of defect pairs. The picture also indicates "walls", which are elongated distortions in the director field.

2.6 Dynamics of Defects

The motion of defects determines to a large extent the flow pattern in active nematics. One can see by comparing the two panels of Fig. 2.12 that the mean separation between defects in the turbulent regime coincides with the typical vortex size. Defects themselves permanently shift their positions due to induced flow, as well as by inhomogeneities of nematic alignment that also depend on their locations. The gradient of active stress, shown in Fig. 2.14, is at its maximum in the vicinity of defects, which suggests that they play a major role in driving the flow.

Asymmetric positive defects are able to propel themselves, as if pushed by the "comet tail", when the activity is tensile ($\zeta > 0$), and in the opposite direction when the activity is contractile ($\zeta < 0$). Symmetric negative defects do not self-propel, but both kinds of defects are driven by active flow induced by other defects as well as by the nematic alignment field transmitting topological attraction of oppositely charged defects and repulsion of defects of the same sign.

Describing the dynamics of active nematic fluids in terms of the motion of defects would bring an intellectual advantage, beyond saving computation effort. A rough analogy would be describing the electromagnetic field through the motion of charged particles. Simulations, such as those in the preceding section, are carried out on a lattice. The lattice should be tight enough to resolve the fine structure of defect cores where not only the alignment but also the deviation from isotropy changes, vanishing at the center of a defect – but keeping the lattice dense elsewhere would be wasteful. Analytical methods of defect dynamics, on the contrary, benefit from this contrast, since it allows for separate treatment of defect cores and the far field.

In the simplest version (Giomi et al, 2013), interaction of defects is literally identified with the Coulomb interaction of electric charges. In 2D, it generates an attractive or repulsive force decaying as the reciprocal of the separation distance. This is not precise: interactions turn out to be dependent also on the velocity of defects, which causes a logarithmic correction to the defect velocity. This has been proven in the theory of interacting vortices (defects of integer charge) through a multiscale expansion matching the alignment distribution and strength in the vicinity of defects and in the far field (Pismen, 1999). The extension to nematic textures with half-

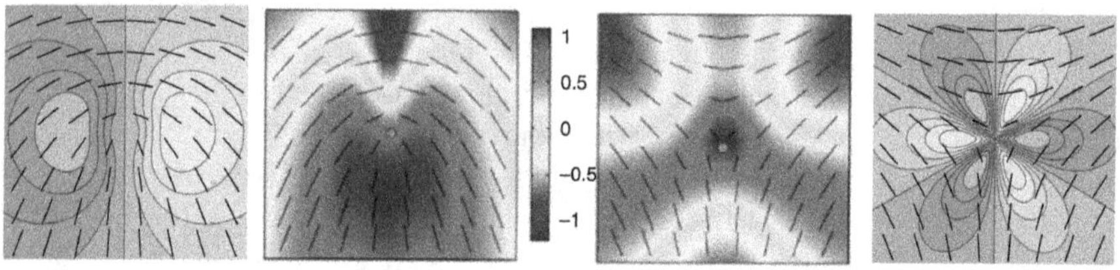

Fig. 2.14 Isotropic stress around $+1/2$ (*left*) and $-1/2$ (*right*) defects. Color maps at the center show the magnitude of the isotropic stress with *blue* and *red* corresponding to mechanical compression and tension, respectively (Doostmohammadi et al, 2018). The corresponding alignment and flow patterns are shown on both sides

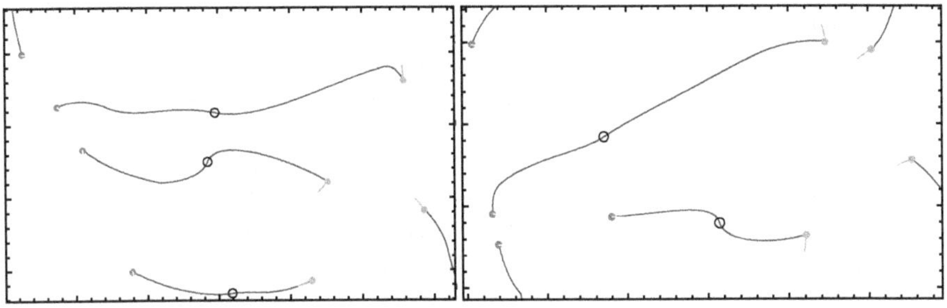

Fig. 2.15 Two examples of trajectories of defects, starting from four positive (*green dots*) and four negative (*red dots*) defects with initially randomly scattered positions in the two halves of the field. The directions of the "comet tails" of +1/2 defects are shown near their initial positions. The annihilation sites are marked by *circles* (Pismen, 2013)

charged defects follows from a complex representation of the nematic tensor (2.2), which formally maps the nematic case onto the polar one (Pismen, 2013).

Topological interactions and self-propulsion of +1/2 defects are the decisive factors in the motion of defects when they are well separated, since active flow generated by defects decays at a faster rate with distance. In the two examples shown in Fig. 2.15, positive and negative defects, initially randomly scattered in the two halves of the field, are either attracted and annihilate or escape out of the computation domain. The result depends on the orientation of the "comet tails", which was held constant, randomly assigned at the outset, in the computation generating Fig. 2.15. The justification was that in the formal expansion in a small parameter (a common feature of analytical methods) the rotation rate is of a higher order and should be neglected; however, the rotation of "comet tails", caused by interactions with all defects on a long journey, as well as by random noise, should be an important factor, as it affects the self-propagation direction. Changing directions due to active and random torques was accounted for in the statistical dynamics of defects at low activity by Shankar et al (2018).

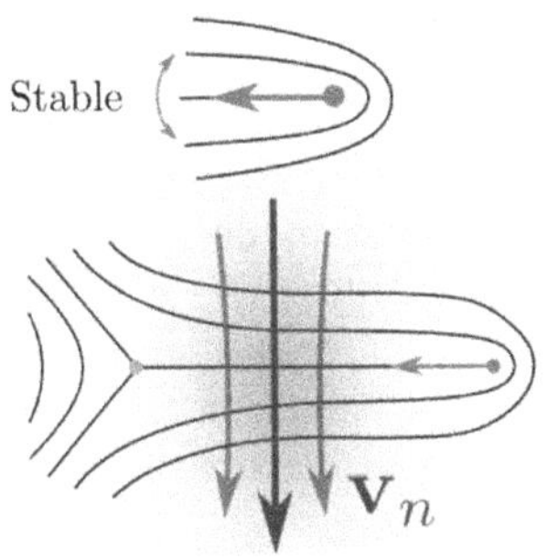

Fig. 2.16 *Lower*: Configuration of a defect pair after its emergence. The *red arrow* shows the direction of the"comet tail", and *blue arrows*, the direction of the phase gradient. *Upper*: Stable location and direction of another +1/2 defect (Shankar and Marchetti, 2019)

Analytical methods, relying on the presence of small parameters that imply large separation and slow nearly steady motion, clarify major features of dynamics, but are limited in many respects. Most importantly, they are not applicable to the most singular events: emergence and annihilation of defect pairs. The latter is not missed: it proceeds fast whenever oppositely charged defects come into a close vicinity. But spontaneous emergence is the crucial moment. In the case of tensile activity, the defect pair should be configured after its emergence, as shown in Fig. 2.16. At a certain separation, decreasing with strengthening activity, pushing by the "comet tail", remaining constant, overcomes the topological attraction, decreasing with distance, and the defect pair is separated. The direction of the "comet tail" should be reversed when activity is contractile.

Simha and Ramaswamy (2002) argued that an ordered state of an active nematic is intrinsically unstable, because a fluctuation strong enough to bring about the critical separation may always occur with a non-zero probability. Strong fluctuations readily happen as extant defects run about in turbulent patterns, such as that shown in Fig. 2.12, but are far less likely in a perfectly ordered state. Shankar et al (2018) argued that fluctuations in the directions of $+1/2$ defects make their motion less persistent, allowing the defect pair to remain bound, so that, counterintuitively, noise may stabilize the ordered nematic phase.

On the other hand, the same authors noticed that an additional $+1/2$ defect may be stationary with respect to a bound pair when positioned as shown in Fig. 2.16. Extending this arrangement in all directions over the plane would lead to a *polar* ordered phase with $+1/2$ defects equally spaced and oriented in parallel. These propositions, ingenuous as they are, are likely to remain virtual, since even attaining a perfectly ordered state of an active nematic, let alone bringing it to the hypothetic ordered polar state, is a difficult task.

Defects moving persistently without annihilating can be realized on a spherical surface where topology requires that defects with total charge two should be present. Commonly, the energy of a passive nematic is minimized by four $+1/2$ defects placed at the vertices of a tetrahedron. However, the flow generated in an active nematic advects defects. At moderate activity, when no extra defects are generated, they were shown to oscillate between the tetrahedral arrangement and a planar configuration on a great circle, as shown in Fig. 2.17.

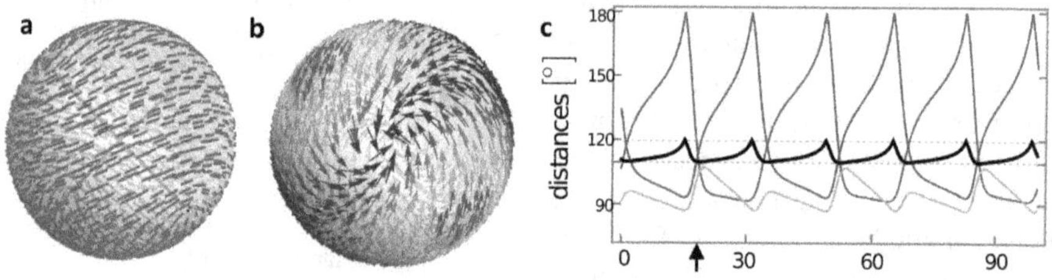

Fig. 2.17 Defects in an active nematic on a sphere. Director (**a**) and flow (**b**) in a tetrahedral configuration. (**c**) Time dependence of pairwise (*colored*) and mean (*black*) angular distances between defects; tetrahedral and planar configurations correspond to 109.5° and 120°, respectively. The *arrow* marks the time when the snapshots in (**a**) and (**b**) were taken (Khoromskaia and Alexander, 2017)

2.7 Active Fluids with Different Symmetries

Although only nematics are featured in the title of this chapter, it is proper to mention here active ordered fluids with different symmetries treated in a similar way. Tjhung et al (2011) and Bonelli et al (2016) combined tensile or contractile activity with self-advection, thereby converting a nematic medium to a polar one. There is a certain freedom in relating vectorial self-advection to the direction of active extending or contracting forces. It has been chosen differently in the cited works, and different patterns can be obtained thereby; patterns also depend on boundary conditions.

When Bonelli et al simulated the system in a doubly periodic 2D box, a marked contrast was observed between tensile or contractile activity (Fig. 2.18). In the case of tensile activity, the flow field **v** exhibits a banded structure, with two bands oriented in opposite directions perpendicular to the averaged polarization **P** (recall the pedestrians filing along a corridor in Fig. 1.12a), but when activity is switched to contractile, the velocity and polarization patterns are distorted and misaligned. Different patterns were obtained when the system was confined between parallel walls. Given the freedom to modify the model, there is much leeway in imitating a variety of patterns seen in real life where colloidal particles or cells may both propel and stress the surrounding fluid.

More exotic patterns were observed in simulations allowing for phase separation between an active polar and a passive isotropic fluid (Bonelli et al, 2019). The patterns depend not only on activity but on anchoring on the interphase boundary. The left panels of Fig. 2.19 show a weakly disordered lamellar pattern at low contractile

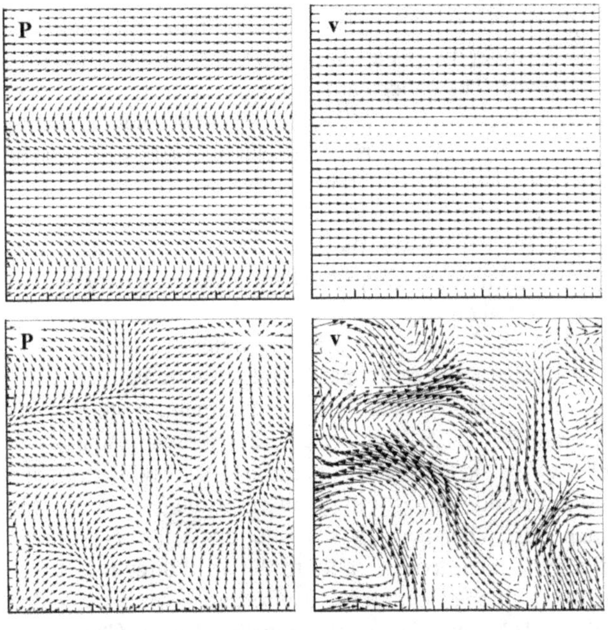

Fig. 2.18 Patterns of polarization **P** and flow velocity **v** in the polar model with tensile (*upper*) and contractile (*lower*) activity (Bonelli et al, 2016)

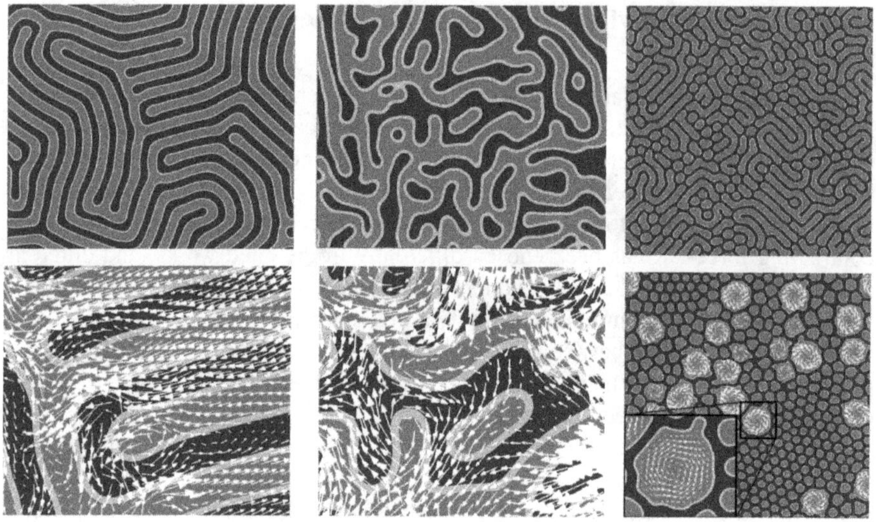

Fig. 2.19 (**a**) Patterns in a 1:1 mixture of phase-separated active polar and passive isotropic fluids. *Left*: Low contractile activity with normal anchoring at the active/passive interface. *Center*: High contractile activity with no anchoring. *Lower panels* show blowups with *white arrows* indicating the direction of the flow field. *Right*: Tensile activity, increasing from above to below, with normal anchoring. The *inset* in the lower picture shows a blowup of a selected droplet, and *spirals* show the polarity pattern in rotating droplets. The active phase is colored *red*, the passive one, *blue*, and a transitional border region, *green* (Bonelli et al, 2019)

activity and normal anchoring at the active/passive interface, with the alignment vector pointing toward the passive phase. Quite surprisingly, the pattern becomes ordered at higher activity, but the order disappears altogether, as shown in the central panels, when anchoring is removed. The pattern at low tensile activity in the upper right panel mixes lamellae with droplets of the active phase, but droplets definitely prevail in the lower right panel as activity increases. Polar droplets with normal anchoring must contain a vortex defect. Its structure is aster-like in small droplets but spiral in big ones, which, accordingly, rotate with an angular velocity increasing with their radius. Given cheap computer power, there is a lot of freedom to modify patterns, changing activity, alignment strength, and the ratio of active and passive phases, multiplying the production of colorful pictures.

Unlike a chimeric combination of nematic stress with polar symmetry, *chiral* active fluids are straightforward analogues of the passive cholesteric liquid crystals briefly mentioned in Sect. 2.1. A picture of cholesteric alignment, more detailed than that in Fig. 2.1, is shown in Fig. 2.20a. A generic hydrodynamic instability of the cholesteric order is *pitch splay*, sketched in (Fig. 2.20b). A splay deformation of the pitch is equivalent to layer undulations with bending of the director field in normal planes along the pitch axis. When activity is tensile, this deformation mode gives rise to active flow along the pitch axis, parallel to the bend direction, which destabilizes the cholesteric order. Pitch orientations have nematic symmetry, and at high distortions develop pairs of 1/2-charged line defects, called λ-lines. Cholesteric

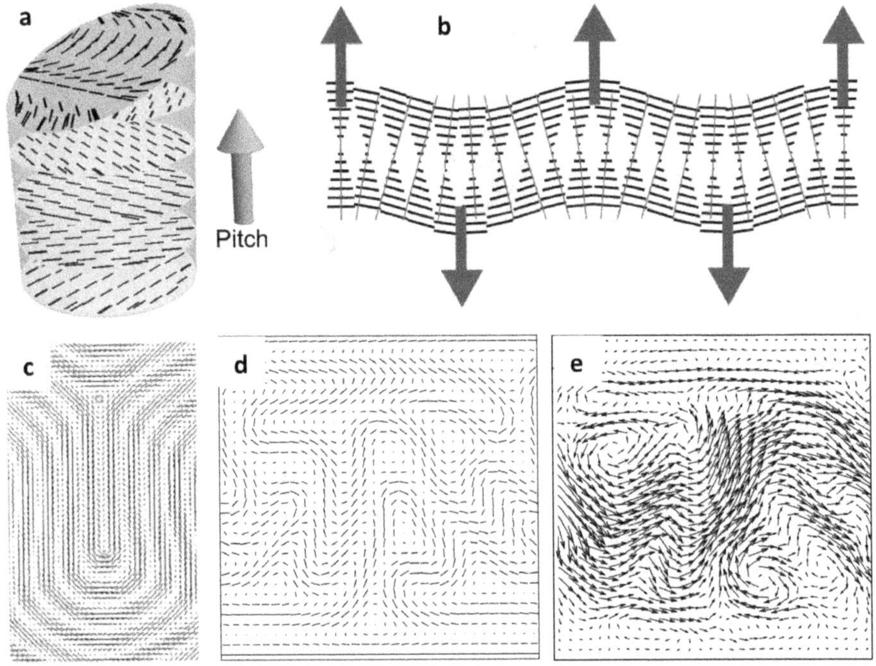

Fig. 2.20 (**a**) Projections of a cholesteric director field onto successive slices normal to the pitch direction. The top inclined slice shows a texture of periodic arcs due to the right-handed twist along the pitch. (**b**) Sketch of the pitch-splay instability. *Black lines* show the projection of the twisted director field onto the plane containing the splayed pitch axis, shown by *red lines*. The *blue arrows* show the active flow direction, which acts to increase the distortion and drives the instability when activity is tensile. (**c**) Cross-section of a pair of line defects of pitch orientation. (**d**), (**e**) Typical snapshots of simulated director alignment and flow fields (Whitfield et al, 2017)

order is essentially three-dimensional, but a cross-section shown in Fig. 2.20c looks the same as a cross-section of a nematic line defect or a point defect in a 2D nematic.

Whitfield et al (2017) analyzed and simulated the dynamics of a cholesteric active fluid. In the case of tensile activity, the induced flow and the related distortions of the ordered state increase gradually as activity intensifies, but when activity is contractile, instability is suppressed until activity exceeds a certain threshold. Typical snapshots of simulated director alignment and flow fields are shown in Fig. 2.20d and e. They are taken in a plane normal to confining flat walls with a homogeneous anchoring of the director at rather high tensile activity, when the pattern is time-dependent. Admittedly, such pictures are not quite specific, as convective vortices in simulations of active fluids with different topology are rather similar.

Like nematic order, cholesteric order can be combined with propulsion (Fürthauer et al, 2012). This description applies to fluids stirred by chiral motors, natural or artificial, such as those sketched in Fig. 2.21. Publications devoted to the dynamics of active fluids typically start by mentioning colloidal and bacterial suspensions, protein

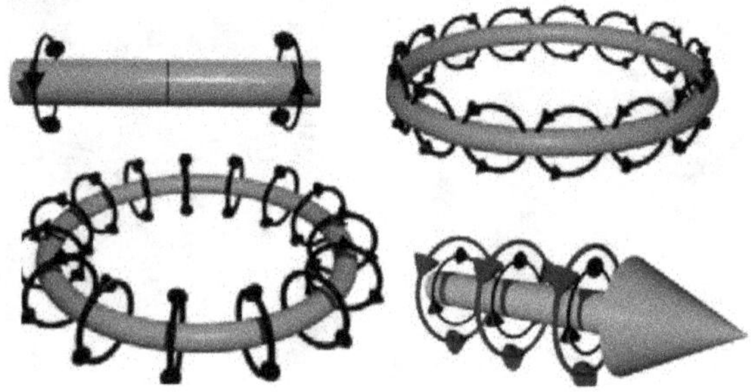

Fig. 2.21 Schematic representation of elementary chiral motors (Fürthauer et al, 2012)

filaments driven by molecular motors, etc., as real systems fitting this description. We will return to a more detailed description of these "wet" active media, approaching their physical and biological realities, in the following chapters.

Chapter 3
Active Colloids

3.1 Irreversible Movements in Reversible Flow

Active fluids are stirred by microscopic agents. We considered above their bulk effect smeared over their multitude together with the surrounding medium, but they also have an individuality that determines how they interact with the fluid they are immersed in. Live organisms commonly move by deforming their body, something inanimate particles cannot do. However, it is possible to swim even without external appendages, and even with no deformation, just by inducing tangential flow along the particle's surface. This kind of a swimmer is called a *squirmer*, further classified as a *pusher* if the surface displacement in the direction of motion is stronger behind, or a *puller* if it is strongest on the forward side. Rather than propelling the swimmer directly, this generates fluid flow, and its back reaction moves the swimmer ahead. A typical flow field around a pusher is shown in Fig. 3.1; in the case of a puller, the picture is reversed with respect to the direction of motion.

Inertia plays no role in the motion of microscopic particles or creatures. Its effect is measured by the Reynolds number: linear dimension times velocity divided by kinematic viscosity. A particle of a few microns in size would have to propel itself as fast as a torpedo to bring the Reynolds number close to unity. The flow particles

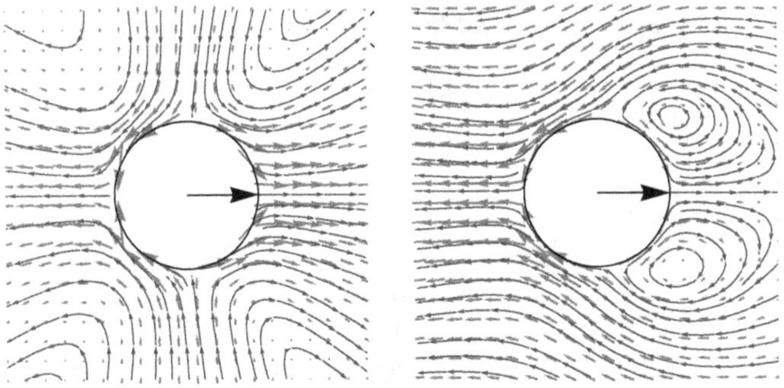

Fig. 3.1 The axisymmetric velocity field (*red arrows*) and stream lines (*blue*) around a pusher in the laboratory frame of reference (*left*) and in the frame moving with the swimmer (*right*), Zöttl and Stark (2018)

L. Pismen, *Active Matter Within and Around Us*, The Frontiers Collection,
https://doi.org/10.1007/978-3-030-68421-1_3

Fig. 3.2 *Left*: Demonstration of the reversibility of Stokes flow (by Ved1123, CC). *Right*: A reversible change of configuration (Qiu et al, 2014)

induce in the surrounding medium is slow as well, as has been taken into account in Sect. 2.5. Viscous (Stokes) flow at low Reynolds number is more amenable to mathematical analysis than bird flight or the swimming of fish. The classical book on swimming organisms by James Lighthill (1975) has not aged, but quite a lot has been added since then, as reflected in a number of reviews on the motion of colloidal particles and bacteria (Lauga and Powers, 2009; Romanczuk et al, 2012; Yeomans et al, 2014; Elgeti et al, 2015; Lauga, 2016; Zöttl and Stark, 2016).

Stokes flow is dissipative but, paradoxically, it is reversible in some respects. A narrow colored strip in a narrow gap between two cylinders spreads out into a formless cloud when the inner cylinder slowly rotates, but gathers back almost precisely (just a bit blurred by diffusion) when the direction of rotation is reversed (Fig. 3.2, three panels on the left). Therefore a swimmer changing its configuration in a reversible way, as a clam would do (Fig. 3.2, right), cannot advance, by the *scallop theorem* due to Purcell (1977). This injunction is lifted when swimming in a viscoelastic liquid (Qiu et al, 2014), or near a deformable surface, or while interacting with other reversible swimmers.

The Stokes equation does not contain the time derivative. This means that the response of the fluid to the motion of an immersed body is instantaneous in a Stokesian world. The equation is linear, and therefore effects of all infinitesimal motions are superimposed: any deformation of the swimmer's body or a tangential shift of the swimmer's surface generates a *stokeslet*, a singular solution diverging at its source and vanishing at infinity. In principle, the entire flow field can be obtained by integrating these stokeslets and their derivatives (force dipoles, quadrupoles, etc.). Otherwise, the flow field around a spherical particle can be computed as a superposition of spherical harmonics, and the flow field generated by motion of several particles, by superposition of their individual contributions. This is still difficult in practice, since applicable boundary conditions should be satisfied at all surfaces. Even if there is a lone swimmer in an infinite expanse of fluid, satisfying boundary conditions on the swimmer's surface, say, no normal velocity and no slip – is a nontrivial task, except for some symmetric shapes and motions, and approximations are necessarily involved.

3.2 Autophoretic Particles

Inanimate active particles, such as Quincke rollers (Sect. 1.7) or grains in a vibrated layer (Sect. 2.3), use external energy sources to power their activity. Such sources can also be found in the surrounding medium when motion along the gradient of some field reduces the overall energy. Those are "phoretic" processes, with the name of a field (thermo-, diffusio-, electro-) attached to this suffix (Anderson, 1989). Colloidal particles moving in this way are not qualified as active, but they are when they create such gradients themselves via surface reactions (Moran and Posner, 2017). Of course, such *autophoretic* particles or droplets are driven in the end either by stored chemical energy or by an external energy source, but, after all, living organisms also depend on nutrients or solar light, although they utilize these resources in a far more sophisticated way.

Autophoretic particles driven by chemical or electrochemical reaction have to be asymmetric, designed in a way that breaks fore–aft symmetry. Any asymmetry, in either surface energy, or activity, or geometric shape, is suitable, but the most straightforward case we will consider is a *Janus* particle, named after the two-faced Roman god. Such a particle may be coated by a catalyst in an asymmetric way or have asymmetric photoactive or heat-adsorbing properties; then ensuing concentration, temperature, or electric potential gradients near the surface cause a local osmotic flow that propels the particle in the opposite direction.

Osmotic flow is driven by gradients of the surface energy due to changes in the state of both the surface itself and the adjacent fluid. The driving force is applied not to the particle itself but to a thin fluid layer within the range of molecular interactions that determine the surface energy. If it was applied on the surface itself, no motion would arise due to the no-slip condition. However, molecular interactions are not concentrated at the surface but extend into the fluid for a nanoscale distance,

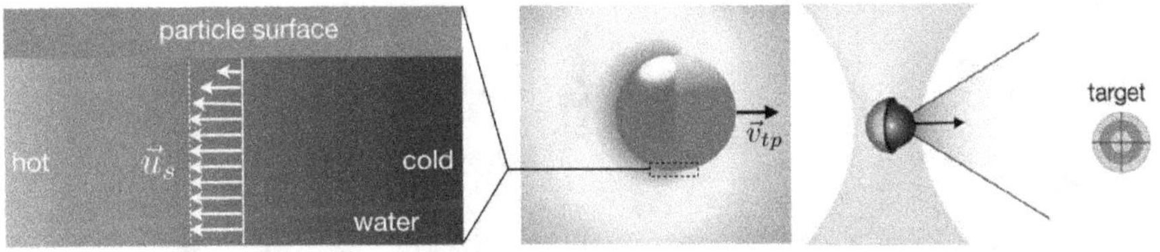

Fig. 3.3 A Janus particle with a heated cap on the left moving towards the colder area by self-thermophoresis. The thermoosmotic flow generating the effective slip is shown on the *left*, and the temperature field color-coded from *blue* (cold) to *red* (hot), in the *center* panel. The *right* panel demonstrates the strategy of driving the particle to a target by switching on the heating only when the particle is suitably oriented (Kroy et al, 2016)

much smaller than the size of a microscopic particle. The velocity of osmotic flow vanishes at the solid boundary and saturates to some effective *slip velocity* at a distance measured by the range of molecular interaction forces, while a tangential gradient driving the osmotic flow does not change over this tiny distance.

Surface slip generates a relative motion between the particle and fluid. As the molecular-scale distance is negligible, the induced flow pattern is the same as if the surface itself was slipping in the direction opposite to the particle's propulsion, as a squirmer's surface does. Ideally, it generates a force dipole, but the precise flow field depends on the tangential distribution of thermodynamic variables affecting the surface energy, and hence, slip velocity. The migration velocity of the particle is equal in magnitude to the area average of the slip velocity over the particle surface.

Self-thermophoresis of a Janus particle can be realized in the simplest way by an external source such as a defocused laser beam differentially heating the metal-covered part of the surface. Thermoosmotic flow can be directed either way and depends, besides the surface energy, on the sign of the thermodiffusion (Soret) coefficient. In the example illustrated in Fig. 3.3 (Kroy et al, 2016), the slip velocity is directed toward the heated part, inducing self-thermophoretic motion in the opposite direction, as shown in the central panel. In this way, the particle can even be driven to a target by switching on the heating source only when the unheated part of the surface is facing the right way.

Self-diffusiophoresis is encountered more often, and can be driven in a more natural way, by a catalytic reaction taking place on the active part of a Janus particle that changes the local surface energy. The induced velocity is proportional to the concentration gradient tangential to the surface or, more precisely, the gradient of chemical potential. In its turn, the concentration distribution around a surface with non-uniform reaction rates is determined by mass transport equations, which generally include advective and diffusional terms. Here another dimensional number comes into play – the Péclet number: linear dimension times velocity divided by diffusivity. In liquids, diffusivity is several orders of magnitude smaller than the

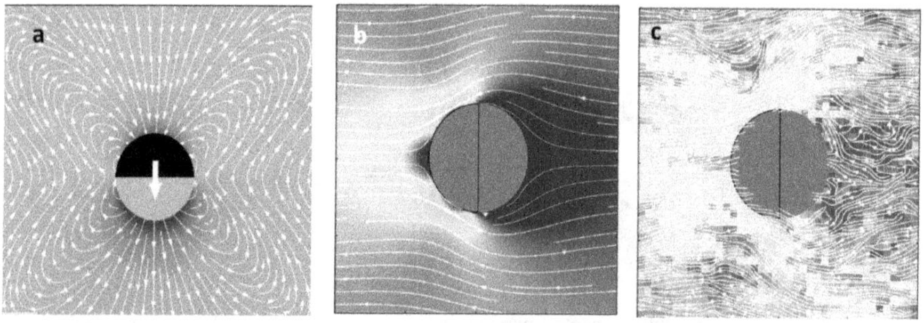

Fig. 3.4 (**a**) A Janus particle with a catalytic cap moving by self-diffusiophoresis. The concentration field of the reaction product is color-coded, increasing from blue to red (Popescu et al, 2018). Simulated (**b**) and observed (**c**) velocity field in the frame comoving with a Janus particle; the magnitude of the velocity is color-coded, increasing from blue to red (Campbell et al, 2019). *White lines* in all panels show flow streamlines

kinematic viscosity entering the Reynolds number. Still, a microscopic particle has to propel itself at several hundred body lengths per second to attain a Péclet number of order unity. A small Péclet number means that advection plays a negligible role in mass transport, compared to diffusion, and therefore the concentration field is determined by solving the linear diffusion equation.

In this way, flow and mass transport are uncoupled, which makes computations far easier. An example of the concentration field computed under this assumption is shown in Fig. 3.4a. Once the concentration distribution is known, this defines the local slip around the particle surface proportional to the concentration gradient, and the flow field can be computed by expanding in spherical harmonics. The flow field, computed in this way (Campbell et al, 2019) and shown in Fig. 3.4b, has been compared with experimental measurements and found to yield a qualitatively similar but noisy picture (Fig. 3.4c). Microscopic colloidal particles experience Brownian motion due to thermal fluctuations in the fluid, which causes their random reorientation. They are therefore called *active Brownian particles* – a term introduced by Schimansky-Geier et al (1995).

Besides the asymmetric chemical activity of Janus particles, asymmetry leading to autophoresis can be *geometrical*. Michelin and Lauga (2015) achieved this by combining two particles with equal and uniform activity but different size. This creates a concentration gradient, such as seen in the left panel of Fig. 3.5, that may drive self-diffusiophoresis. Of course, each particle would move as as result in its own way, and they have to be kept at a constant distance by an inflexible link, but asymmetry required for self-propulsion can be attained by asymmetric clustering of a larger number of symmetric particles (Varma et al, 2018).

The dependence of the velocity of the tied pair on their size ratio and the gap d_c between them is shown in the right panel of Fig. 3.5. There should be no fluid flow and no motion when the particles are of the same size; on the other hand, if the size of one of the particles goes down to zero, it has no effect, and there will be no motion

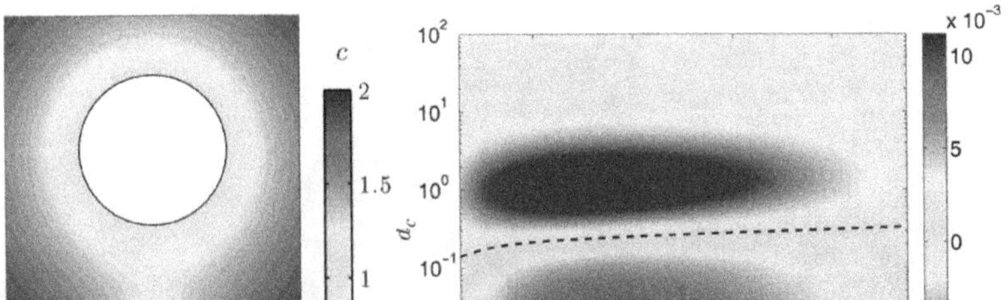

Fig. 3.5 *Left*: Example of the concentration distribution around a well separated active pair. *Right*: Dependence of the self-propulsion velocity on the size ratio R_2/R_1 and the gap d_c between the two particles. The *dashed line* corresponds to no propulsion (Michelin and Lauga, 2015)

either. The nontrivial result is that the direction of motion reverses as the separation increases from direct contact $d_c = 0$ to a distance of the same order of magnitude as the radius of the larger particle.

All the above kinds of behavior can also be induced by an electrochemical reaction on the active surface of a Janus particle that causes a change of ion concentration in the solution and, if the particle is conductive, electron transfer between the active and passive parts. A surface charge induces an electrical double layer with a dense layer of counter-ions and a diffuse layer of ions of the same charge. The surface energy depends on the thickness and composition of the double layer. Again, its gradient induces electroosmotic flow saturating to a slip velocity at a small distance of the same order of magnitude as the thickness of the double layer, and leading to self-electrophoretic propulsion toward the region where the energy is lower. These effects are often hard to separate from self-diffusiophoresis due to composition changes caused by the same reaction (Kuron et al, 2018).

3.3 Surface Effects

The direction of autophoretic motion is influenced by the immediate environment of colloidal particles, which allows one to steer them by applying external gradients or bringing other particles or droplets into their vicinity. The particles interact both chemically and hydrodynamically, which influences their motion and mutual orientation. The most immediate effect is geometrical. Bounding walls affect the motion of Janus particles through both the induced flow pattern and the concentration distribution, caused, respectively, by the no-slip and no-flux boundary conditions at the wall. The change in the solute distribution leads to changes in the phoretic slip at the particle surface, which translate into changes in the flow induced by the particle. On the other hand, flow induced by the particle is reflected by the wall, creating a

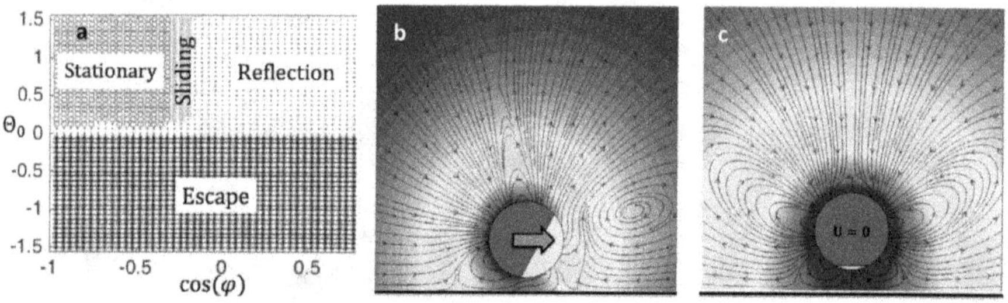

Fig. 3.6 (**a**) Phase diagram for a Janus particle near a wall. Θ_0 is the initial orientation; $\Theta_0 = \pi$ when the catalyst covers the "upper" side up to the latitude φ and the motion is directed towards the wall; full coverage corresponds to $\varphi = \pi$. (**b**) Sliding motion. (**c**) Stationary hovering state. *Black lines* indicate streamlines; the concentration field of the reaction product is color-coded, increasing from blue to red (Ibrahim and Liverpool, 2016)

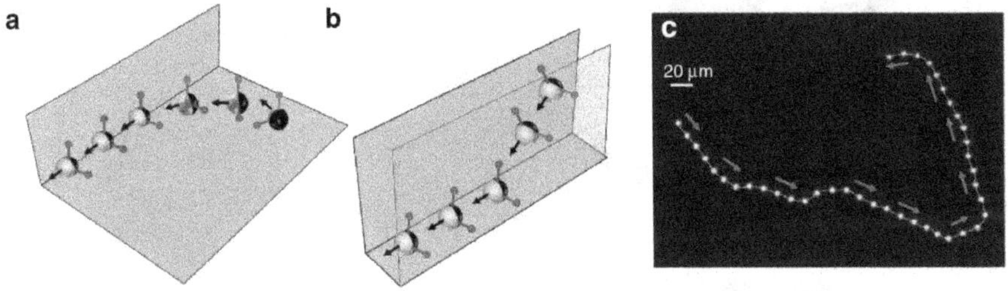

Fig. 3.7 Janus particle sliding toward a vertical wall (**a**) and within a rectangular groove (**b**). The allowed rotation axis is shown in *green* and the blocked one, by *red*. (**c**) Wandering path along a horizontal plane (Das et al, 2015)

feedback loop between the chemical and hydrodynamic sensing and response (Uspal et al, 2015).

This results in two specific modes of behavior near a wall: a "sliding" state, in which a particle translates along the wall at a constant elevation and orientation (Fig. 3.6b) and a stationary "hovering" state when a particle remains motionless with its axis of symmetry oriented perpendicular to the wall, while the induced flow persists (Fig. 3.6c). Ibrahim and Liverpool (2016) mapped the distinct modes of behavior, dependent on the orientation and extent of the catalyst-covered part (Fig. 3.6a). The particle with the surface facing the wall covered by the catalyst and moving away from the wall never comes into its vicinity, and is reflected from the wall when the covered area is small; sliding sets in when more than half of the area is catalytic, and hovering requires still greater coverage.

A sliding particle is free to rotate around the axis normal to the wall (marked green in Fig. 3.7a), thereby changing its propagation direction; the path, such as the one shown in Fig. 3.7c, can be affected by transient concentration inhomogeneities. Rotation is suppressed when the particle comes near another wall, such as the vertical wall in Fig. 3.7a or the bottom of a rectangular groove in Fig. 3.7b. In this way, it can be steered both by chemical inhomogeneities and by geometric constraints.

A special situation arises when a Janus particle is confined to a liquid surface, as in the experiment by Dietrich et al (2017), where they concentrated at the oil/water interface. The wetting properties of both Janus faces come into play here. The catalyst-coated part is slightly hydrophobic, and the uncoated part is somewhat more so. Due to the low wettability contrast between the two hemispheres, either side can be exposed to the oil phase. As shown in Fig. 3.8a, particles split into two populations, with either hemisphere mostly in the liquid or water phase, and, since the difference between the contact angles is not large, the two populations have rather similar immersion depths. However, they will move only if a part of each hemisphere is immersed in water, with the speed, as usual, proportional to the concentration gradient shown by the color contrast in Fig. 3.8b and c. The directions of motion in both these cases are opposite, and the speeds are different as well.

What is characteristic to this situation is that rotations causing a change in the direction of motion are even more restricted at the liquid interface than near a solid

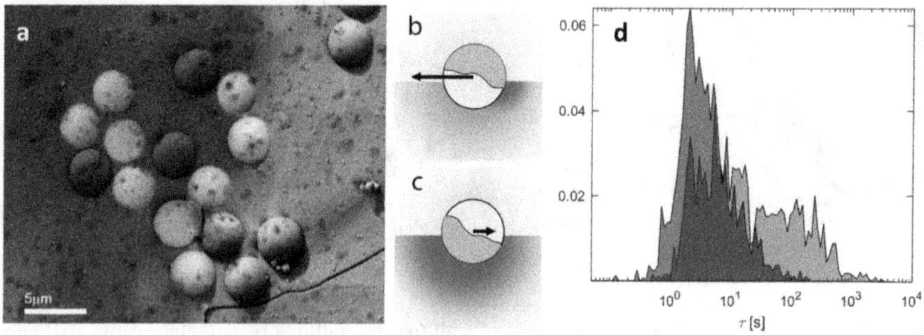

Fig. 3.8 (**a**) View from the oil side, showing two populations of Janus particles, with either the catalyst-coated (*bright*) part or the uncoated (*dark*) part prevailing. (**b**), (**c**) The particle velocities, dependent on the position and area of the catalyst-coated part. The fuel concentration increases from blue to red; the size of the *arrow* is proportional to the velocity. (**d**) Distributions of the rotation time τ for particles at the oil/water interface (*red*) and near a solid surface (*blue*) (Dietrich et al, 2017)

surface. The distributions of the rotation time τ in the two cases, obtained in the same setting, are compared in Fig. 3.8d. The rotation times are both much larger and spread more widely at the oil–water interface (red) than near a solid surface (blue), where they are still much larger than the characteristic rotation times in the bulk.

3.4 Collective Effects

A great variety of patterns of motion, compared by Saha et al (2019) to *waltzing*, arise due to pair interactions of autophoretic particles. The problem of hydrodynamic interactions between two spherical squirmers has been solved exactly (Papavassiliou and Alexander, 2017), but Saha et al (2014, 2019) concentrated upon diffusional interactions. Each particle creates a chemical field characterized by its activity α, and responds to a chemical gradient via its effective mobility μ, which is controlled by interfacial interactions. The concentration gradient due to another particle breaks the axial symmetry of a Janus swimmer, and its polar axis, which defines its self-propulsion direction, rotates in proportion to its vector product with the gradient

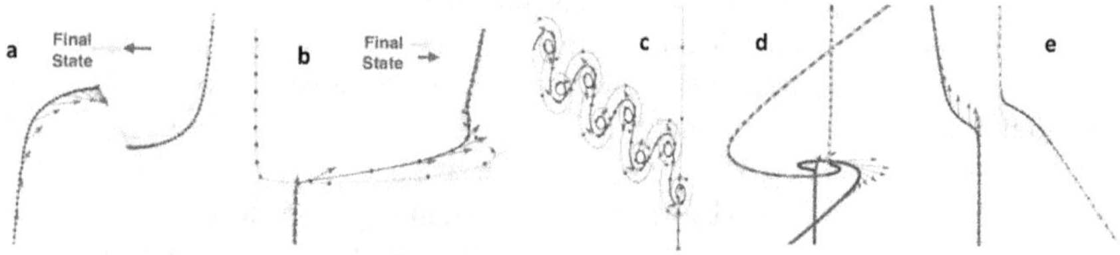

Fig. 3.9 Examples of trajectories of interacting particles (Saha et al, 2019)

direction. This leads to different "dancing" figures, depending on the activity of both particles and the sign of the response to an external gradient, called *chemotactic* or *antichemotactic* when the self-propulsion axis rotates, respectively, to point parallel or antiparallel to an imposed chemical gradient.

Different kinds of trajectories, depending on the design of the catalytic coverage of an interacting pair, are sketched in Fig. 3.9. A pair of identical, mutually attractive chemotactic swimmers form a stationary dimer (Fig. 3.9a). In panel (b), the swimmers chase one another, arriving at a final state of an active dimer with both polar axes pointing in its direction of motion. Panel (c) shows a looping trajectory of a rotating active dimer. Both chemotactic (d) and antichemotactic (e) particles may scatter off one another when self-propulsion is stronger than attraction.

Nasouri and Golestanian (2020) mapped regimes of pair interactions as a function of the ratios of the activities $\alpha_1/|\alpha_2|$ and mobilities $\mu_1/|\mu_2|$ of the two particles. The velocity of each particle, proportional to its activity, is perturbed by the velocity component induced by interaction with its partner: for particle #1, it is proportional to $\mathbf{n}\mu_1\alpha_2$, where $\mathbf{n}$ is the vector directed from the center of particle #1 to the center of particle #2. For the latter, the induced velocity is proportional to $-\mathbf{n}\mu_2\alpha_1$. The interaction strength decays in inverse proportion to the distance between the centers of the particles.

Each coefficient can be attributed either sign. Inverting the sign of α is equivalent to interchanging the positions of its Janus faces, and inverting the sign of μ turns a chemotactic pair into an antichemotactic one. There are four possibilities for the relative motion of two chemically active spheres: the two particles may collapse onto one another (regime I), move away and separate (regime II), reach a stable bound dimer with a certain gap size (regime III), or develop a critical gap size above which they move apart and below which they aggregate (regime IV). Regimes III and IV are located between regimes I and II, suggesting that the transition from the fully

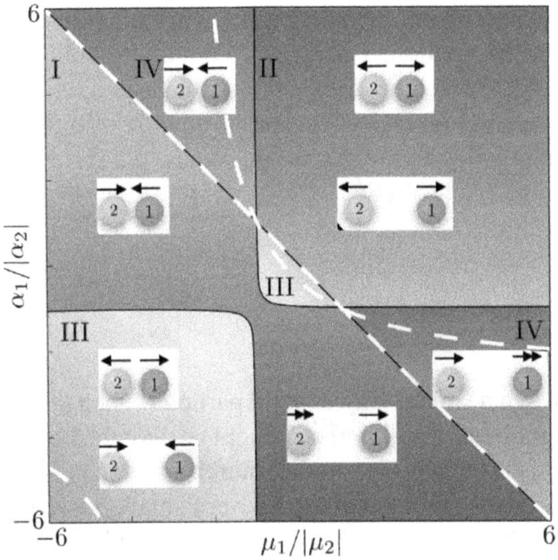

Fig. 3.10 Diagram showing the regimes of pair interactions of active particles as a function of the ratios of their activities $\alpha_1/|\alpha_2|$ and mobilities $\mu_1/|\mu_2|$ at $\alpha_2/\mu_2 > 0$. *Insets* show the directions of velocities due to interactions in each regime at small and large distances; *double arrows* indicate a greater speed. *Dashed lines* show the regime boundaries when hydrodynamic interactions are neglected. If $\alpha_2/\mu_2 < 0$, regime I changes to regime II, regime III changes to regime IV, and vice versa (Nasouri and Golestanian, 2020)

attractive to the fully repulsive mode is not abrupt (Fig. 3.10). Oscillatory "waltzing" in Fig. 3.9c requires a special design of the catalytic coat.

According to Fig. 3.10, taking into account hydrodynamic interactions does not change the picture in a qualitative way, but the theory has another weak point: it views the concentration distribution as quasistationary at current particle positions. This approach is justified for hydrodynamic interactions, as we mentioned at the end of Sect. 2.1, but diffusion in liquids is very slow, and quasistationarity, while valid in nanoscale layers that define the slip velocity (Sect. 3.2), may already become questionable on the microscale. Of course, solving the time-dependent diffusion equation would make the theory far more complicated.

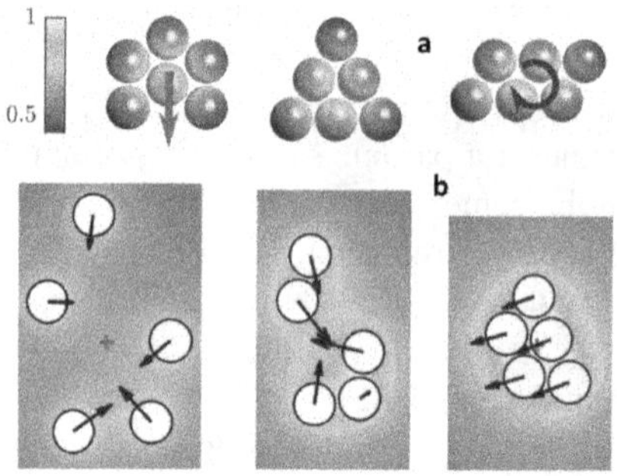

Fig. 3.11 (**a**) *From left to right*: Propagating, stationary, and rotating clusters of symmetric chemically active particles. (**b**) Self-assembly of a propagating cluster (Varma et al, 2018)

Even symmetric chemically active particles can form stable mobile clusters with an asymmetric reactant concentration distribution within their interstices. Examples of propagating, stationary, and rotating clusters are shown in Fig. 3.11a. Such clusters can self-assemble in a natural way. Two symmetric particles may attract because of a different fluid composition in the space between them; the doublet formed by two identical particles is symmetric and stationary, but several particles attracted in this way may form an asymmetric mobile composite, as shown in Fig. 3.11b. It is still easier for symmetric particles to join a mobile cluster by nucleating around a Janus particle. Various modes of assembly, either spontaneous or manipulated by illumination, are reviewed by Popescu (2020).

Generally, motility of active particles depends on other particles in their surroundings, which affect the concentration distribution. This works similarly to "quorum sensing" by microorganisms (Sect. 4.4), which involves more sophisticated internal mechanisms. The dynamics of active particles also affects passive particles through hydrodynamic interactions when both kinds are mixed; all this leads to a variety of separation and clustering effects (more on this in Sect. 3.7).

3.5 Autophoretic Droplets

Droplets of immiscible liquids move in a similar way to solid particles when driven by an external force or an imposed gradient. The difference might be twofold. First, a droplet might deform. This effect is measured by the capillary number: velocity times dynamic viscosity divided by surface tension. It is commonly very small, rendering the deformation negligible. Second, a flow may be induced within the droplet as well

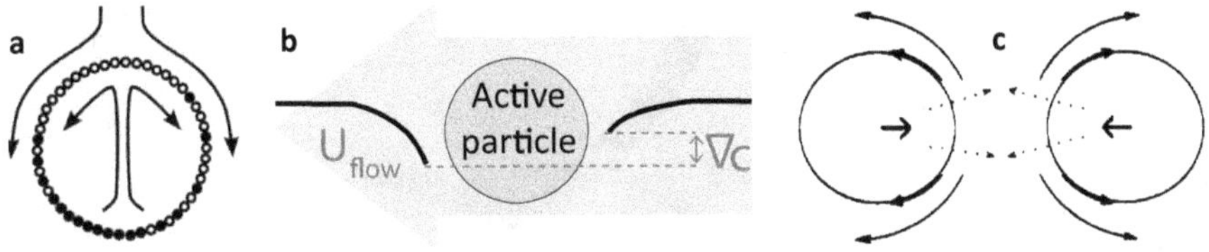

Fig. 3.12 (**a**) Flow pattern (shown in the rest frame of the droplet) due to inhomogeneous surfactant coverage. (**b**) Self-propulsion due to coupling between the advective flow and the surfactant gradient (Moerman et al, 2017). (**c**) Attraction between two droplets due to a self-induced surface tension gradient (Golovin et al, 1995)

as outside, as the no-slip condition is replaced by the condition of equality of the internal and external velocities at the interface. The inner flow may be suppressed when the interface is so densely covered by a uniform surfactant layer that it becomes immobile; in this case, there will be no difference between the phoretic motion of a solid particle or a droplet.

A distinct mechanism, present only in droplets or bubbles, is the surface motion of adsorbed surfactants (Fig. 3.12a). If it is induced by fluid flow around a droplet, this can cause a considerable increase in the drag force (Levich, 1962). A chemical reaction taking place on the surface of a droplet may modify the surfactant, thereby changing the surface tension locally and inducing *Marangoni flow* – surface flow toward a high surface tension area. On the other hand, the motion of the droplet creates a non-uniform distribution of the surfactant on its surface (Golovin and Ryazantsev, 1990). This feedback loop may cause self-propulsion to emerge as a dynamic instability without an imposed Janus-like asymmetry (Fig. 3.12b) and induces attraction between two like droplets, as in Fig. 3.12c (Golovin et al, 1995).

Thutupalli et al (2011) observed this spontaneous motion in experiments with aqueous droplets submerged in the oily phase in a confined flat layer. A chemical reaction at the interface modified the surfactant in a way increasing surface tension, and spontaneous symmetry breaking triggered by random local inhomogeneities of the reaction rate led to consolidation of high and low surface tension areas, which triggered the self-propulsion instability as described above. A moving droplet is a

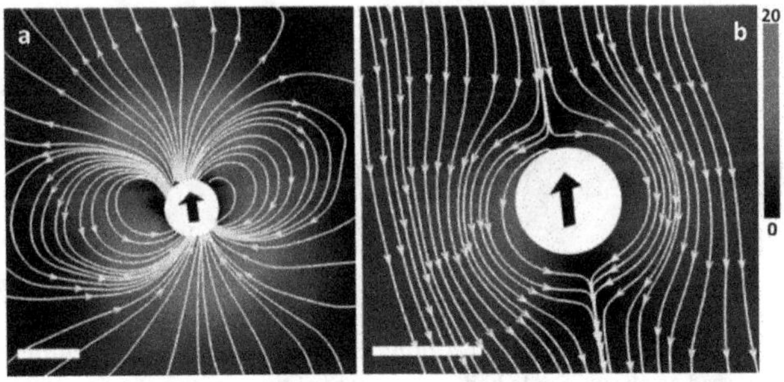

Fig. 3.13 Streamlines and the magnitude of the velocity (color-coded, scale in microns per second) around a self-propelling droplet in the laboratory reference frame (**a**) and in the comoving reference frame (**b**). Scale bar 100 μm (Thutupalli, 2014)

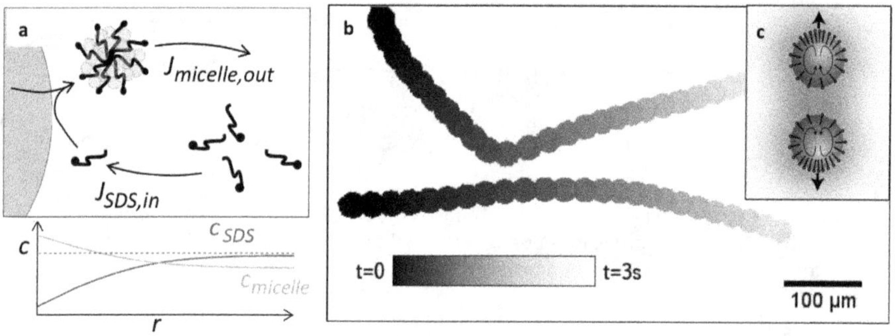

Fig. 3.14 (**a**) Scheme of micelle and surfactant (SDS) transport (*top*) and the radial concentration gradients (*bottom*). The *dashed line* shows the bulk concentration. (**b**) Two oil droplets swimming in given initial directions repel one another due to the induced concentration gradients. *Circles* map their trajectories over time (Moerman et al, 2017). (**c**) Surfactant concentration distribution and the direction of Marangoni flow in repelling droplets (Meredith et al, 2020)

squirmer, similar to self-phoretic solid particles, but the induced flow pattern shown in Fig. 3.13 is different due to the internal circulation.

Another factor causing spontaneous motion is the transfer of surfactant into micelles within a continuous phase. In the experiments by Moerman et al (2017), oil slowly dissolved in the water phase and accumulated in swollen surfactant micelles (Fig. 3.14a). This gave rise to radial concentration gradients due to depletion of dissolved surfactant and accumulation of micelles near a droplet. When two droplets come close together, surfactant depletion, and hence an increase in surface tension on the sides facing each other, induces Marangoni flow towards this area (Fig. 3.14c), in the direction opposite to that in Fig. 3.12c, leading to repulsion, as shown in Fig. 3.14b.

Meredith et al (2020) experimented with two *different* microscale oil droplets, one of which, colored red in Fig. 3.15a, was dissolving and transferring its content to micelles much faster than another one, colored blue. This made the interactions asymmetric and non-reciprocal. The surfactant concentration in the faster dissolving

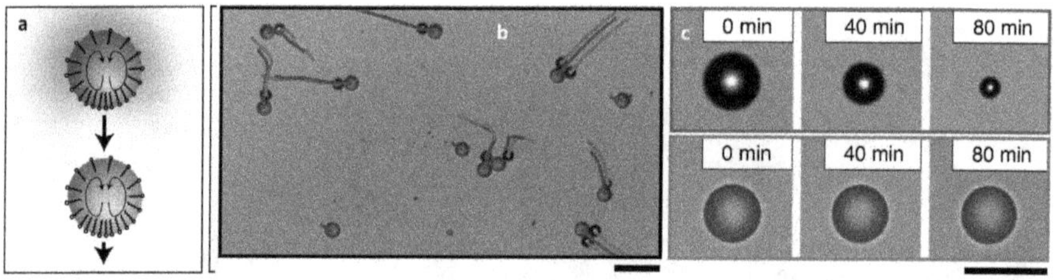

Fig. 3.15 (**a**) Surfactant concentration distribution and the direction of Marangoni flow in nearby faster (*red*) and slower (*blue*) dissolving droplets and the resulting chasing motion. (**b**) Trajectories of chasing and paired droplets, colored as above. Scale bar 250 μ. (**c**) Shrinking of the chasing (*top*) and swelling of the escaping (*bottom*) droplets. Scale bar 100 μ (Meredith et al, 2020)

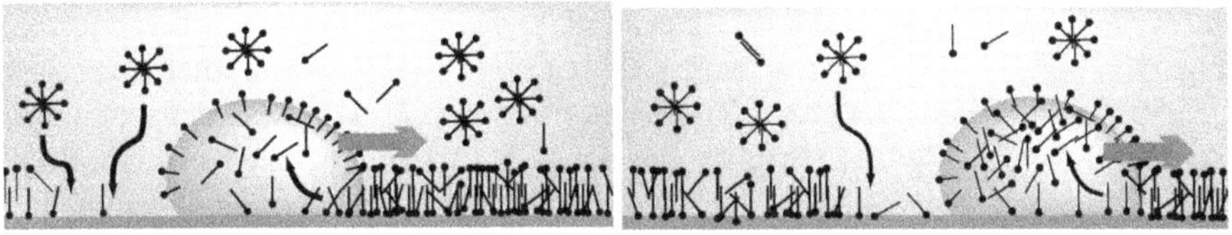

Fig. 3.16 Oil droplet crawling on a substrate in the direction of denser surfactant coverage, which is itself created due to its absorption by the moving droplet (Nagai et al, 2007)

droplet increases on the side facing its more slowly dissolving counterpart, where the effect is opposite, as shown in Fig. 3.15a. As a result, the induced Marangoni flow is oriented in the same way in both droplets, causing them to move in the same direction. The "red" droplet runs faster and catches up, whereupon the two droplets continue to translate as a pair (Fig. 3.15b). Meredith et al liken this chase to a prey–predator interaction, but in this case the"predator" rather than the "prey" is eaten, as its content migrates into micelles and further dissolves in its counterpart, whence the "predator" droplet shrinks and the "prey" swells, as shown in Fig. 3.15c.

In a droplet placed on a solid support and therefore "crawling" rather than swimming, motion can be induced by a wettability gradient. Nagai et al (2007) observed spontaneous motion of an oil droplet placed on a glass substrate in an aqueous phase containing surfactant, which tends to adsorb both on the substrate and on the droplet surface. The droplet better wets the substrate surface when it is more densely covered by the surfactant, and if an inhomogeneity arises in the coverage, moves in this direction, as shown in Fig. 3.16. The surfactant dissolves within the droplet, while the surfactant, present in abundance in the water phase in the form of micelles, adsorbs in the bared area behind the droplet. The crawling motion is fed by a disequilibrium in the surfactant distribution between the droplet and the aqueous phase. It arises as a randomly triggered dynamic instability and persists for a considerable time, till equilibrium is restored.

Similar erratic motion due to surface tension inhomogeneities can be induced on an interface between two immiscible liquid phases. In experiments by Sumino

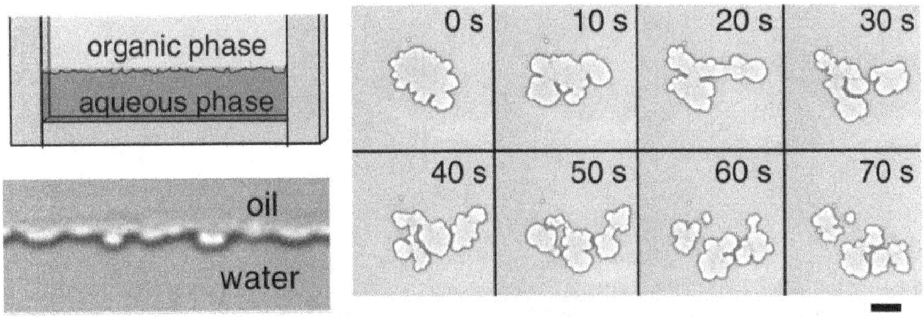

Fig. 3.17 *Left*: Experimental setup (*top*) and blebbing oil–water interface (*bottom*). *Right*: Erratic motion of the oil droplet on the water surface. Scale bar 10 mm (Sumino and Yoshikawa, 2014)

and Yoshikawa (2014), it caused interfacial blebbing on the oil–water interface (Fig. 3.17 left), which became more pronounced when the oil droplet floated on the surface of the aqueous phase. The droplet spontaneously moved, deformed, and split, as shown in Fig. 3.17 (right). Interfacial blebbing continued for about 100 seconds until the surfactant distribution over the interface came to equilibrium. The droplets in blebbing and crawling experiments are in the millimeter range, far above colloidal sizes, but the principal actors in their motion are nanometer-range surfactant structures.

3.6 Imitated Cells

Some aspects of autonomous droplet dynamics are reminiscent of the motion of living cells. Precluding the more realistic description in Chap. 5, a droplet filled by an active nematic fluid, like the one considered in Sect. 2.5, may be viewed as a rough model of a cell with cytoskeletal filaments stressed by molecular motors. A contractile nematic fluid is generically unstable (Simha and Ramaswamy, 2002). When there is no distortion, the active forces, shown by blue arrows in Fig. 3.18a, balance. A splay leads to an imbalance creating flow to the right that increases splay, thereby forming a destabilizing positive feedback loop. However, at low activity, the droplet just deforms, contracting in the alignment direction, and a critical activity level is necessary to set the droplet in motion, as sketched in Fig. 3.18b. Both stationary and moving droplets induce flow in a surrounding passive fluid (Fig. 3.18c and d).

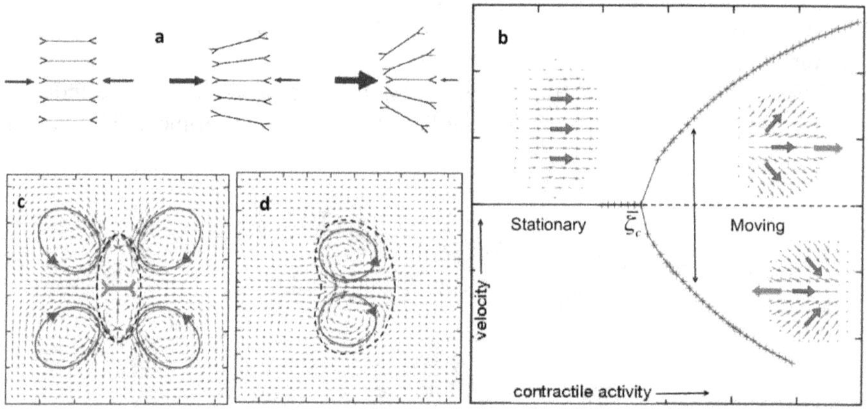

Fig. 3.18 (**a**) Instability of the ordered phase of a contractile active nematic. (**b**) Bifurcation diagram showing the transition from a stationary to a translating droplet at a critical contractile activity $\bar{\zeta}_c$. (**c**), (**d**) Flow field in a passive isotropic fluid surrounding stationary and translating active droplets, respectively (Marenduzzo, 2016)

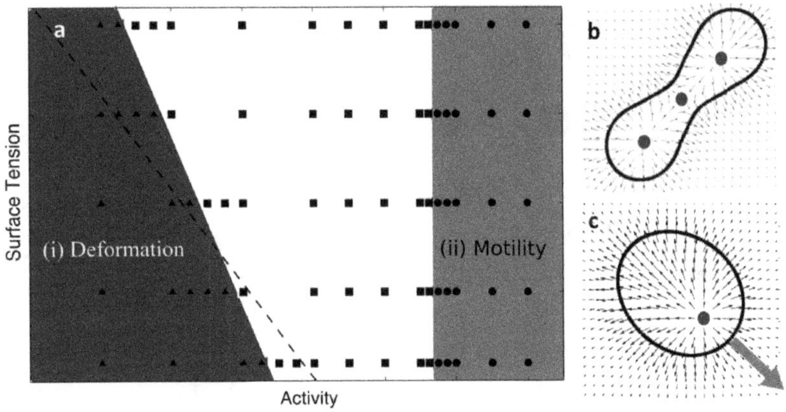

Fig. 3.19 Instability of a contractile polar droplet. (**a**) Diagram showing the domains of stationary (*white*), deformed (*blue*) and mobile (*red*) states, as a function of the level of activity and surface tension. *Squares*, *circles*, and *triangles* mark the results of computations used to produce this diagram. The *dashed line* shows the analytical result of linear stability analysis. (**b**), (**c**) Polarization field (assumed to decay gradually outside the droplet). *Blue dots* mark the positions of defects. The *red arrow* shows the propagation direction (Whitfield and Hawkins, 2016)

A similar transition takes place in a droplet with polar activity. In computations leading to the diagram in Fig. 3.19a, surface tension was varied alongside contractile activity, and, besides a transition to directed motion, a stationary deformed state was observed (Whitfield and Hawkins, 2016). A 2D polar droplet with normal anchoring on its surface must contain a vortex defect, but deformation is accompanied by nucleation of an extra pair of oppositely charged vortices, as can be seen in Fig. 3.19b.

The most essential property of living cells is their ability to multiply and perpetuate themselves. An early hypothesis about the origin of life (a question still unsettled) was that it had started from a self-replicating droplet containing a chemically active mixture of proteins (Oparin, 1924). Zwicker et al (2017) reproduced this scenario in an amazingly simple metabolic model (Fig. 3.20). The droplet material B is supposed to degrade into "waste" A, which leaves the droplet and is recycled in its environment with the help of "fuel" C (degraded to C′ in the process), returning to B, which diffuses into the droplet.

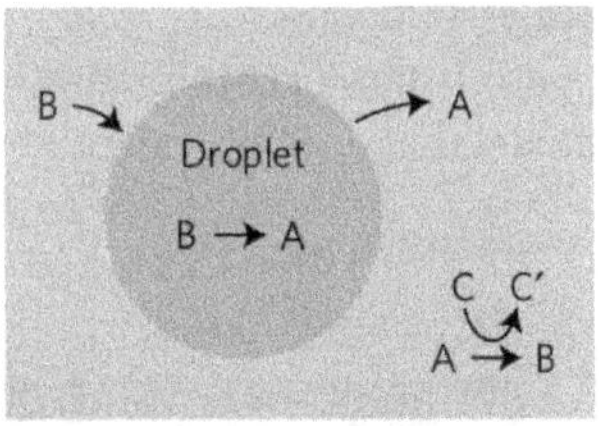

Fig. 3.20 Reaction scheme of the self-replication model (Zwicker et al, 2017)

The essential element of the model is the deformation of the spherical shape of the droplet caused by mass transfer. The explanation (hidden in the paper's supplementary material) relates this phenomenon to the well known Mullins–Sekerka (1963) instability. When the growth of an aggregate is limited by diffusion of material from an outside source, a protrusion coming closer to this source grows faster; this causes, in particular, the elaborate shapes of snowflakes formed in air supersaturated by water vapor. Linear stability analysis indicates the

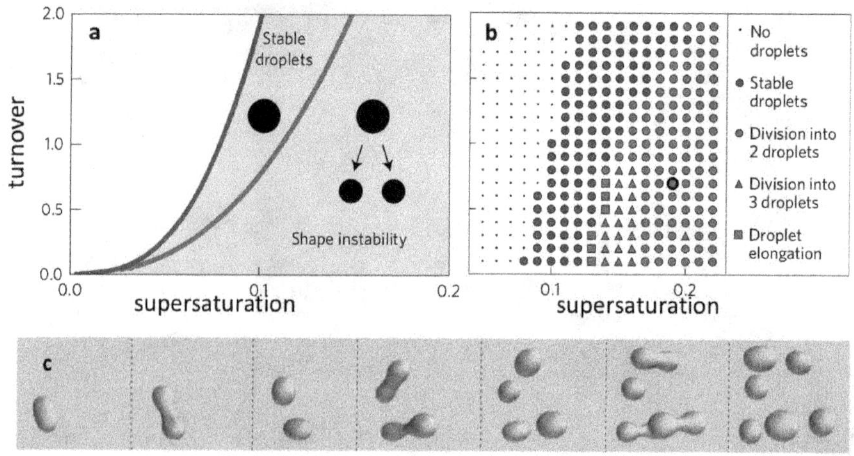

Fig. 3.21 (**a**) Stability diagram of active droplets as a function of supersaturation and turnover of droplet material. Droplets either dissolve and disappear (*white region*), or they are spherical and stable (*blue region*), or they undergo a shape instability (*red region*). (**b**) The same diagram obtained as a result of simulations. (**c**) Cycles of growth and divisions of chemically active droplets (Zwicker et al, 2017)

instability limit (at a fixed surface tension), depending on the turnover of the droplet material and its supersaturation in the background fluid (which depends in turn on the rate of the recovery process), as shown in Fig. 3.21a. Droplets lose their spherical shape, elongate, and eventually divide: thus, chemical activity overcomes the coarsening tendency, leading instead to cycles of growth and divisions (Fig. 3.21c). Computations roughly reproduce analytical predictions (Fig. 3.21b). However, it should be noted that pinching is a singular phenomenon involving vanishingly small scales (Cohen et al, 1999), and its details are sensitive to a computation grid in hydrodynamical simulations. Such fine points are, of course, irrelevant for a basic model that does not account for fluid flow.

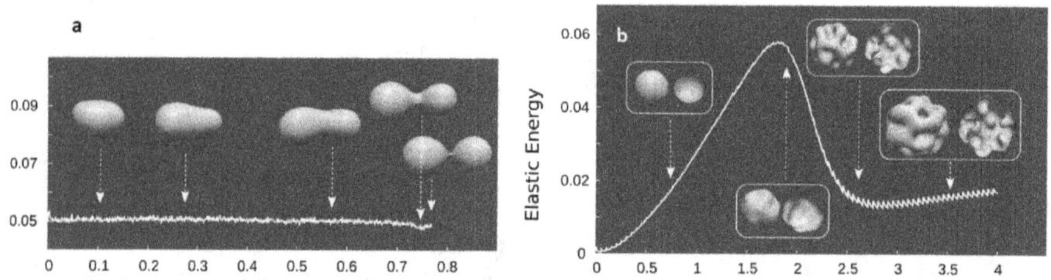

Fig. 3.22 Scaled elastic energy and vesicle shapes as a function of scaled time for two different growth modes: vesicle division at a high mechanical relaxation rate and positive preferred curvature (**a**) and multiple sites of inward vesiculation following the buildup of elastic energy in the case of slow mechanical relaxation and negative spontaneous curvature (**b**). The shapes along the curve show both the exterior and the interior of the vesicles (Ruiz-Herrero et al, 2019)

A requisite element of Oparin's hypothesis is a membrane keeping the content of protocells intact. Such a barrier separating replicating biomolecules from the environment is essential for specialization and competition between cells. It could have emerged quite naturally at the dawn of life, built of surfactant molecules present in a chaotic primordial soup of organic matter. Ruiz-Herrero et al (2019), instead of focusing on a basic chemical metabolism, built up a model that couples membrane growth and fluid permeation. The surface of a vesicle may rapidly increase by addition of surfactant molecules, but its volume could grow slowly, as it is limited by the permeability of the membrane and may be counteracted osmotically, depending on the balance of solute concentrations inside and outside the micelle. In this model, further evolution depends on two factors: the rate of mechanical relaxation and the preferred spontaneous curvature. It unfolds in a similar way to the evolution in the model of Zwicker et al when the mechanical relaxation is fast and the preferred curvature is positive: as the surface to volume ratio increases, the vesicle elongates, becomes mechanically unstable, and breaks into smaller vesicles (Fig. 3.22a). There is also a technical problem here related to the division event. This behavior may be irrelevant for the question of the origin of life (where Oparin's hypothesis is no longer viewed as viable) but it is consistent with observations of cell deformation and division, such as the sequence shown in Fig. 3.23, which involves dynamical imbalances in the surface area to volume ratio, either due to excess membrane growth or low permeability.

At low mechanical relaxation rates, droplets do not break up but acquire a variety of complicated shapes, which are particularly fancy in the not quite realistic case of negative spontaneous curvature (Fig. 3.22b).

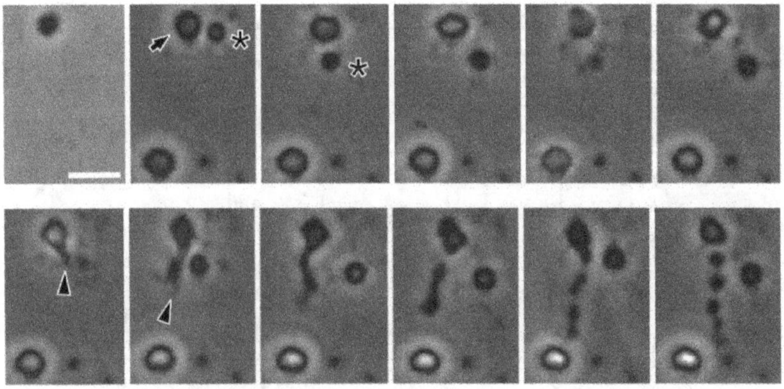

Fig. 3.23 A sequence of cell division. The cell marked with an *arrow* begins to form a protrusion (marked with an *arrowhead*), which resolves into a string of six cells visible in the last picture. Another cell (marked with an *asterisk*) drifting into the field of view was not the result of a proliferative event. Scale bar 5 μm (Leaver et al, 2009)

3.7 Active Suspensions

Interactions between colloidal swimmers carried by the induced flow turn their assembly into a correlated flock and convert the entire system into an active medium. It can be described on a continuum level by effective mass transport and hydrodynamic equations (Jülicher and Prost, 2009), leading eventually to the simulations of the dynamics of active fluids described in Sects. 2.5 and 2.7. The next level is a statistical description taking into account fluctuations of colloidal particles (Romanczuk et al, 2012). Long-range hydrodynamic interactions make theoretical description of dense suspensions and emulsions very difficult, even in the absence of activity. Passive suspensions are treated as a fluctuating continuous medium (Batchelor, 1976), in the spirit of one of the *annus mirabilis* papers by Einstein (1905). The theory of active suspensions and emulsions may look unassailable when we recall that short-range interactions make even "dancing" movements of paired active swimmers so elaborate and variegated, as we have seen in Sect. 3.3, but a statistical description abstracts from such intricate details.

Experiments (as well as simulations) of active suspensions are commonly carried out in 2D, either in a shallow cuvette or allowing heavy particles to settle on the bottom. In this way, Thutupalli et al (2011) measured space-dependent correlations between autophoretic droplets. The flock shown in the left panel of Fig. 3.24 appears at first sight do be disordered. But the plot on the right shows the averaged correlation of alignment angles C_ϑ as a function of the distance between droplets. It decays quite fast with separation, which suggests that only droplets within denser clusters are correlated. An interesting feature is a damped oscillation with period just above the

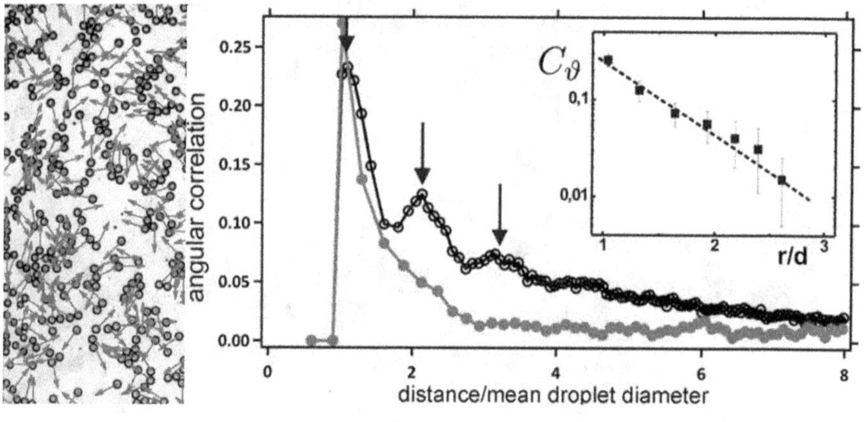

Fig. 3.24 *Left*: Flock of active droplets with their velocity vectors shown by *red arrows*. *Right*: Correlation C_ϑ of droplet propulsion directions as a function of the scaled distance between droplet centers r/d. The *red* and *black curves* correspond to the droplet densities 0.46 and 0.78, respectively. *Black arrows* mark the correlation peaks. *Inset*: Semi-logarithmic plot of the decay of the correlation function (*dashed line*). A decaying oscillation has been superimposed to remove the peaks on the *black curve* and fit the data (Thutupalli, 2014)

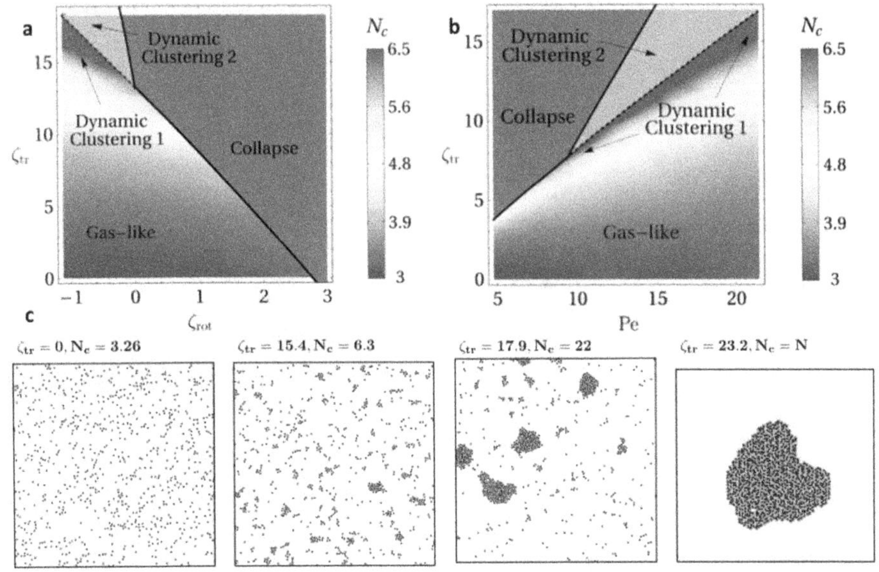

Fig. 3.25 (**a**) Clustering regimes as a function of the translational and rotational chemotactic response parameters ζ_{tr} and ζ_{rot}. (**b**) Clustering regimes as a function of ζ_{tr} and the Péclet number. The mean cluster size N_c is color coded, with the collapsed state set in *dark gray*. (**c**) Representative snapshots of clustering regimes. From *left to right*: gas-like, dynamic clustering 1 and 2, and collapse to a single cluster. N is the total number of particles in the simulation (Pohl and Stark, 2014)

droplet diameter, as seen on the black curve, which corresponds to a higher average density. The correlation peaks on this curve hint at some kind of local layering due to short-scale interactions, like molecular layering in a fluid close to a solid boundary.

A common feature of active suspensions is the *clustering* that we have already encountered in the more abstract setting of the Vicsek model in Sect. 1.2 and the dynamics of rod-like particles in Sect. 2.2, not to mention crowds of scared people in Sect. 1.6 and racing Quincke rollers in Sect. 1.7. The common cause, irrespective of detailed mechanisms, is motility-induced phase separation. Being a universal phenomenon, clustering can be reproduced by any kind of simulation. Thus, Buttinoni et al (2013) claimed that their observations could be captured qualitatively even by a minimal model without any alignment interactions and neglecting hydrodynamics. Ishikawa and Pedley (2008) demonstrated that aggregation, mesoscale spatiotemporal motion, and band formation in 2D can be generated by purely hydrodynamic interactions. Hydrodynamic simulations leading to clustering were also extended to 3D (Blaschke et al, 2016).

On the other hand, Pohl and Stark (2014) reproduced all varieties of clustering patterns based on chemotactic interactions alone. They compiled the two diagrams shown in Fig. 3.25a and b, mapping the clustering regimes represented by the snapshots of simulations in Fig. 3.25c. In the first diagram, the regimes depend on two response parameters ζ_{tr}, quantifying the chemotactic velocity response of

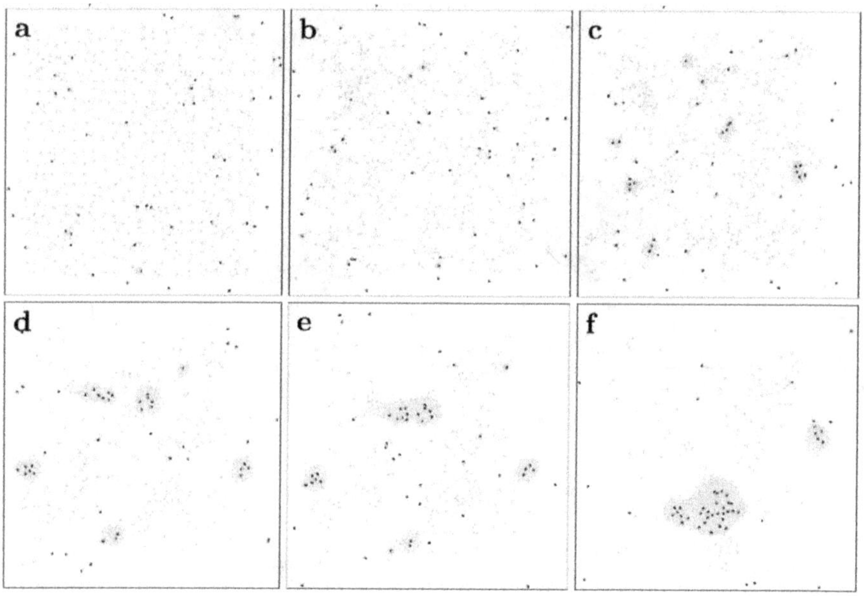

Fig. 3.26 (**a**)–(**f**) Time series of snapshots showing the crystallization of passive particles (*green*) around active colloids (*yellow* and *black*). (Stürmer et al, 2019)

particles, and ζ_{rot}, quantifying the rate of rotation of the polar axis of a Janus particle in response to a concentration gradient. The diagram is drawn at a constant value of the Péclet number Pe proportional to the self-propulsion velocity. The route from the gas-like state to collapse into a single cluster is followed as both ζ-parameters grow, but intermediate dynamic clustering regimes are realized when high translational motility is balanced by negative rotational response. The value $\mathrm{Pe} = 19$ adopted here is quite high: advection prevails over diffusion at $\mathrm{Pe} > 1$, but still higher Pe values are commonly used in purely hydrodynamic simulations. The second diagram

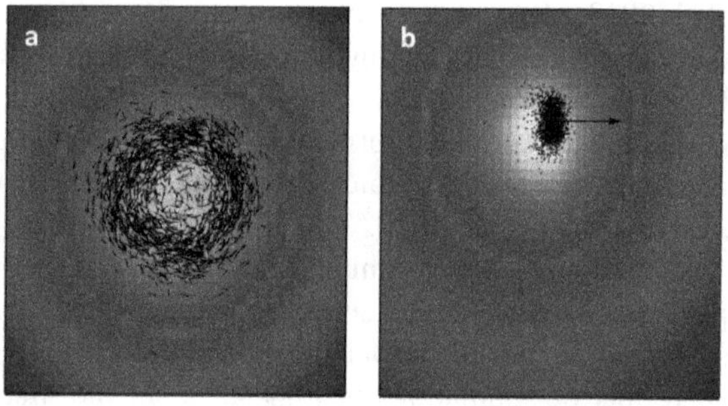

Fig. 3.27 Active particles with chemotactic coupling and velocity alignment. The attractant concentration is color coded, increasing in brighter regions. (**a**) Collective rotation at an intermediate velocity alignment strength. Each particle is shown by its velocity vector. (**b**) A cluster moving in the direction shown by the *arrow* at strong velocity alignment (Romanczuk et al, 2012)

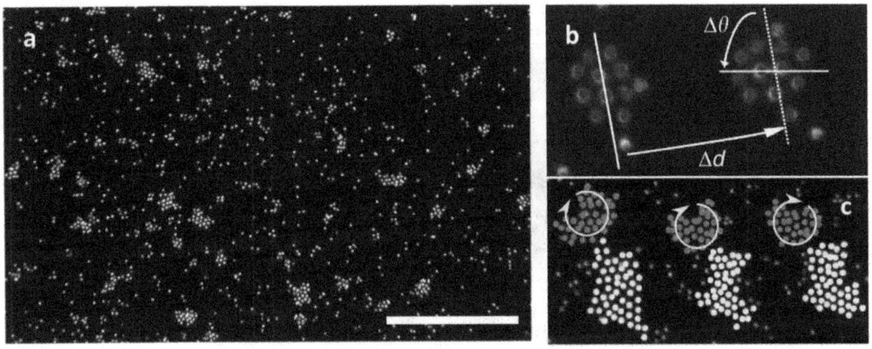

Fig. 3.28 (**a**) Snapshot of sedimented Janus particles; scale bar 40 μm. (**b**) Two overlaid snapshots of the same cluster shifting by a distance Δd and rotating by an angle $\Delta\theta$. (**c**) Series of snapshots of a cluster rolling over another cluster (Ginot et al, 2018)

shows that a gas-like state prevails at high Pe values and not too strong a chemotactic response.

Chemotactic clustering may also involve passive particles (Stürmer et al, 2019). With increasing diffusiophoretic intensity, the system evolves from the gas-like state to intermittent clustering to collapse into a single cluster. Passive particles tend to gather around active ones, and the fledgeling cluster stabilizes if it reaches some critical size. A ripening process may follow, as clusters merge, as shown in Fig. 3.26, which may eventually lead to a total collapse.

Although either kind of interaction may induce clustering, the results are different. When only chemotactic interactions are present, there is no collective motion within a collapsed cluster, which performs at most slow diffusive motion (Fig. 3.27a). This changes dramatically if, in addition to the chemotactic coupling, velocity alignment is introduced (Romanczuk et al, 2012). Increasing the alignment strengths leads first to collective rotation within a stationary cluster. A further increase in alignment strength enables the particles to escape collectively out of the stationary maximum of the chemical attractant. As the particles keep on producing the attractant, they drag its cloud of chemoattractant around them, creating the compact chemically-bound moving cluster shown in Fig. 3.27b.

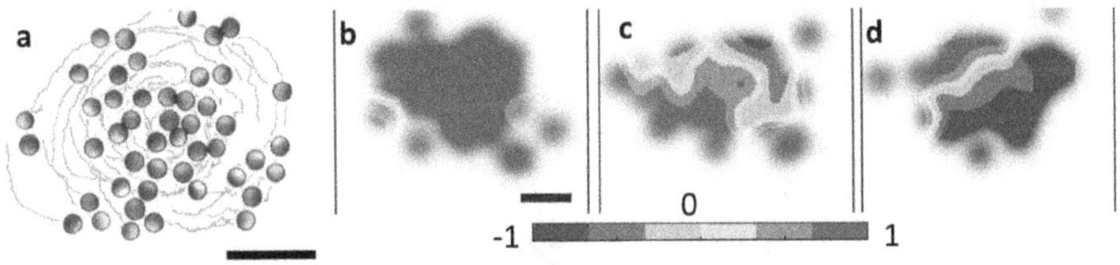

Fig. 3.29 (**a**) Rotating cluster of active particles with trajectories shown by *green lines*. (**b**)–(**d**) Spontaneous reversal of rotation direction (color coded). Scale bar 30 μm (Bäuerle et al, 2020)

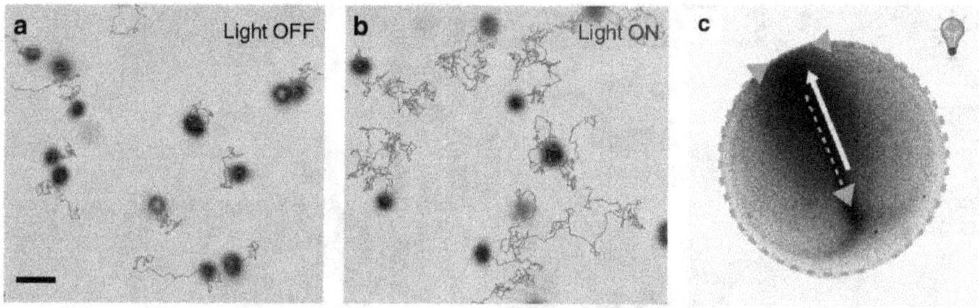

Fig. 3.30 (**a**) Brownian motion of passive colloidal particles sedimented at the bottom of a sessile drop. Scale bar 5 μm. (**b**) Enhanced diffusion due to self-propulsion upon illumination with UV light. (**c**) Self-organized vortical flow (directed as indicated by *orange arrows*) formed upon illumination. The *white arrow* pointing towards the region of increased particle density indicates the emerging polarity of the drop (Singh et al, 2020)

Fine details of clustering patterns are revealed in dedicated experiments. Cluster shapes are dynamic, since particles constantly join and leave them. At the same time, clusters move as distinct units. Figure 3.28a shows a snapshot of a clustering pattern of sedimented Janus particles immersed in a bath of hydrogen peroxide fuel. The clusters, while slightly modifying their shape and occasionally losing or gaining individual particles, translate and rotate, as shown in Fig. 3.28b and c. Rotating clusters, such as the one shown in Fig. 3.29a, may retain their identity, while occasionally changing their direction of motion, as in the sequence Fig. 3.29b–d.

In a suitable macroscopic setting, a global self-organized pattern may emerge. In the experiments by Singh et al (2020), photocatalytic colloidal particles sedimented at the bottom of a sessile drop showed enhanced diffusion due to self-propulsion upon illumination with UV light (Fig. 3.30a and b). At higher particle densities, the inhomogeneities self-organized into vortical flow patterns, as shown in Fig. 3.30c. The authors infer that, since particles do not possess a defined polarity and only a fraction of them exhibit significant self-propulsion, as is evident from the wandering trajectories in Fig. 3.30b, the observed self-organized collective motion is caused by chemicals emanating from the particles, rather than by their hydrodynamic interactions. Concentration inhomogeneities, coupled to the density inhomogeneities seen in Fig. 3.30c, cause surface tension gradients driving Marangoni flow, and this in turn sustains the uneven distribution of active particles in the vortex encompassing the entire drop.

Chapter 4
Motion of Microorganisms

4.1 Prokaryotic Flagella

Microorganisms, with their moving appendages and the ability to deform their body, display a much wider variety of propulsion modes than colloidal particles. The most common propulsion aid is a *flagellum* ("whip" in Latin). Prokaryotic bacteria have invented it as an advanced rotational motion device; their helical filaments do not actually deform in a whiplike way at all. This design principle was not taken over by eukaryotes, which developed more whiplike flagella (Sect. 4.2), and it was left to humans to reinvent the propeller.

The structure of the motor has even been claimed by advocates of intelligent design to be too sophisticated to have evolved by blind mutations, but certain features point to its descent from excretion openings in the cell membrane. A flexible helical filament is fastened to a "hook", a short flexible polymeric beam which serves to regulate the filament's alignment relative to the cell's body. The hook is attached to a shaft protruding through the membrane, supported by a stator composed of protein rings, which act as a kind of a clutch that allows the rotation to be switched on and off (Fig. 4.1). The protein units are activated

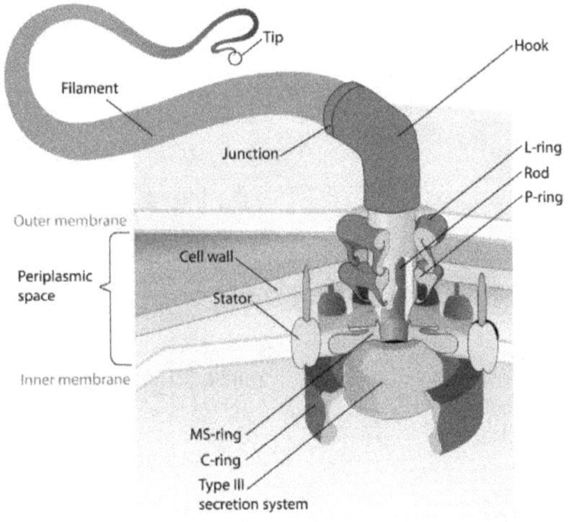

Fig. 4.1 Structure of the bacterial flagellum. More detail is shown than we need, but the helical shape of the filament is not clear in the picture (Public Domain)

L. Pismen, *Active Matter Within and Around Us*, The Frontiers Collection,
https://doi.org/10.1007/978-3-030-68421-1_4

by steps in the direction of rotation. The motor is powered electrochemically, using energy from either sodium or hydrogen ion flux through the membrane – this is the way things usually work in prokaryotic cells, taken over by eukaryotes in their mitochondria. Experimentalists have succeeded in following the stepped activation sequence of protein units, and have been able to reduce the motive force by lowering the ion concentration, which led to a decrease in both the number of stator units and the speed per unit (Sowa et al, 2005).

A helical filament is a cylinder formed by 11 protofilaments built of a stack of *flagellin* protein monomers, which can assume either left-hand or right-hand (L-R) conformations differing in length. Each protofilament contains a single type of monomer, and therefore L- and R-protofilaments have different lengths. Mixing them in the flagellar filament produces bending which, combined with an intrinsic twist of the filament, leads to a helical configuration with the pitch decreasing as the difference between the number of protofilaments with different conformations is reduced (see the inset of Fig. 4.2). The shape of a filament can also be changed by applying a force. Darnton and Berg (2007) demonstrated changing conformations by pulling the two ends of a flagellar filament apart by optical tweezers. These transitions play an important part in bacterial movement strategy, as we shall see presently.

Bacteria, in particular the intensively studied *E. coli*, move in a *run and tumble* manner, alternating straight-line motion with random changes of direction. This allows them to explore their surroundings and find the way to nutrients or to avoid peril. Bacteria are *not* diffusiophoretic: they are too small to feel concentration gradients, but they have a kind of a short-term memory, which allows them to measure the spatial gradient by sensing a change in time as they move. Engelmann

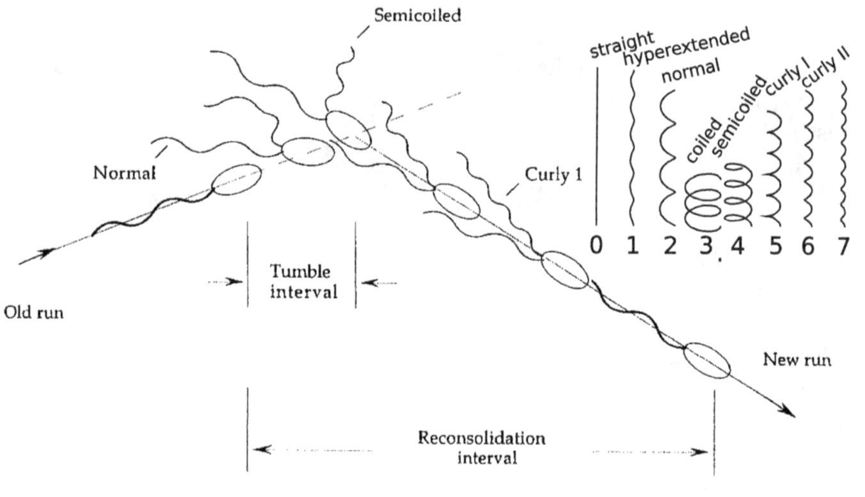

Fig. 4.2 Changing the filament's conformation and the direction of motion at a tumble (Berg, 2004). *Inset*: Shapes of helical filaments with different numbers of R-protofilaments (Vogel and Stark, 2010)

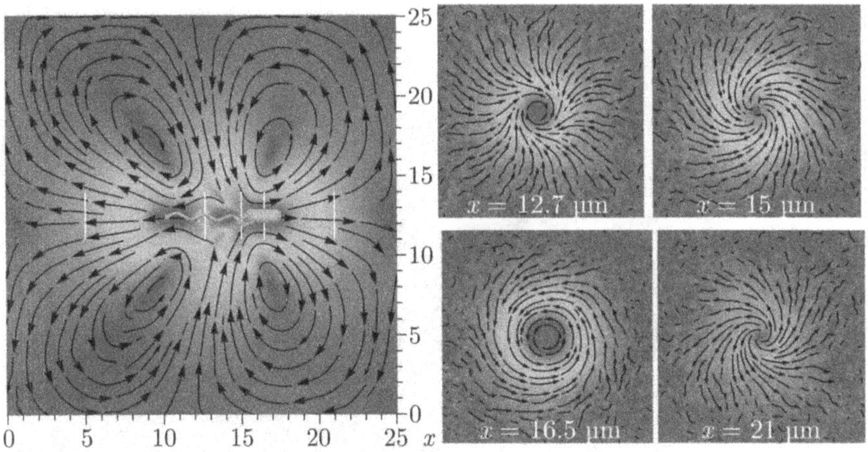

Fig. 4.3 Simulated time-averaged flow field near a bacterium of length 8 μm. *Left*: Axisymmetric flow field in a plane of motion. *Right*: Swirling flow in planes normal to the direction of motion at the positions marked by *white lines* in the left panel. The magnitude of the velocity is color-coded, increasing from blue to red (Hu et al, 2015)

(1883) proved this by switching illumination on and off in a homogeneous solution, and bacteria reacted by backing up and stopping to steer their path. If a change in time is beneficial in some respect, bacteria increase the duration of runs and, respectively, decrease the frequency of tumbles. This results in a net motion along the gradient of an attractant or against the gradient of a repellent.

During a run, usually lasting about one second, a flagellum rotates counterclockwise, propelling the cell forward. The motion of the helix is irreversible in time, so that the "scallop theorem" is not applicable, even if the filament does not change its shape. The simulated flow field in a plane of motion shown in the left panel of Fig. 4.3 is close to the theoretical flow field of a pusher in Fig. 3.1.

When a tumble is signaled, the rotation direction is reversed, which causes the filament to change its conformation from "normal" (helical) to "semicoiled" or "curly", as sketched in Fig. 4.2. Following the tumble interval of about 0.1 second, as counterclockwise rotation is resumed, the helical conformation is restored but the new run is directed randomly at some angle to the preceding one. Buckling instabilities play an important role in reconfigurations in the course of a tumbling event (Vogel and Stark, 2012). Detailed records of trajectories (Son et al, 2013) exhibit alternations between reversals and large reorientations with a broad angular distribution centered about 90°, shown, respectively, by green circles and red squares in Fig. 4.4. As is evident from the right panel of this figure, where bacteria are attracted to a dead copepod, run-and-tumble motion is not at all effective in minimizing the path to the target.

Most bacteria have several flagella, some of them *lophotrichous*, with multiple flagella located at the same spot on their surface, and many others (including the "model bacterium" *E. coli*), *peritrichous*, with flagella projecting in all directions

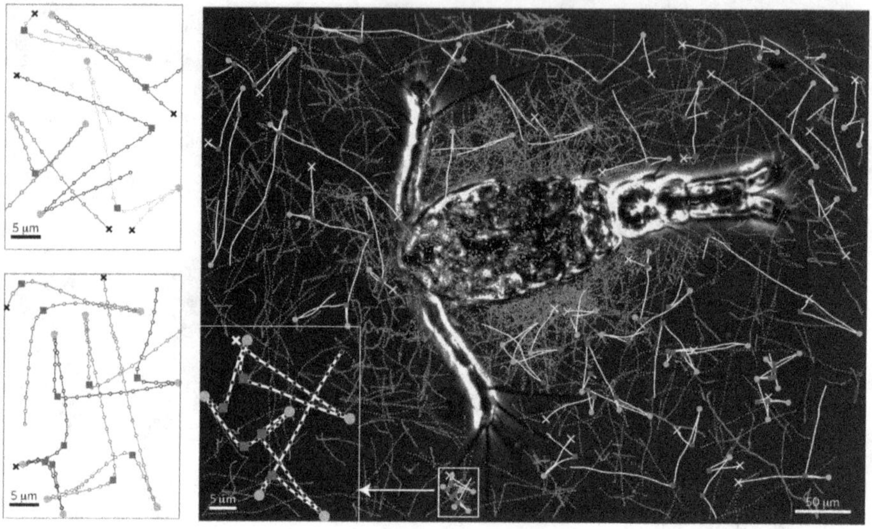

Fig. 4.4 Examples of run-and-tumble trajectories. In the *right* panel, bacteria are attracted to a dead copepod (Son et al, 2013)

(Fig. 4.5). Having multiple flagella facilitates tumbling, since reversing the rotation direction of a single motor is sufficient for breaking up the bundle and stopping the run. If flagella are distributed around the entire surface of the cell, there would be no propulsion were the filament axes directed randomly. However, flexible hooks allow filaments to gather into a bundle behind the cell and work collectively, propelling the cell as a "pusher" swimmer and thereby increasing the speed several times over. Computations by Riley et al (2018) indicated that collective action arises as a result of an elasto-hydrodynamic instability when the stiffness of hooks decreases beyond a certain threshold.

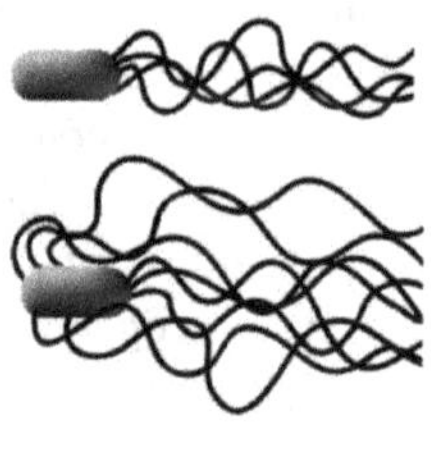

Fig. 4.5 Lophotrichous (*above*) and peritrichous (*below*) bacteria (CC)

The dynamics of multiple flagella is extremely complicated, as it involves nonlocal hydrodynamic interactions between filaments, short-range steric and electrostatic interactions, and elastic deformations of the filaments and hooks. Lauga (Gompper et al, 2020) uses the simile of long hair gathering behind the head of a swimmer at the pool as a reason for passive flagellar assembly (although inertial effects dominate in the case of a swimmer). He also notices another passive effect: a flagellum's rotation is compensated by counter-rotation of the cell body, which also causes flagella to wrap around each other. In addition, the gathering of multiple flagella is facilitated by long-range hydrodynamic interactions (Riley et al, 2018). The propulsive force created by each flagellum pushes the fluid away from the cell body and thus creates a stream along its surface that causes mutual attraction of flagellar filaments. Moreover, swirling flow due to flagellar rotation, as shown in the right panels of Fig. 4.3, leads to their wrapping. Nevertheless, the filaments do not get tangled up since all of them coil in the same direction.

4.2 Eukaryotic Flagella and Cilia

Eukaryotic flagella do not rotate, but they can beat in various fashions; sperm also propels itself in this way. Their core is a microtubule-based cytoskeletal structure called an *axoneme*. More variegated beating patterns are possible because molecular motors, called *axonemal dyneins*, are distributed along the axoneme rather than being concentrated in prokaryotic rotors. The dyneins power the beat by generating sliding forces between adjacent doublet microtubules, which causes the flagellum to bend, as sketched in Fig. 4.6a and b. Feedback control is carried out by dyneins detaching under the action of excessive curvature or a force applied along or normally to the filament (Fig. 4.6c–e).

Flagella and cilia (except non-motile ones) are similar in their structure and function, although the distinction is prominent in their Latin names, translating first, as *whips* and second, as *eyelashes*. This suggests that cilia are shorter (which they commonly are, although not necessarily), more numerous, and less forceful. Both flagella and cilia propel by oscillatory beating, but the oscillations may contain multiple harmonics and a variety of spatial forms. By the rules of Stokesian flow, the beating cannot be reversible, and the scallop theorem of Sect. 3.1 is avoided by a circular deformation pattern through a traveling wave propagating along the flagellum, as shown in Fig. 4.7a. Vilfan (2012) classified filamentary beating patterns by their symmetry, as illustrated in Fig. 4.7b–d.

Various spatio-temporal beating patterns are driven by the activity of molecular motors distributed along the filament, which obey signaling cues adjusted to the en-

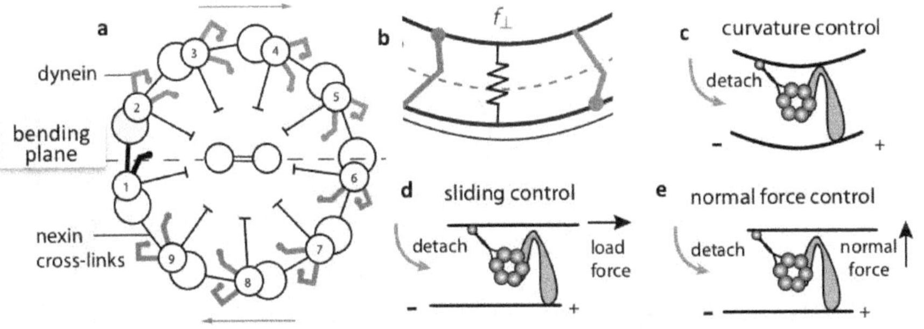

Fig. 4.6 (**a**) Cross-section of an axoneme, as seen from the basal end looking towards the distal tip, with numbered doublet microtubules. The attached dyneins, colored *green* and *blue*, bend the axoneme, respectively, to the right and to the left; their combined action creates the normal bending force $f_\perp$ (**b**). (**c**)–(**e**) Modes of feedback control for dynein attachment (Sartori et al, 2016)

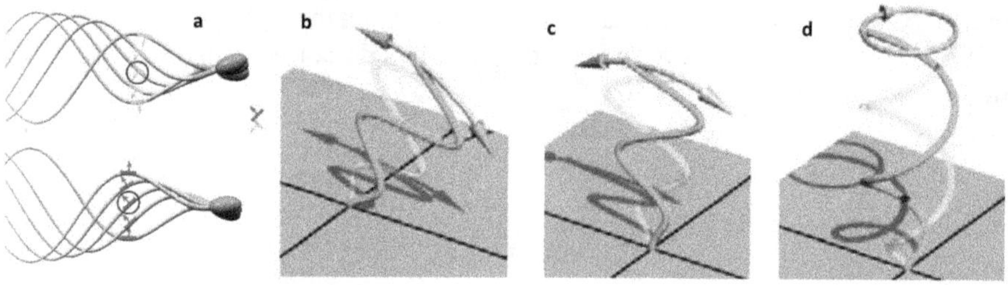

Fig. 4.7 (**a**) Illustration of irreversibility of motion through a propagating wave: forces applied during two half-cycles (*green* and *red*) are oppositely directed (Kaupp and Alvarez, 2016). (**b**)–(**d**) Beating patterns according to their symmetry properties. In each figure the *faint line* shows the filament half-a-period later. (**b**) A planar beating pattern. (**c**) A planar pattern that is symmetric with respect to reversal of the horizontal coordinate and the time shift by half a period. (**d**) A pattern that is symmetric with respect to rotation around the vertical axis and a simultaneous time shift (Vilfan, 2012)

vironment. The filament's deformations obey the laws of elasticity of incompressible slender bodies, reducible to a single longitudinal dimension (Landau and Lifshits, 1986) and dependent on the bending and torsional elastic moduli, but the drag force by the surrounding fluid brings about additional complications (Cox, 1970). The common way to compute flows generated by slim bodies is to approximate them as a sequence of touching beads.

Various shapes and their time evolution can be analyzed by expanding them in Fourier modes, both in time and along the filament's length, most commonly, in the form of traveling waves $\sin(ks - \omega t)$, where k is a wavenumber along the arc length s of the filament and ω is a frequency. Symmetric waveforms result in straight-line net motion, as in the upper left-hand panel of Fig. 4.8. Asymmetry, leading to

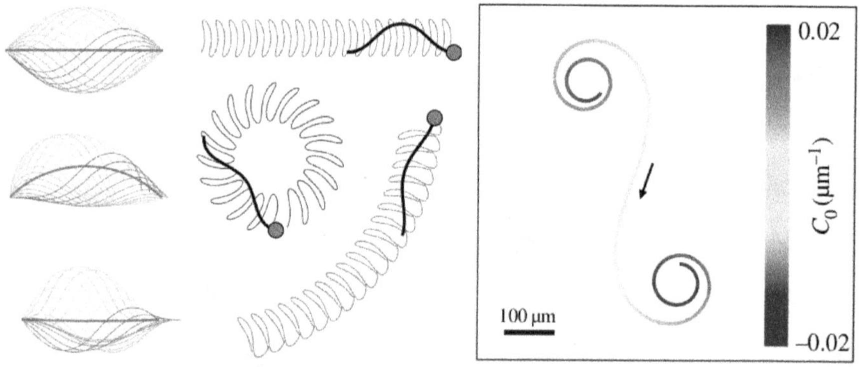

Fig. 4.8 *Left*: Stroboscopic images of beating patterns (with the color changing from blue to yellow in the course of an oscillation period) and the corresponding trajectories; *straight* or *curved red lines* indicate, respectively, a zero or non-zero average flagellar curvature. *Right*: Trajectory with a color-coded variable average flagellar curvature (Gong et al, 2019)

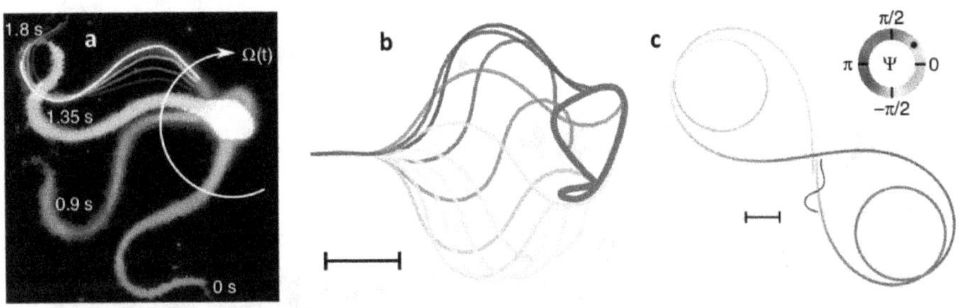

Fig. 4.9 (**a**) Snapshots of a tethered sperm rotating clockwise with the angular velocity $\Omega(t)$ taken at the indicated time. *Gray lines* below the blue curve show the tracked flagellum at 2 μs intervals. (**b**) Simulation of the flagellar beat with an admixture of the second harmonic. Scale bar 10 μm. (**c**) Simulated sperm trajectory resulting from slowly changing the phase difference between the two harmonics, color-coded along the trajectory. Scale bar 20 μm (Saggiorato et al, 2017)

curved swimming paths, can be arranged by maintaining a constant average flagellar curvature or by mixing a symmetric traveling wave, as in the lower left-hand panels of Fig. 4.8 (Gong et al, 2019). Increasing the magnitude of the average curvature of the flagellum causes a more sharply bent trajectory. This is demonstrated in the right-hand panel of Fig. 4.8, showing the trajectory traced when the curvature changes as $C = C_0 + C_1 \sin(ks - \omega t)$ with a variable C_0 and constant C_1.

Asymmetry can also be attained by adding a second temporal harmonic of form $\sin(ks - 2\omega t)$. Saggiorato et al (2017) observed that the second temporal harmonic of flagellar beating causes rotation of sperm around the tethering point (Fig. 4.9a). The other two panels of this figure show a typical simulated beating pattern of the sperm with an admixture of the second harmonic (Fig. 4.9b) and a trajectory with the period-averaged curvature changing as the phase difference between the two harmonics changes slowly (Fig. 4.9c).

Changes in the curvature of the path of a microswimmer are important because they influence the way it approaches a target. Similar to prokaryotes (Sect. 4.1), but through periodic motion rather than tumbling, the sperm's signaling system detects the attractant's gradient translated into a temporal cue. This has been convincingly demonstrated by studies of sperm chemotaxis in 2D (Kaupp and Alvarez, 2016). In the absence of stimulation, sperm swim in circles (colored green in Fig. 4.10a) in a shallow observation chamber. The female sex hormone causes Ca^{2+} ions to enter the sperm and this in turn produces a periodic modulation of the swimming path curvature, whereupon the resulting looping path, marked in yellow, guides the sperm along the gradient.

While swimming along a looping path, the cell is exposed to the attractant concentration, which varies periodically with the frequency of circular swimming. The local maxima, indicated by yellow dots in Fig. 4.10b, cause maxima of curvature, which follow with a phase shift ϕ. The time derivative of the Ca^{2+} concentration (red) and the curvature of the swimming path (magenta) shown in Fig. 4.10c are strongly correlated. Saggiorato et al (2017) have also detected an increasing second-harmonic intensity and enhanced rotation caused by stimulation.

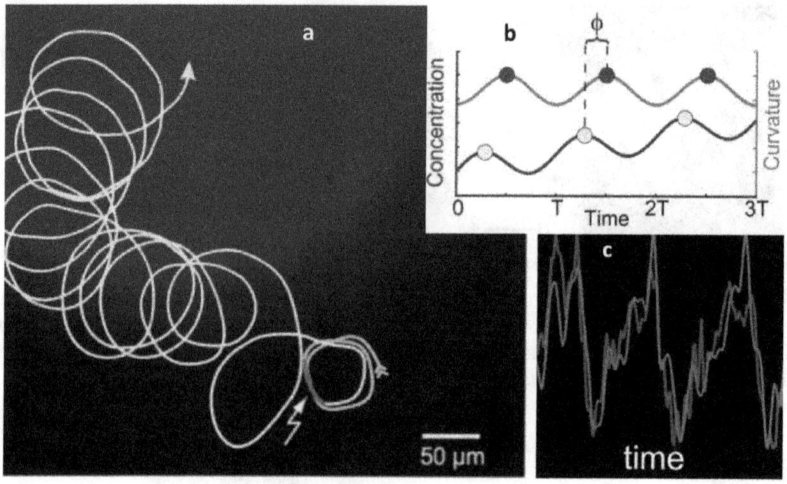

Fig. 4.10 (**a**) Sperm trajectory before stimulation (*green*) and in the attractant gradient (*yellow*), switched on during the interval marked in *red*. (**b**) The periodic modulation of the attractant concentration (*blue curve*) is translated into the modulation of the curvature of the path (*red curve*) with a phase shift ϕ. (**c**) Time derivative of the Ca^{2+} concentration (*red*) and the curvature of the swimming path (*magenta*) as functions of time (Kaupp and Alvarez, 2016)

Many eukaryotes follow this strategy of swimming up or down a gradient. The single-cell alga *Euglena gracilis* swims by asymmetric beating of a single anterior flagellum, which produces helical trajectories accompanied by body rotations. Its phototactic behavior is based on switching flagellar beating patterns and the

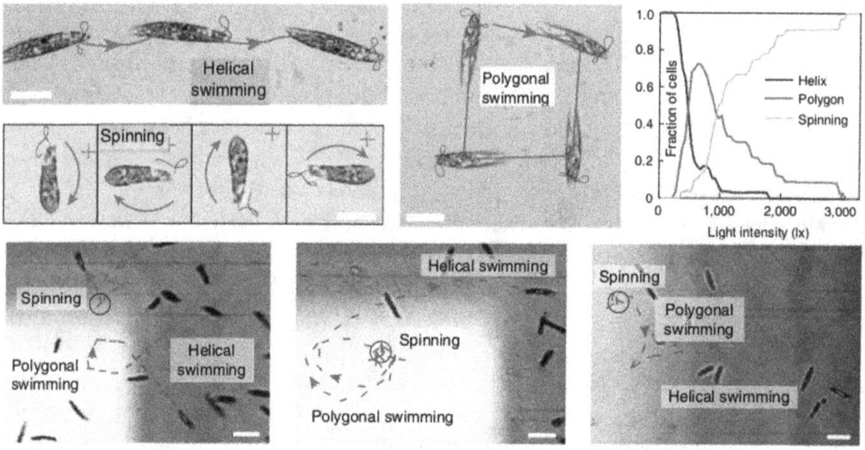

Fig. 4.11 *Top*: Swimming modes of *Euglena gracilis* (*left*) and the dependence of their probability on light intensity (*right*). *Bottom*: Switching swimming modes when encountering a sudden step-up in light intensity (shown by *shading*) (Tsang et al, 2018)

corresponding swimming modes (which include unusual polygonal movements) in response to illumination gradients, as illustrated in Fig. 4.11 (Tsang et al, 2018). These motions change the cell's spatial orientation, which in turn affects the detected light signal, allowing the alga, depending on circumstances, either to enhance photosynthesis or to avoid ultraviolet damage.

4.3 Synchronization

Experiments with sperm may have been favored not only because of its availability and ease of handling but due to relative simplicity of navigation with the help of a single flagellum. Having several flagella brings about the additional problem of synchronizing them. A model organism serving to elucidate this problem is as simple as possible, as it has just two flagella. This is the unicellular green alga *Chlamydomonas*, swimming in a breast-stroke manner, which necessitates synchronizing the flagellar beats: they should be in-phase for efficient straight-line propulsion, while finer coordination is necessary for turns.

The problem is superficially related to the general problem of synchronization of oscillators (Pikovsky et al, 2001), but is substantially more complicated, since not only oscillations in time but also beating patterns have to be coordinated, and there is no evidence for a chemical master oscillator that would force molecular motors in the two flagella to coordinate their actions. Sir Geoffrey Taylor (1951) himself, the most prominent researcher in fluid mechanics ever, took charge of this problem, attributing synchronization to hydrodynamic interactions between the flagella. He found that the dissipation rate is minimal when the waves down the neighboring "tails" (i.e., filaments) are in phase, and, moreover, viscous stress in the fluid between the flagella tends to force synchronization of the two wave trains. Taylor also extended this mechanism to cells with a single flagellum, like sperm swimming close to one another in the same direction, in agreement with observations by Lord Rothschild (1949) whom he thanked for bringing this problem to his attention.

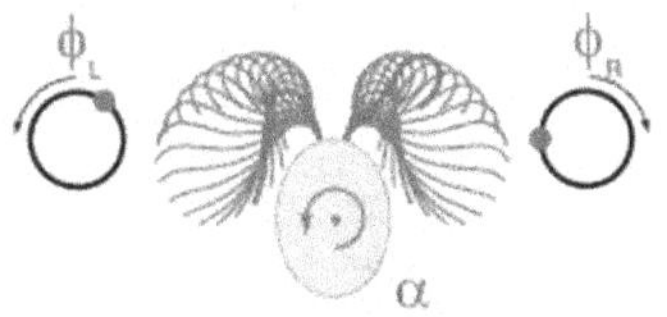

Fig. 4.12 The phases of the beating wave of two flagella change in the opposite directions in proportion to the cell's rotation rate (Friedrich, 2016)

The hydrodynamic synchronization mechanism has been supported and refined by computations and high-speed tracking experiments (Geyer et al, 2013), demonstrating that a deviation from the synchronized state causes rotational motion of the cell body. The ensuing hydrodynamic friction force feeds back on the phases $\phi_{R,L}$ of the beating waves of the two flagella, $\sin(ks - \omega t + \phi_{R,L})$, shifting them in opposite directions in proportion to the cell's rotation rate (Fig. 4.12), and thereby restoring the synchronized state. Sartori et al (2016) explored additional regulation mechanisms due to the feedback control of dynein attachment by changes in the filament's curvature and normal or tangential forces applied to its surface, as sketched in Fig. 4.6c–e.

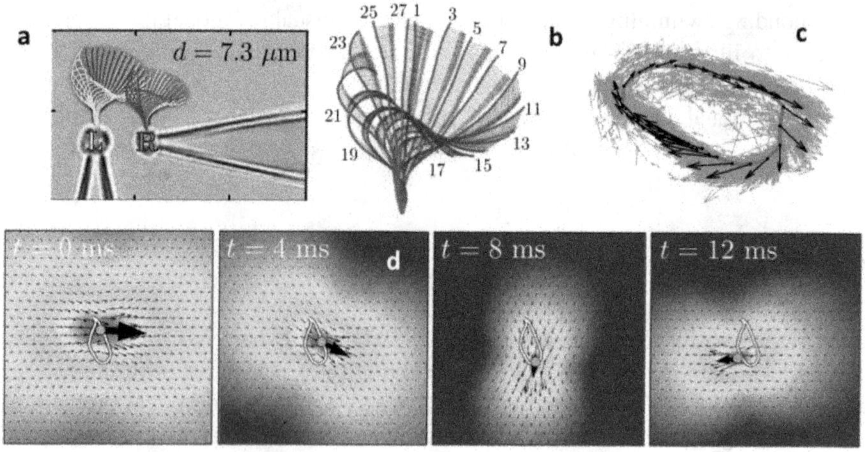

Fig. 4.13 (**a**) A pair of hydrodynamically interacting cells held by tweezers. (**b**) Instantaneous velocity distribution along the flagellum during a complete beat cycle (indexed by the numbers of frames at 2 ms intervals). (**c**) Vector force integrated along the flagellum, shown at a centre of mass position at a certain time during a beat cycle (*red*: per frame, *black*: averaged over about 100 beats). (**d**) Approximated instantaneous velocity field during the first half of a representative flagellar beat (Brumley et al, 2014)

Brumley et al (2014) carried out detailed observations of the flow field generated by flagella, aimed at proving that hydrodynamic interactions are sufficient for synchronization. For this purpose, they held cells at a certain distance using tweezers, as in Fig. 4.13a, to exclude all interfering factors. However, to attain quantitative results, the flow field generated by a single flagellum had to be traced in detail, which is interesting in its own right. Similar to the way we swim, the beating cycle includes a power stroke and a recovery stroke. During the power stroke (frames 1–11 in Fig. 4.13b), the filament is stretched out straight and moves rather fast in one direction, while during the recovery stroke (frames 13–27), it bends and slowly retracts. The force integrated along the flagellum and the fluid velocity field are shown in Fig. 4.13c and d at various stages of the beat. These data are sufficient, in principle, for estimating the mutual influence of the two flagella; however, they do not differentiate between interactions due to viscous friction and those caused by dynein detachment.

A contrasting example is the synchronization of a great number of flagella or cilia. Some microorganisms have their entire body covered with short cilia (Fig. 4.14a), and propel themselves by waving them in a coordinated way, thereby inducing tangential flow. Waving a thin hairy cover has the same effect as inducing slip in a thin layer adjacent to the surface, similar to a colloidal squirmer (Sect. 3.2). A particularly interesting feature of the beat pattern of large coordinated arrays is the formation of *metachronal* waves when the beating is not synchronous, but follows a pattern resembling a wheat field in the wind (Fig. 4.14b).

Brumley et al (2015) studied hydrodynamic interactions responsible for the formation of metachronal waves as thoroughly as synchronization of two flagella. The chosen "model organism" was the multicellular alga *Volvox carteri*. A *Volvox* colony

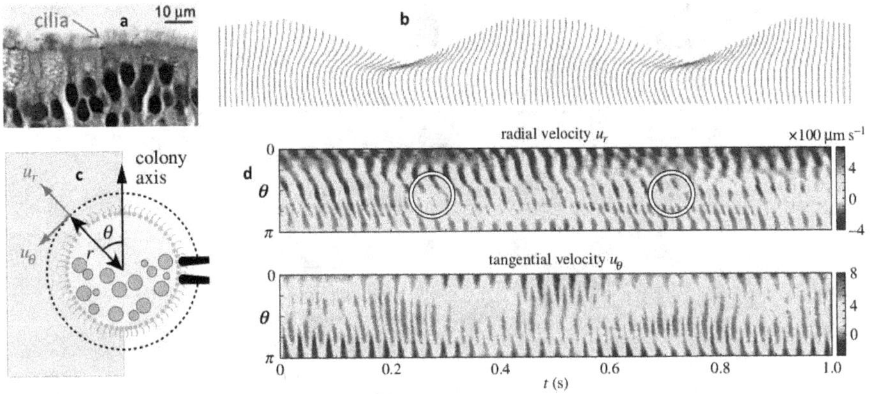

Fig. 4.14 (**a**) Cell covered by multiple cilia (Cicuta, 2020). (**b**) Sketch of a metachronal wave (by Shyamal, CC). (**c**) *Volvox* colony; the *dashed line* shows the location of the velocity measurement. *Green disks* show immature daughter colonies irrelevant for this experiment. (**d**) Dependence of radial and tangential velocities on time t and latitude θ; *circles* surround phase defects (Brumley et al, 2015)

comprises thousands of flagellate somatic cells embedded within a spherical extracellular matrix shell (Fig. 4.14c) and beating their flagella towards the colony's posterior, approximately along meridian lines. As in Fig. 4.13a, the alga was held in place by tweezers, and the dependence of the flow field on time t and latitude θ, shown in Fig. 4.14d, was measured along the circle shown by the dashed line in Fig. 4.14c.

The picture clearly demonstrates metachronal waves interrupted, as wave patterns commonly are, by recurrent phase defects indicated by circles in Fig. 4.14d. The authors attribute the synchronization leading to this characteristic pattern to hydrodynamic interactions, but the influence of the elastic resistance of the body of the colony may also play a role.

4.4 Bacterial Suspensions

Bacteria are habitually modeled as chemotactic or phototactic pushers or pullers, and their suspensions as analogues of colloidal flocks or as continuous active fluids with a symmetry suitable to their mode of locomotion. Many bacterial species are oblong, and, in a similar way to the rod-like particles in Sect. 2.2, they align either in parallel or antiparallel upon collision, as seen in the upper panels of Fig. 4.15. They also show a tendency to cluster, and colliding clusters tend to arrange in a similar manner, as in the central and lower rows of Fig. 4.15. Two co-propagating clusters typically merge upon colliding, but bacteria in counter-propagating clusters are able to penetrate through the opposing cluster.

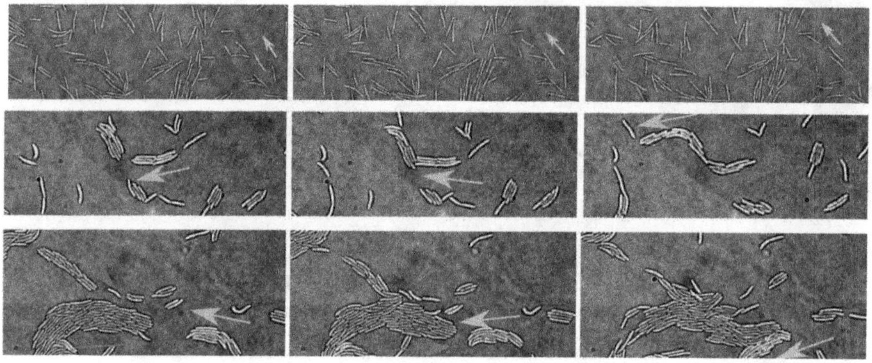

Fig. 4.15 *Top row*: Collisions of individual bacteria. Two counter-propagating myxobacteria become anti-parallel. *Middle row*: Collision of two small clusters moving in opposite directions and passing each other. *Bottom row*: Merging of two clusters (Harvey et al, 2013)

Further clustering may lead to the formation of bands propagating along a chemical gradient, similar to polar (Sect. 1.2) or nematic (Sect. 2.2) active particles. The model put forward by Keller and Segel (1971) couples a diffusion–drift equation for the bacterial density to a reaction–diffusion equation for the nutrient to arrive at a traveling band solution. The model has acquired a "classical" status, and found wide applications in studies of bacterial chemotaxis.

Yet, bacteria are more sophisticated than that. Just as people, while belonging to the same species *H. sapiens*, have different physical (and other) abilities, so too do bacteria. They use run-and-tumble motion to detect chemical gradients (see Sect. 4.1), but the ratio between the durations of the run and tumble periods, what is called the *tumble bias* (TB), varies among individuals. Procrastinating fellows with a high TB waste more time tumbling and diffuse more slowly in a uniform environment (Fig. 4.16a). This makes them less sensitive to chemical gradients, and

Fig. 4.16 (**a**) Trajectories of bacteria with different tumble bias (color coded) in a uniform environment. Scale bar 0.2 mm. (**b**) Simulated density profiles of migrating bacteria with different tumble bias. The *black curve* shows the nutrient concentration distribution (Fu et al, 2018)

this in turn is apt to reduce their chemotactic velocity and make them lag behind a traveling band. However, this does not actually happen. And while human society has instituted social services that help those less able or less fortunate to keep up, bacteria do it in a simpler way.

Experiments by Fu et al (2018), supported by simulations, show self-organized *sorting* of bacteria in traveling bands, with the average TB increasing from front to back (Fig. 4.16b). This has a simple explanation. As an attracting chemical is not only perceived but consumed by the bacteria, its gradient increases from front to back before flattening down. Bacteria with a higher TB, which don't run fast enough in a shallow gradient, lag behind, until they fall into the region where their velocity equalizes with that of individuals with the lowest TB at the front. Only those with the highest TB unable to sustain the collective velocity at the maximal gradient, fall into the gray area in Fig. 4.16b and drop out of the band as weaklings out of a herd.

Moreover, bacterial chemical communications are far more sophisticated than in active colloids. They are capable of producing a full repertoire of signals and respond to a wide variety of chemicals in their environment. Rather than moving to reduce free energy as Janus particles do or even just running towards a source of food as the Keller–Segel model assumes, they react to signals sent by other bacteria of the same or different species. This *quorum sensing* enables organisms to measure their local population density and to regulate their response accordingly, as well as to exchange information about environmental conditions. Warnings of adverse circumstances may trigger aggregation in bacterial suspensions, independently of either cell division or hydrodynamic interactions. This enables a multicellular community to defend against predators or harmful environmental factors.

Bacteria emit and detect signaling molecules called *autoinducers*. At a low cell density, when the autoinducer concentration is below a certain threshold, gene expression programs that benefit individual bacteria are active, but exceeding this limit switches on gene expression programs beneficial for the community and drives clustering. Jemielita et al (2018) proved that quorum sensing is capable of driving aggregation by experimenting with two strains of *V. cholerae* cells, of which only one was sensitive to autoinducer signaling. The snapshot sequence in Fig. 4.17, where

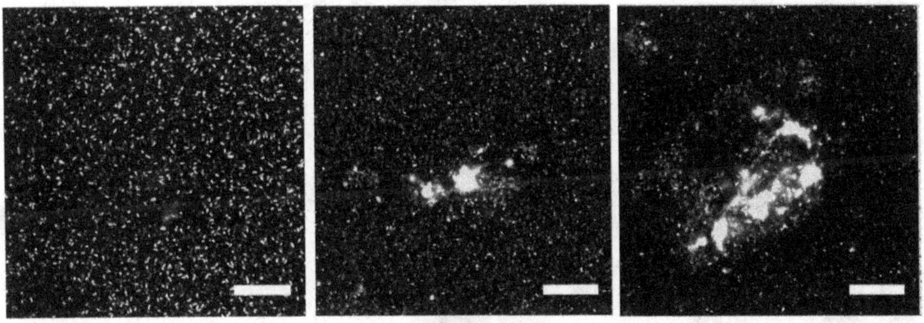

Fig. 4.17 Clustering sequence of bacteria containing a gene responsible for sensitivity to quorum sensing signals (colored *green*); scale bar 250 μm (Jemielita et al, 2018)

bacteria belonging to this strain are marked green, shows their rather rapid (on the scale of half an hour) aggregation from a dilute suspension.

Crowding in dense suspensions inhibits chemotactic sensing, leaving bacteria less space for runs that would allow them to feel the gradient, although the swimming speed may increase slightly due to collective entrainment. As the cell is swimming, it monitors the change in chemoattractant concentration within a few seconds, to decide whether to tumble. If during this time the direction in which the cell swims has changed significantly due to steric interactions with close neighbors, the decision becomes less relevant, thereby making the bacterial chemotaxis strategy inefficient (Colin et al, 2019).

Nearly close-packed populations of swimming bacteria form a collective phase, such as shown in Fig. 4.18a, labeled "zooming bionematic" (Cisneros et al, 2007). It exhibits long-range orientational order, analogous to the molecular alignment of nematic liquid crystals, and a flow field of turbulent appearance (Fig. 4.18b) with strong spatial and temporal correlations of velocity and vorticity, measured in the cited paper. Indeed, dynamics of this kind has prompted a continuous description of dense suspensions as active nematic fluids (Sect. 2.5). In this regime, collective motion leads to an increase in the mean cell velocity with growing density (Sokolov et al, 2007), up to a jamming limit. Notably, a bacterial suspension may display a "superfluid-like" behavior with vanishing viscous resistance to shear (López et al, 2015). This happens when the activity of swimmers organized by shear fully overcomes dissipative effects due to viscous losses.

Patteson et al (2018) observed a more realistic evolution of boundaries between active and passive domains than the simulated sequence in Fig. 2.19. They used ultraviolet light exposure to selectively block cell motility in a dense suspension of *S. marcescens* and create compact domains of passive bacteria within swarms. Post-exposure, the boundaries separating motile and immotile cells reshaped and eroded due to emergent collective flows, resulting in the dissolution of passive domains within active swarms, as demonstrated by the snapshots in the upper panels of

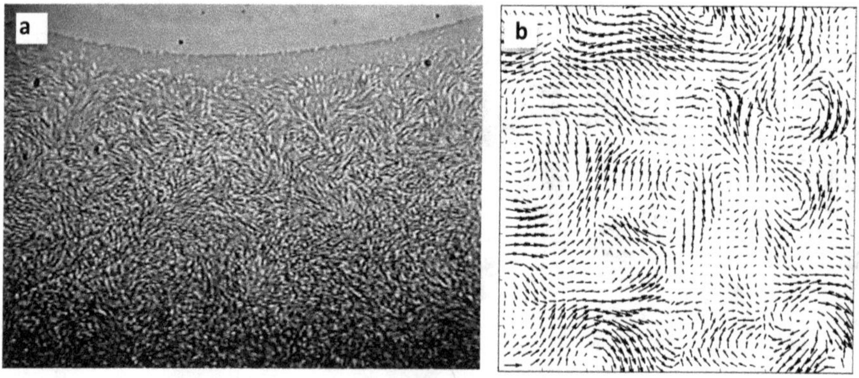

Fig. 4.18 (**a**) *B. subtilis* cells concentrated at a sloping water/air interface. (**b**) Snapshot of the bacterial swimming vector field estimated by particle imaging velocimetry (Cisneros et al, 2007)

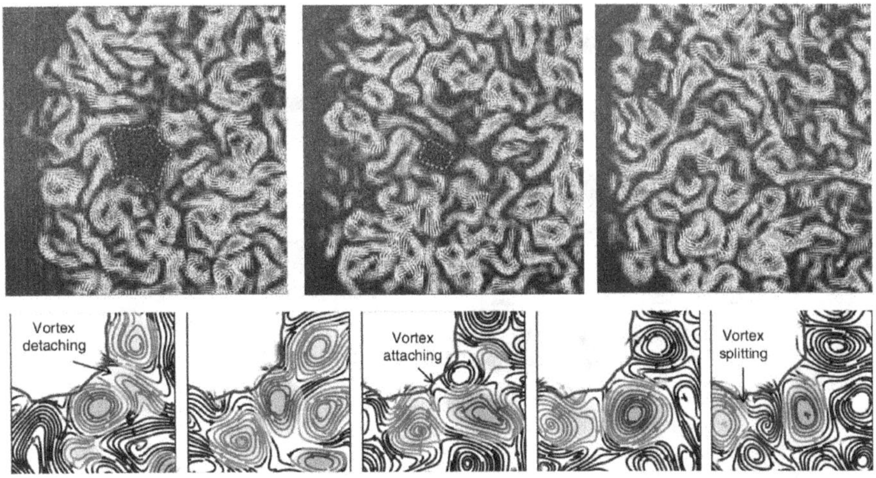

Fig. 4.19 *Top*: Series of snapshots demonstrating the dissolution of the 50 μm star-like passive domain in the *left panel*. The magnitude of the velocity is color coded, increasing up to 50 μm/s from blue to dark red. *Bottom*: Dynamics of vortices near the active/passive interface (Patteson et al, 2018)

Fig. 4.19. The sustained erosion of the interface is facilitated by vortices that form in its vicinity. A montage of the flow streamlines in the lower panels of Fig. 4.19 reveals the motion of vortices (labeled by color) near the active/passive interface marked by the blue line. Vortices starting in the bulk can collide and attach to the interface (e.g., the brown vortex); some vortices at the surface detach and move away (green, orange), while others fade away (purple) or split (blue).

4.5 Bacterial Circus Arena

The sensitivity of bacterial motion to external signals and geometric constraints has prompted experimentalists to "train" them in specific motion patterns. Nishiguchi et al (2018) rectified the seemingly chaotic motion of *B. subtilis* cells seen in Fig. 4.18 by confining them within a 2D periodic array of microscopic vertical pillars. The bacteria self-organized in this setting into a lattice of hydrodynamically bound vortices with a long-range antiferromagnetic order of alternating left- and right-rotating vortices controlled by the pillar spacing (Fig. 4.20). The patterns were most stable and maintained a nearly perfect order when the pillar spacing was comparable with the typical vortex size of unconstrained bacterial turbulence, such as that in Fig. 4.18b.

In other experiments involving the same group, bacteria in a turbulent suspension were induced to move heavy loads. They complied, bringing edges with parallel

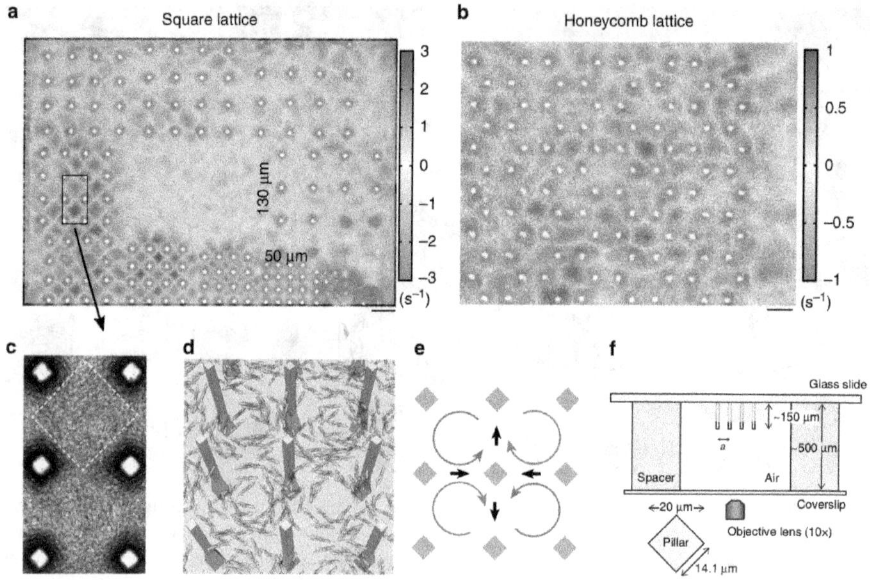

Fig. 4.20 (**a**) Snapshot of bacteria swimming between square lattices of pillars overlaid by a color plot of the average vorticity magnitude. Pillars are arranged in nine arrays with different lattice constants increasing clockwise in 10 µm increments from 50 to130 µm. Scale bar 100 µm. (**b**) Distribution of the average vorticity magnitude in a honeycomb lattice. Scale bar 50 µm. (**c**) Close-up of the rectangular area shown in (**a**). *Arrows* indicate instantaneous velocities. The *yellow dashed line* depicts a single region of interest. Scale bar 50 µm. (**d**) Artistic representation of bacteria swimming between 3D-printed micropillars. (**e**) Swimming bacteria self-organized in a lattice of counter-rotating vortices. *Black arrows* indicate the direction of bacterial flow between the vortices. (**f**) Experimental setup. For clarity, only one set of pillars is shown (Nishiguchi et al, 2018)

orientation (controlled by the applied magnetic field) closer together, and moving wedges oriented antiparallel further apart (Fig. 4.21).

Bacteria can self-organize in a special way in liquid crystals. Most of these materials are toxic, but bacteria are capable of growing in *lyotropic chromonic* liquid crystals (LCLC) found in some food and textile dyes. Nematic LCLC molecules are formed by stacking flat units containing aromatic rings with attached acidic groups and sodium atoms (Fig. 4.22a). Sodium ions dissociate in water solutions leaving

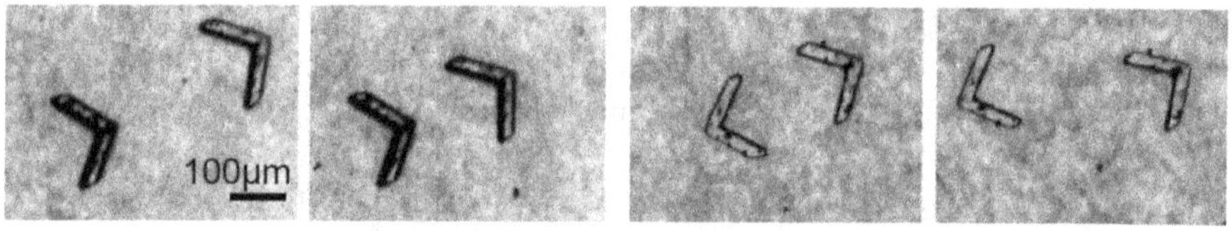

Fig. 4.21 Illustration of the attraction and repulsion of wedge-like carriers in a turbulent bacterial bath (Kaiser et al, 2015)

the stack negatively charged (Fig. 4.22b). The nematic order of these elongated aggregates is enhanced at lower temperatures, when the stacks grow longer. The experiments are carried out in a shallow cuvette where nematic order can be set by rubbing the confining plates (Fig. 4.22c).

Lavrentovich and coworkers have taken advantage of this medium to explore interactions of bacterial suspensions with liquid-crystalline patterns in what they call *living liquid crystals*, setting the aim of controlling the chaotic behavior of bacterial suspensions (Zhou et al, 2014; Peng et al, 2016). When rod-like *B. subtilis* cells are dispersed in a liquid crystalline environment with spatially varying orientation, they tend to align with the director and move along these lines in either direction. At the same time, they recognize subtle details in liquid crystal patterns. This makes it possible to control the distribution of bacteria and the geometry of their trajectories by arranging nematic alignment patterns, which can be created on demand by anchoring the nematic director with the help of a light-sensitive lithographic design of confining plates (Guo et al, 2016). In this way, it is possible, for example, to arrange a periodic pattern of defects that induces bacteria to self-organize into a lattice of co-rotating vortices (Peng et al, 2016).

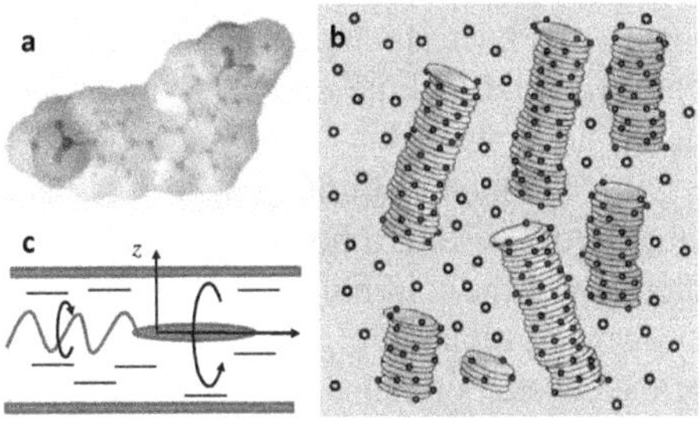

Fig. 4.22 (**a**) The structure of an LCLC unit. The acidic group is colored *magenta*, and sodium atoms, *pink*. (**b**) Nematic stacks. *Small circles* show sodium ions. (**c**) Sketch of the experimental setup, showing the director and a moving bacterium (Zhou, 2017)

Interactions between nematic and bacterial alignments are mutual. While bacteria orient along the director, the nematic order is distorted by their activity (Fig. 4.23b), leading eventually to nucleation of paired defects (Fig. 4.23c). Bacteria differentiate between topological defects, heading toward defects of positive topological charge and avoiding those of negative charge. This is seen in both experimental (Fig. 4.24b)

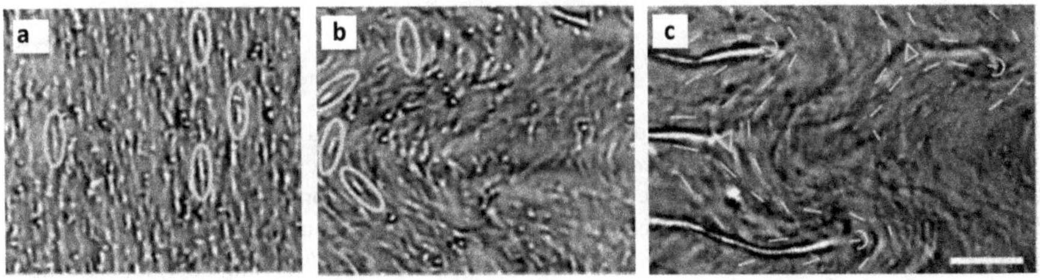

Fig. 4.23 (**a**) Inactive bacteria (highlighted by *ellipses*) in an ordered nematic align along the director. (**b**) Active bacteria distort nematic order. (**c**) Nucleating positive (*semicircles*) and negative (*triangles*) half-charged defects. *Yellow dashes* show bacterial orientation (Zhou, 2017)

and simulated pictures (Fig. 4.24e and f) of patterns in a free-standing liquid crystal film unconstrained by the director anchoring (Genkin et al, 2017). It is explained by the arrangement of bacterial trajectories that follow nematic alignment lines, as sketched in Fig. 4.24c and d. The incoming trajectories converge at the core of the positive defect, so that bacteria accumulate in its core and depart while swimming in the opposite direction. In contrast, the negative defect creates a nematic configuration that expels bacteria, independently of their orientation. Bacteria close to the defect core (in the dark blue region) swim away, whereas bacteria swimming towards the defect are deflected.

The common swimming pattern of bacteria along the nematic director is impossible in the case of a homeotropic alignment when the director is oriented perpendicularly to the confining plates. In this case, bacteria can either spin along their long axis, while remaining parallel to the imposed director but not moving in the plane of the cuvette, or swim perpendicularly to the director at a certain distance from the plates. A swimming bacterium produces a symmetric quadrupolar pattern of direc-

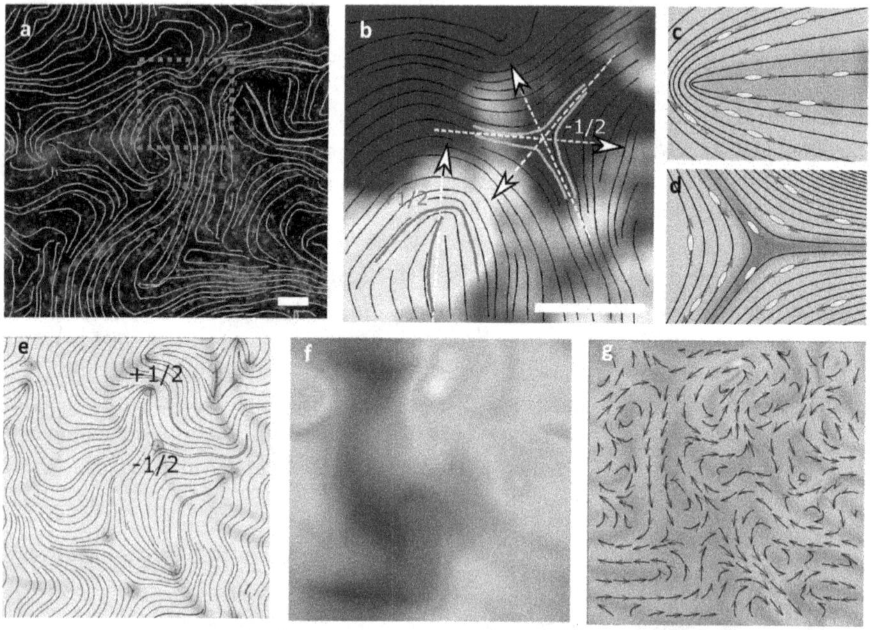

Fig. 4.24 (**a**) Swimming bacteria suspended in a free-standing liquid crystal film in the regime of chaotic motion. Scale bar 50 μm. *Green lines* show the nematic director orientation reconstructed from the bacterial orientation. (**b**) Bacterial concentration distribution in the area indicated by the *blue dashed box* in (**a**). (**c**) Bacteria accumulate in the *yellow region* near a +1/2 defect due to the convergence of their trajectories. (**d**) Bacteria escape from the *dark blue region* near a −1/2 defect, independently of their initial orientation. (**e**)–(**g**) Results of numerical modeling of a film with no director anchoring, showing nematic orientations and the magnitudes of the order parameter (**e**), concentration fields (**f**), and flow velocity magnitudes and streamlines (**g**). In all pictures, the respective magnitudes increase as the color changes from dark blue to yellow (Genkin et al, 2017)

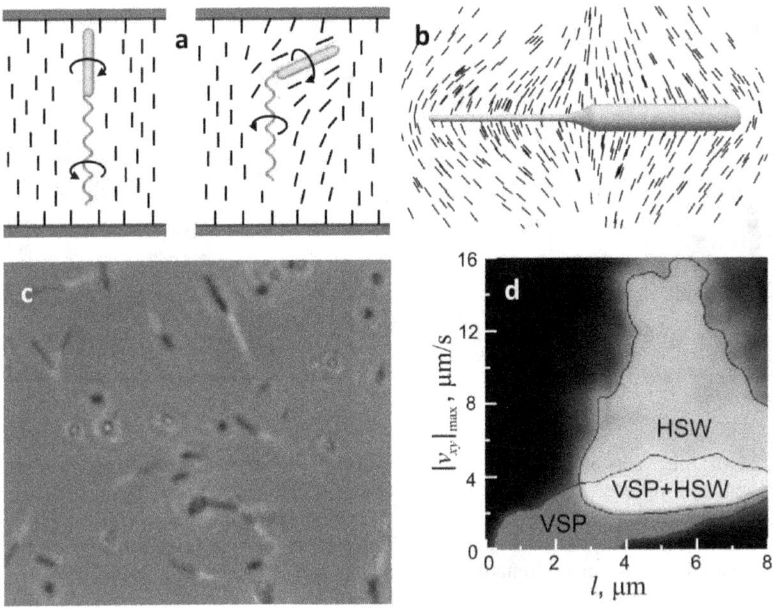

Fig. 4.25 (**a**) Homeotropic director alignment around a rotating bacterium and its distortion with a change in the incline. (**b**) Director profile generated by a bacterium swimming horizontally. (**c**) Phase contrast microscopy shows horizontally moving (HSW) bacteria as dark rods with white traces produced by the flagella bundles; rotating (VSP) bacteria appear as dark dots. (**d**) Diagram of VSP, HSW, and mixed regimes, depending on the length l of the bacteria and their maximum swimming velocity $|v_{xy}|_{max}$ (Zhou et al, 2017)

tor distortions around itself (Fig. 4.25a and b), and this is minimal if they sustain a horizontal orientation. Bacteria behaving in either way are seen in a microscopic snapshot in Fig. 4.25c. They may occasionally change their swimming direction, either by reversing it and backtracking, or through a vertical flip-flop. The prevailing mode of behavior depends on the length of the bacteria and their maximum swimming velocity, as shown in the diagram in Fig. 4.25d. At higher concentrations, swimming bacteria create large areas of strong director distortions that attract other bacteria and generate areas of horizontal director orientation.

4.6 Swarms and Colonies

Besides swimming through liquid media, bacteria can use flagella to crawl on solid surfaces. When confined to a surface, a bacterial suspension turns into a *swarm*. Bacterial swarming may take the form of mass migration, with multitudes of cells spreading collectively to colonize surfaces. Swarms containing elongated cells retain a tendency to arrange themselves in a nematic order and gather into clusters. Be'er et al (2020) carried out a series of experiments, observing swarms of *B. subtilis*

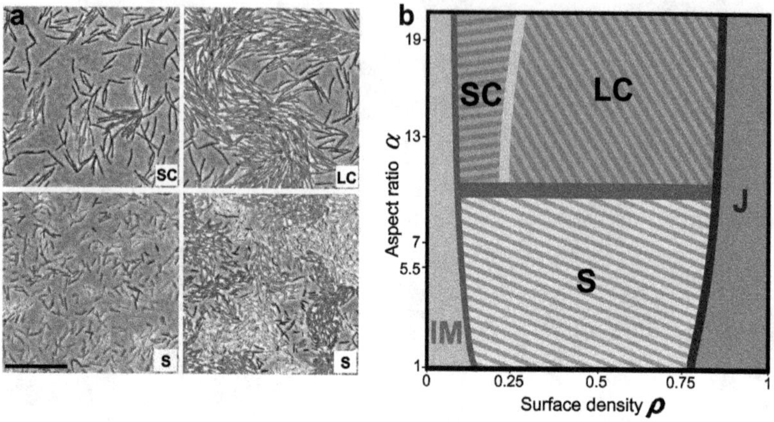

Fig. 4.26 (**a**) Snapshots of monolayer bacterial swarms with different aspect ratios and densities, representative of different motile phases: small clusters (SC), large clusters (LC), and swarming (S) at low or high densities. Moving clusters are *colored*, while immobile cells are *gray*. Scale bar 50 μm. (**b**) Phase diagram as a function of the aspect ratio α and surface density ρ, showing, besides swarming phases in (**a**), immotile (IM) and jammed (J) states (Be'er et al, 2020)

mutants with different aspect ratios, all of them attaining a state with a homogeneous density and no preferred direction of motion. As in suspensions (Sokolov et al, 2007), the mean cell speed tends to increase with density, showing that cells cooperate to produce faster motion.

Typical swarming regimes are shown in Fig. 4.26a. Based on their observations, Be'er et al compiled a phase diagram (Fig. 4.26b) showing the dependence of qualitatively distinct swarming regimes on the aspect ratio α and surface density ρ of a bacterial population. They found that clustering occurs only above a critical aspect ratio (about 10), and that, naturally, a transition from small to large clusters takes place with increasing density. Within the S region, swarming statistics were not sensitive to the density, or to small changes in the aspect ratio, save for the immobile and jamming states at the two extremes.

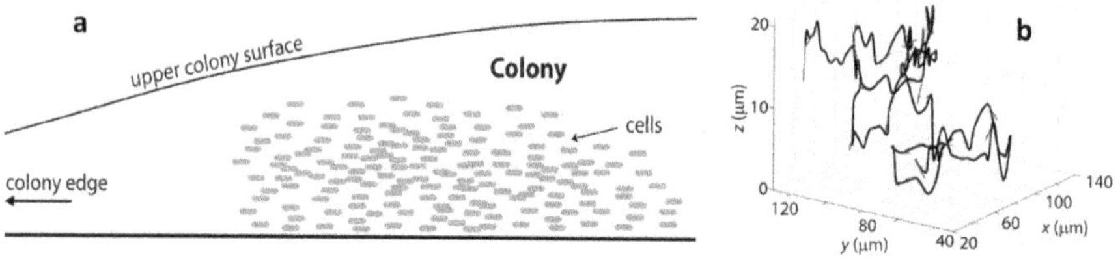

Fig. 4.27 (**a**) Sketch of the cross-section of the 3D colony near the edge. (**b**) 3D trajectory of a single cell. *Red arrows* indicate instantaneous normal vectors to the trajectory, with length proportional to the local curvature (Partridge et al, 2018)

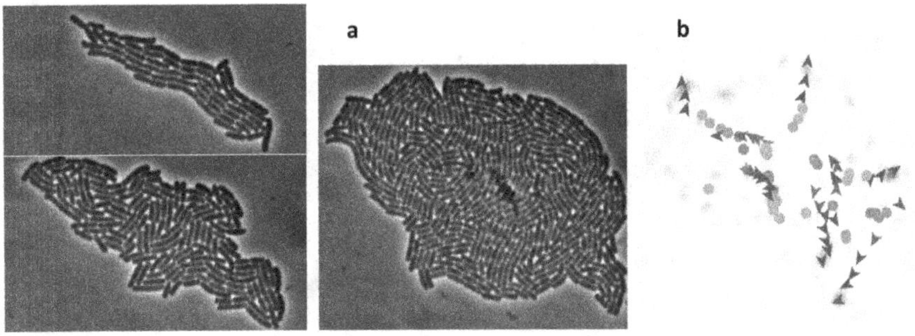

Fig. 4.28 (**a**) Snapshots of a growing bacterial colony. (**b**) Trajectories of positive (*red*, with *arrows* indicating their polarity) and negative (*green*) topological defects within a colony. The markers are increasingly faded for earlier times (Dell'Arciprete et al, 2018)

A swarm can also extend to 3D, as in Fig. 4.27a. In this setting, Partridge et al (2018) recorded bacterial trajectories like the one shown in Fig. 4.27b. Bacteria move freely within the swarm, without being affected by chemical gradients, displaying an intricate swirling motion where hundreds of dynamic bacterial clusters continuously form and dissociate. This vivid mobility distinguishes a 3D swarm from more consolidated biofilms, to be discussed in the next section. Cells tend to move in relatively straight lines in the bulk, and change their direction when coming close to the boundaries.

As oblong cells divide and a swarm grows into a bacterial colony, as shown in Fig. 4.28a, it retains nematic order in what Dell'Arciprete et al (2018) call a "Hubble expansion" of their little universe. Half-charged defects form and move as the colony expands. Negative defects are just advected by the expansion, perhaps aided by elastic interactions. In contrast, positive defects show marked directed motion, recorded in Fig. 4.28b, being pushed by their "comet tails", as in Sect. 2.6.

Many bacterial species are not normally free-living, but attach to surfaces and transform from swimmers to crawlers or "stickers" through changes in gene expression. Surfaces provide a degree of stability in the growth environment. This has prompted the hypothesis by von Nägeli (1884) that life may have emerged in a safer environment of adsorbed layers, rather than in a "prebiotic broth". Prebiotic evolution

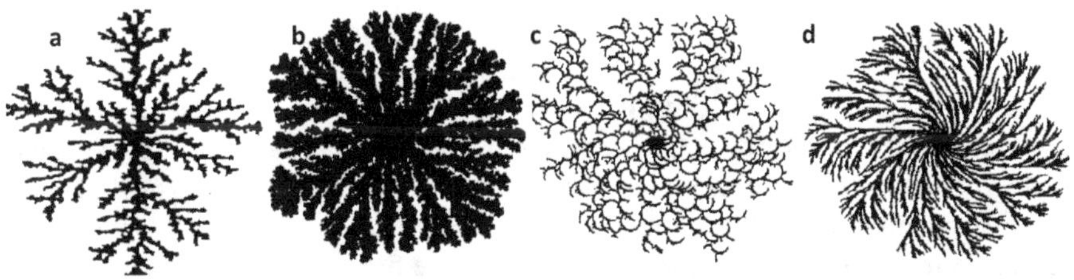

Fig. 4.29 Different shapes of bacterial colonies: branched (**a**), (**b**) and chiral (**c**), (**d**), Ben Jacob et al (1995, 1997)

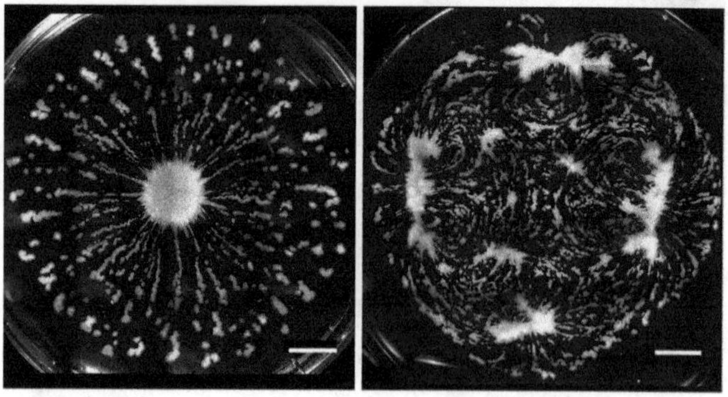

Fig. 4.30 Fingering (*left*) and fragmented (*right*) structure of a bacterial colony growing on a substrate with the viscosity within the range from 300 to 450 Pa s and from 50 to 300 Pa s, respectively. Scale bars 10 mm. (Atis et al, 2019)

on catalytic surfaces, supported by volcanic or geothermal energy, is now viewed as more viable than Oparin's (1924) hypothesis of protocells (e.g., Skoblikow and Zimin, 2018).

Surface-bound bacterial colonies grow on a macroscopic scale in a variety of *morphotypes*, depending on environmental conditions and interactions between the bacteria. The shapes of the colonies in Fig. 4.29a and b resemble branched structures emerging in non-living systems due to the Mullins–Sekerka instability (Sect. 3.6). The observed and simulated branched or "tip-splitting" patterns (Ben Jacob et al, 1995, 1997) are denser and more compact at higher nutrient levels or easier spreading, as would occur on a softer substrate. Bacterial colonies can develop a chiral morphology with the colony consisting of twisted branches, growing into vortex-like spirals, all with the same handedness (Fig. 4.29c and d). Ben Jacob et al attributed the observed macroscopic chirality to the microscopic chirality of the flagella magnified via orientation interactions among bacteria.

The shape of a colony is also sensitive to the substrate viscosity. A colony growing on a solid or a very viscous liquid substrate is compact, but, when the viscosity of the liquid substrate is lowered, it develops first fingering and then fragmentation, as shown in Fig. 4.30. Growth of the colony induces flow in the liquid substrate (Atis et al, 2019).

4.7 Biofilms

Dense growing swarms expanding to 3D develop into *biofilms*. An individual founder cell, expanding from the outset into a surface-bound colony, as in Fig. 4.28, may give rise to a dome-shaped biofilm, containing thousands of cells. Such a "verticalization" involves both disruption of nematic order and mechanical stresses, and is triggered by both the compressive and peeling instabilities of the flat layer caused by the growth of the colony (Beroz et al, 2018). A sharp change in the orientation of the cells is necessary to trigger expansion in the vertical dimension; it initiates a reorientation wave spreading outwards, and eventually the biofilm develops a roughly circular core of nearly vertical cells surrounded by a halo of horizontal cells. The threshold surface

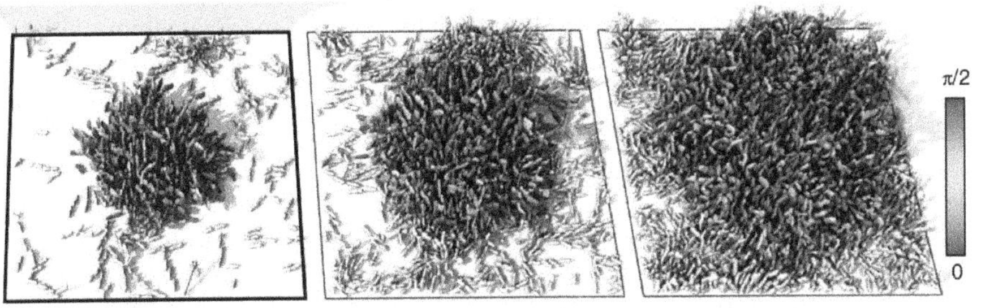

Fig. 4.31 Biofilm growth sequence. Each cell is colored according to its alignment with the vertical, with horizontal cells shown in *gray* (Hartmann et al, 2018)

density for verticalization increases with the cell length, whence biofilms composed of longer cells maintain a wider periphery of horizontal cells, and therefore yield more rapidly expanding, flatter biofilms.

Two complementary close-up views of this process are presented in Figs. 4.31 and 4.32. Part of the observed biofilm growth series in Fig. 4.31 shows thousands of individual cells, digitally colored according to their orientation. Figure 4.32 presents

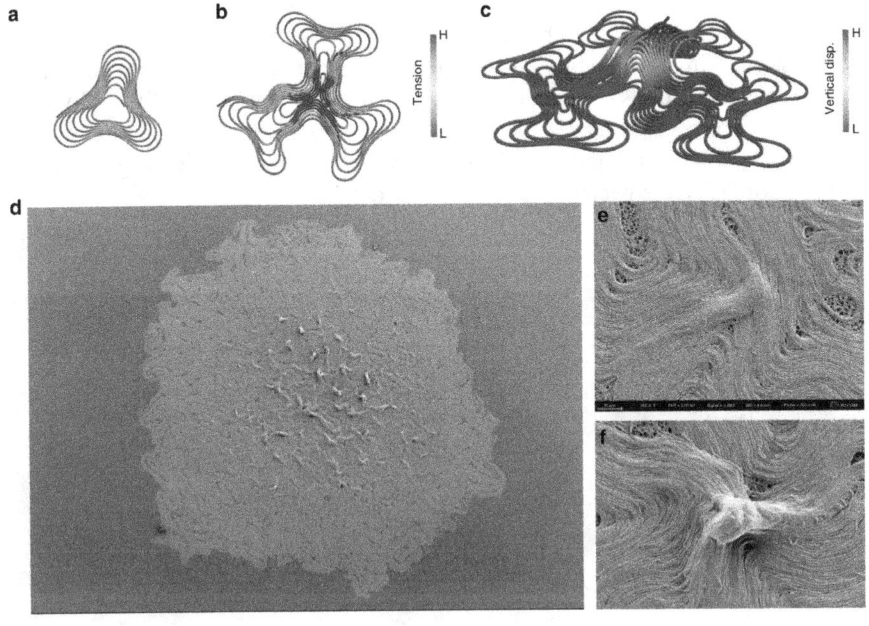

Fig. 4.32 Accumulation of stress and vertical lift-off. (**a**)–(**c**) Simulation results of a growing multilayered circular colony folding after reaching a critical radius. The stress accumulated in the red-colored region results in vertical lifting of the biofilm. (**d**) Overall view of the observed biofilm. (**e**) and (**f**) Accumulation of stress causes vertical lift-off. *Red arrows* indicate the directions of the driving forces (Yaman et al, 2019)

a continuum view on a larger scale, illustrating the effect of mechanical instabilities that generate the verticalization nuclei shown in Fig. 4.32d. These then grow as can be seen in the other panels of this figure, both simulated and experimental. The authors try to tie the vertical lift-off, driven by compressive stress, to propagation of nematic defects, but this approach is questionable, since the topology of the defects changes when the medium becomes three-dimensional (see Sect. 2.3).

Mature biofilms develop variegated morphologies, which are determined not only by microbial gene expression programs, but also by contributions from mechanical forces (Yan et al, 2019). The mismatch between an expanding biofilm and a non-growing substrate constrains the biofilm expansion, and the compressive stress first causes wrinkling and then delamination, as illustrated in Fig. 4.33. The interfacial energy of the biofilm is identified as the key force driving the morphogenesis, because it dictates the ability of the interface to expand as the biofilm buckles and wrinkles. The adhesion energy between the biofilm layers and the substrate and the substrate stiffness are additional contributing factors. Delamination is more "expensive", because new interfaces are generated, and it occurs in systems with film–substrate adhesion energies that are much smaller than the energy of elastic deformations.

In complex biofilms, microcolonies of different species coexist, compete for diffusing nutrients, and cooperate through chemical signaling. The integrating mechanism is *quorum sensing*, already mentioned in Sect. 4.4. Biofilms often include different microbe species, and bacteria are able to differentiate between signals sent by kin or non-kin. Cooperation between bacteria, either genomicaly related or not, involves metabolic costs of signaling, which are rewarded by the benefits of division of labor, which leads to more efficient proliferation and collective defense against invaders or antibiotics. Interaction between bacterial species may even go as far as exchange of DNA, the bacterial analogue of sex.

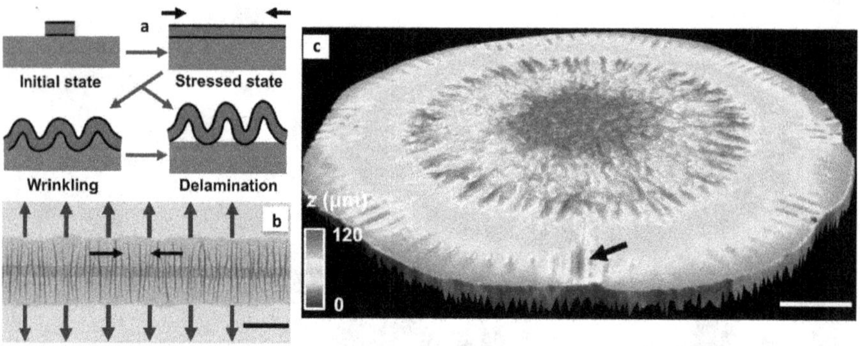

Fig. 4.33 (**a**) Sketch of wrinkling and delamination due to compressive stress. (**b**) Cross-section of the biofilm. *Blue arrows* show the expansion directions, and *black arrows*, the tangential direction of compressive stress. Scale bar 5 mm. (**c**) The overview of the biofilm at the onset of the wrinkling to delamination transition. The *arrow* indicates a blister. Scale bar 2 mm. (Yan et al, 2019)

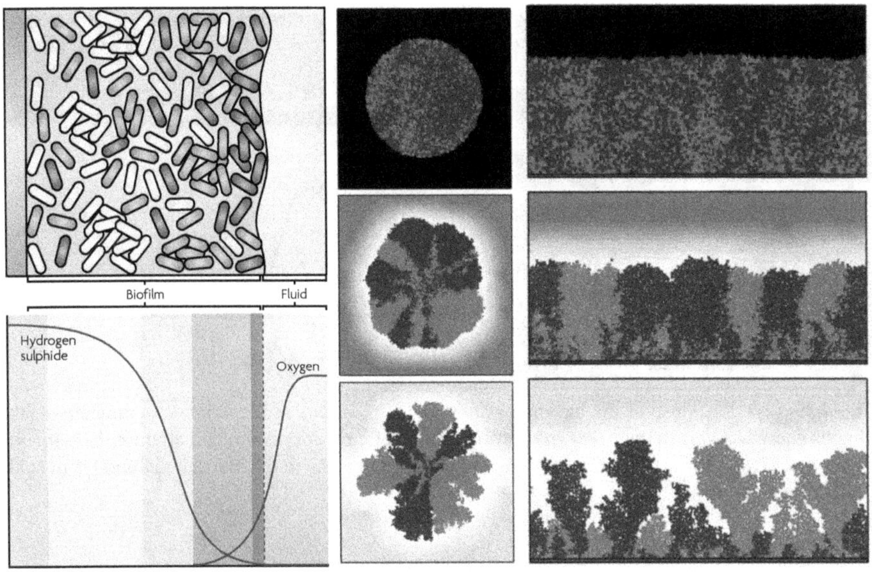

Fig. 4.34 *Left*: Three groups of microorganisms with different metabolisms are distributed within a biofilm, concentrating in areas that suit their needs – and match their color (Stewart and Franklin, 2008). The simulated segregation pattern between cooperators and cheaters (*red* and *blue*) under conditions of radial (*center*) and vertical (*right*) growth and high (*upper row*), moderate (*middle row*), and low (*lower row*) growth substrate availability (Nadell et al, 2010)

Different species may be unevenly distributed within a biofilm when they concentrate in locations most suitable to their abilities and needs. In a simple example schematized in the left panels of Fig. 4.34, hydrogen sulfide diffuses from the left and oxygen from the right. Accordingly, the abundance of aerobic bacteria, colored brown in the picture, rapidly decays from right to left, while the density of sulphate-reducing bacteria, colored yellow, decreases in the opposite direction, in parallel to the decay of concentrations of, respectively, oxygen and hydrogen sulfide, shown in the lower panel. Sulphide-oxidizing bacteria, colored blue, prefer the blue area in the lower panel where both of the chemicals they need are present.

Segregation may be social, driven by motives not unfamiliar to us humans. Bacteria often depend on insoluble substrates for growth, and secrete digestive enzymes, which degrade such substrates into soluble units. These soluble, nutrient-rich products can be captured by neighboring cells, which do not invest themselves in the production of digestive enzymes. Nadell et al (2010) built up a biofilm growth model imitating such a situation. It includes two cell lineages: exploitative that devote all resources to growth, and cooperative that secrete a diffusible compound benefitting all other cells in the vicinity. The cell's growth rate is proportional in this model to the local substrate concentration when it is low, and saturates at high concentration, according to the common Michaelis–Menten kinetics. Once a cell reaches a maximum size, it divides, and cells move passively due to the forces exerted by their neighbors

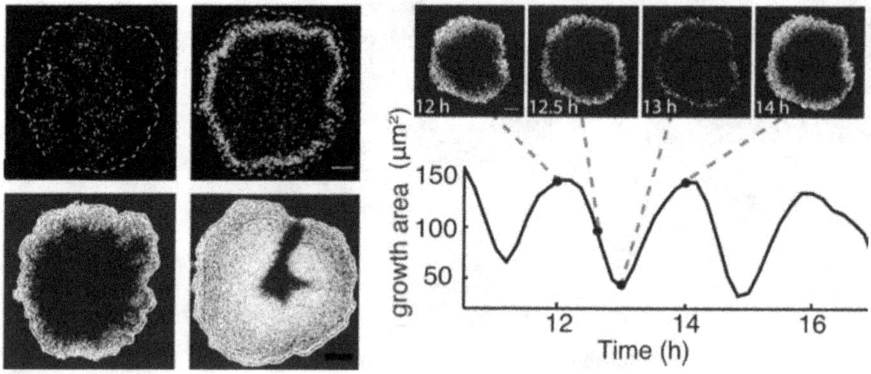

Fig. 4.35 *Left*: Response of the biofilm to an attack. *Green dots* in the *upper row* indicate death of cells and *white dots* in the *lower row*, proliferation before (*left column*) and after (*right column*) the attack. *Right*: Oscillations of the growth area (shown in *white* in the relevant pictures). Liu et al (2015)

as they grow and divide as well. Cooperative cells are at disadvantage when they are mixed with cheaters, which exploit the public goods without themselves paying a cost; accordingly, they benefit by segregating in space and preferentially interacting with each other. This is still irrelevant when the growth substrate is supplied in abundance, so that the entire community prospers, but segregation becomes a more and more attractive choice as the economy deteriorates. This leads to segregation of the lineages, becoming more pronounced with growing scarcity, as shown in the central and right-hand panels of Fig. 4.34. The egoistic lineages do not die away but spread more slowly without public benefits.

Darwinian evolution favors communities of a different kind, where cohabitation of different species is mutually beneficial. In mixed biofilms, cells of metabolically cooperative species may benefit by growing together, but each microcolony must define its optimal size and suppress cell division by quorum sensing when a physiologically optimal size has been reached.

Division of labor may also develop among bacteria of the same strain. In the experiments by Liu et al (2015), bacteria near the outer fringe of the biofilm had better access to externally supplied nutrients, while those in the interior were starved. However, peripheral bacteria are vulnerable to outside threat. When attacked, they die, protecting their kin further in. The latter get access to food and proliferate, replacing peripheral cells, as illustrated in the left panels of Fig. 4.35. The division of labor between well-fed defenders and undernourished folk inside increases resilience to outside threats – but the bacterial community also contrives to prevent mass starvation in quiet times. Outside cells periodically stop proliferating, allowing more nutrients to penetrate inward, which leads to growth oscillations shown in the right panels of Fig. 4.35. Cooperation among cells is the bacterial alternative to integrating into a multicellular organism, and might have been the precursor of the transition to multicellular life.

Chapter 5
Eukaryotic Cells

5.1 The Most Wonderful Machine

Going further down from the scale of microorganisms, we now coming to the inner workings of the most sophisticated machine ever invented: the *eukaryotic cell*. It is the most wonderful device that Nature has evolved, more complicated than anything we have or will ever devise, more complicated than a planet or a star, more complicated than all the physical laws of the Universe, those we know and those we don't know yet or will never know. If there is anywhere in the Universe a finer device, it is sustaining alien life.

Some might claim that we are more sophisticated than our cells, but all our actions and thoughts come from their synergy, and we ourselves came from a single cell containing the development software. Think of a microscopic device containing a command center (nucleus) with a universal digitally coded program (DNA) that is able to direct an all-round chemical factory synthesizing proteins on demand and breaking them after they have served their purpose. A device bound by a shrewd membrane, able to receive signals affecting the decisions made in the command center, and to send signals to a network of cooperating machines, and to block invasion by unwelcome guests. A device having its own power plant (mitochondrion), a designated system of energy conversion among several chemical agents, and an internal communication network serving to distribute material and energy to dedicated destinations. A device capable of reproducing itself at the proper moment and altruistic to the degree that it is ready to commit suicide (apoptosis) when higher powers demand it. The exterior of the nucleus is the *cytoplasm*, a maze of cytoskeletal filaments immersed in watery *cytosol*, where all the organelles are located.

The amazing structure of the cell has been honed for eons by unicellular organisms; for them, it is a matter of survival. Animals and plants may not fully use it in all their cells and all the time – but Nature is conservative, she does not abandon solutions once they have been found to work. Nature cares very much about this structure, dedicating to it a substantial fraction of genomes, highly conserved further up the evolutionary tree; it is estimated that between 2 to 5% of the *Homo*

L. Pismen, *Active Matter Within and Around Us*, The Frontiers Collection,
https://doi.org/10.1007/978-3-030-68421-1_5

sapiens genome codes proteins taking part in the assembly of actin filaments alone. Researchers care no less about it, uncovering minute details from the molecular to the mechanical level.

Armies of biologists, biochemists, and biophysicists, more generously funded than their colleagues in any other branch of science, study various modes of operation of the cells in the innards of healthy or sick individuals or along the conveyor belt of organism development. Thanks to these studies, we live on the average some twenty years longer than hundred years ago, but we still know much less than we don't know. In this book, we cannot delve deeply into the workings of the cell. Its intricacies are hidden in its chemistry, but for us, here as in diffusiophoresis or bacterial growth, chemistry provides an unspecified source of energy driving the mechanics of an active medium. Yet, we need to view this medium here in finer detail than up to now. Even though we restrict in this chapter to mechanical aspects of the cell, we arrive at the peak of complexity midway through the book, based on observations and experiment rather than theories and simulations.

5.2 Filaments and Motors

The mechanical integrity of a cell is sustained by a network of filaments built of protein units and stressed by molecular motors. The sturdiest filaments of this kind, microtubules, rigid tubular structures that hardly bend over their length, are assembled from dimers of the *tubulin* protein, curling into a tight helix 25 nanometers wide around a hollow center. As the strongest structural element of the cytoskeleton, they play an important role in the process of cell division (Sect. 8.2). The most numerous of all are *actin* filaments, also called *microfilaments*, assembled from

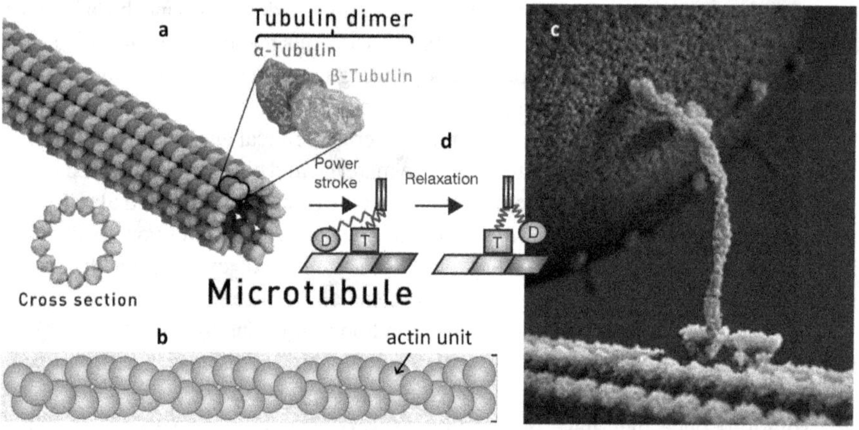

Fig. 5.1 (**a**) Structure of a microtubule. (**b**) An actin filament. (**c**) Kinesin motor carrying protein cargo along a microtubule (CC). (**d**) Power and recovery strokes of a molecular motor (Karsenti et al, 2006)

actin monomers (Fig. 5.1b). They are about 6 nm in diameter, and more flexible than microtubules, but they still bend only slightly over their length. The least prominent kind are *intermediate* filaments, about 10 nm thick and far more flexible.

Both microtubules and actin filaments, but not intermediate ones, are polar. The polarity defines both their direction of growth and the direction of motion of the associated *molecular motors*. The latter's functions are different. *Kinesin* motors (Fig. 5.1c) carry protein cargo along microtubules. They have a dimer structure, with each unit having a globular *head*; the attached long strands are intertwined in a *stalk*. The dimer is structured in such a way that it literally walks upright on its track, moving its two heads alternately, as we move our feet (Fig. 5.1d).

The required energy comes from the common cellular currency – ATP. The energy is released when one of its three phosphorous groups is hydrolyzed, ATP $\rightarrow$ ADP (ATriP to ADiP, P standing for phosphate and A for adenosine, the name of the carrying nucleotide), and is recharged when the missing phosphorus group is joined back in the mitochondrion. Each step involves a change of conformation, and requires one ATP molecule to be released. The track is unidirectional: as a rule, kinesin motors walk towards the "positive" end of polarized microtubules, although some are able to switch directionality. Actin filaments are associated with *myosin* motors, structured, walking, and powered in a similar way, but their main function is to tie up and stress the filaments, as they are integrated into the cytoskeletal network.

Myosin motors walk from the negative (*pointed*) to the positive (*barbed*) end of an actin filament, but they may be connected into dipoles or longer filaments of their own and attach their heads to nearby actin filaments. In this case, their motion is

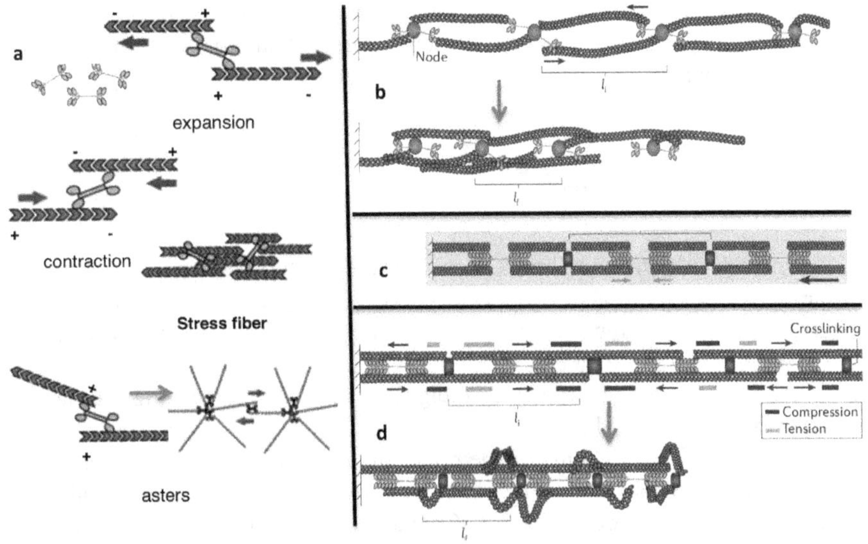

Fig. 5.2 Interactions of myosin motors (*green*) with actin filaments (*red*). (**a**) Structures formed with the help of myosin dipoles (Koenderink and Paluch, 2018). (**b**) Compressive quasi-sarcomeric structure. (**c**) Sarcomere. (**d**) Buckling of a disordered assembly (Murrell et al, 2015)

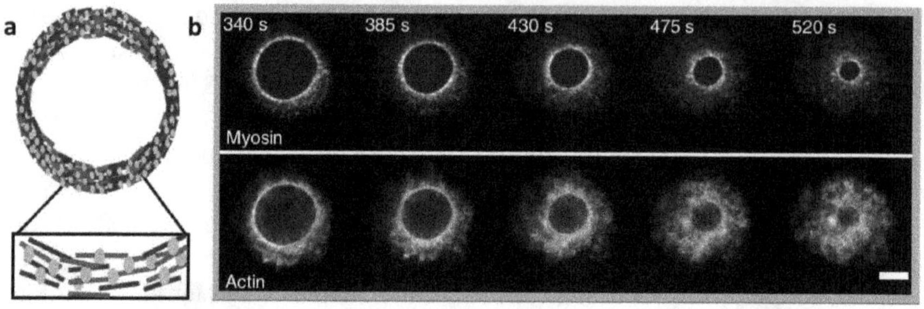

Fig. 5.3 (**a**) Structure of a cytokinetic ring. The filaments are colored *red* and *blue* for myosin clusters (shown by *green dots*) moving clockwise and anticlockwise, respectively (Pachong and Müller-Nedebock, 2017). (**b**) Evolution of the myosin and actin distributions during contraction of a cytokinetic ring (Wollrab et al, 2016)

restricted, and instead they exert a force, which may be contracting or expanding, dependent on the filaments' mutual orientation, as sketched in Fig. 5.2a. Several actin filaments can integrate into a *stress fiber*, which is usually consolidated by bundling proteins. A myosin dipole walking towards the barbed ends of two actin filaments causes them to be joined at their barbed ends, and other filaments may join to form an *aster* structure.

Several actin and myosin filaments can combine into a contractile quasi-sarcomeric structure (Fig. 5.2b). True sarcomeres, well organized and strengthened by long actin filaments, sketched in Fig. 5.2c, are the principal components of muscle cells. Sarcomeric structures also form the basis of contractile *cytokinetic rings* (Fig. 5.3), which effect cell division following the duplication and separation of the genetic material. Disordered assemblies may include both contractile and tensile segments, and buckle when stressed by myosin filaments, as shown in Fig. 5.2d.

5.3 Branched Structure

The eukaryotic plasma membrane sacrifices the mechanical strength of sturdy prokaryotic cells for the sake of recognition, signaling, and transport functions. The integrity of the cell is supported by the *cytoskeleton* (Fig. 5.4a) which, as is already clear from the name, holds the cell together. The network of actin filaments is most dense near the plasma membrane, forming the cell *cortex*, enhancing mechanical strength where it is needed most.

What makes the cortex tough is the branching and interconnection of actin filaments. Special proteins nucleate their branching at attachment points (Fig. 5.4b). For steric reasons, branches are directed at 70° to the mother filaments, but this still does not fully determine their direction, and the entire network turns out to be quite disordered. Filaments going in various directions come close together at some points, and are fastened there by bundling proteins. The network is stressed by myosin molecular

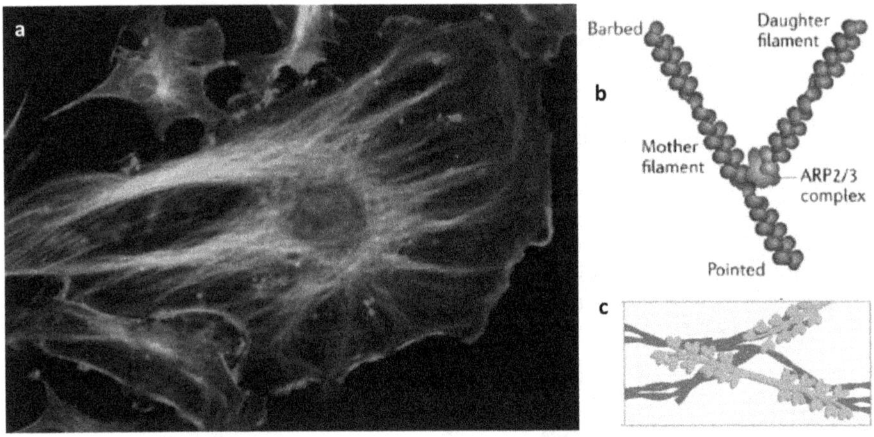

Fig. 5.4 (**a**) Eukaryotic cytoskeleton. Actin filaments are shown in *red*, microtubules in *green*, and nuclei in *blue* (CC). (**b**) A branched actin filament. The *blue blob* shows the protein complex facilitating branching (Goley and Welch, 2006). (**c**) Actin filaments tied by myosin motors (CC)

motors (Fig. 5.2), which can also facilitate crosslinking by attaching their heads to nearby actin filaments (Fig. 5.4c).

Cytoskeletal filaments are live; they continually grow, dissolve, and break. Actin monomer units are attached at the *barbed* end, treadmill along the filament, and detach at the opposite *pointed* end (Fig. 5.5a), unless protected by *capping* proteins. An entire filament can also rapidly depolymerize and shrink. Even sturdy microtubules live on the average only five to ten minutes, and a network of actin filaments can "fluidize" under excessive stretching.

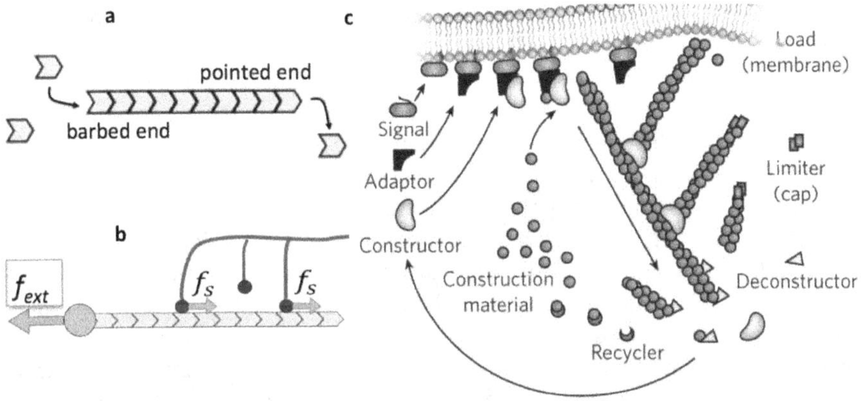

Fig. 5.5 (**a**) Attachment and detachment of actin monomers. (**b**) Stressing an actin filament to enable insertion of an actin monomer at the barbed end. (**c**) Schematic overview of the recycling of an actin filament (Fletcher and Mullins, 2010)

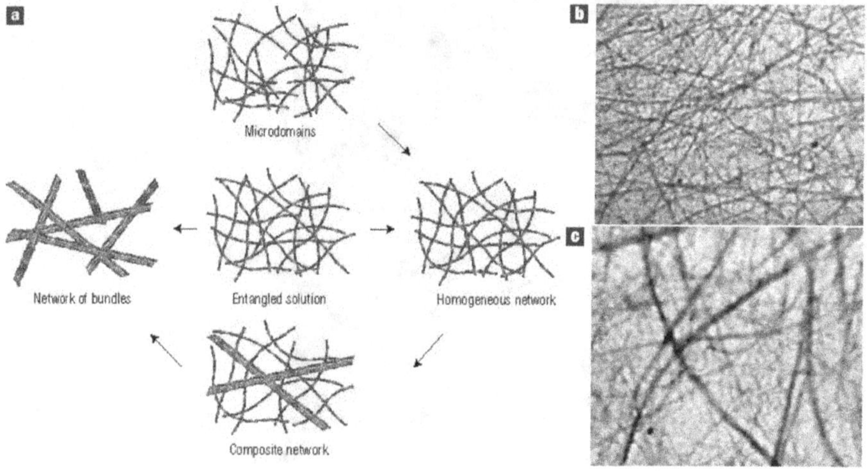

Fig. 5.6 (**a**) Transitions between different cytoskeletal structures, showing actin filaments and thick bundles. Crosslinker proteins are shown by *red dots*. (**b**) and (**c**) Electron micrographs of an entangled actin network (**b**) and a composite network (**c**) with bundles embedded in an isotropic background (Bausch and Kroy, 2006)

In a filament at rest, the barbed end is attached to another filament or to the plasma membrane, and should be detached for a moment to enable another monomer unit to be inserted, so that the treadmill could keep moving. Stressing the filament, either by external force or with the help of attached myosin motors (Fig. 5.5b), alleviates insertion and makes the filament grow faster. An overview of the recycling of an actin filament, aided by deconstructor, constructor, and capping (limiter) molecules, is sketched in Fig. 5.5c.

Depending on the specific type of crosslinker proteins, different structures, sketched in Fig. 5.6, are predicted and observed. Bundles may be embedded in a continuous isotropic background network. The principal cytoskeletal components, actin filaments and microtubules, "cross-talk" in a number of ways (Dogterom and Koenderink, 2019). Dynamic links attach actin bundles to the plus ends of growing microtubules, thereby guiding their mutual alignment. On the other hand, the actin cortex near the plasma membrane anchors or imposes a physical barrier on the growth of microtubules, preventing them from hitting the plasma membrane (Fig. 5.7). Conversely, actin filaments nucleate at the ends of microtubules.

Why should Nature prefer dynamic structures, continuously rebuilding themselves, while remaining stationary in the long run? Treadmilling of actin monomers costs energy, and so does fluctuating stress due to attachment and detachment of myosin motors: this is why we get tired even holding a hand horizontally in the air without doing any work. The benefit is flexibility, readiness to reshape and move at any moment following external inputs or internal needs. A crude analogy is the advantage of movable type, Gutenberg's invention, over woodblock printing. The cell combines amino acids as letters to the lines of proteins, which can be broken

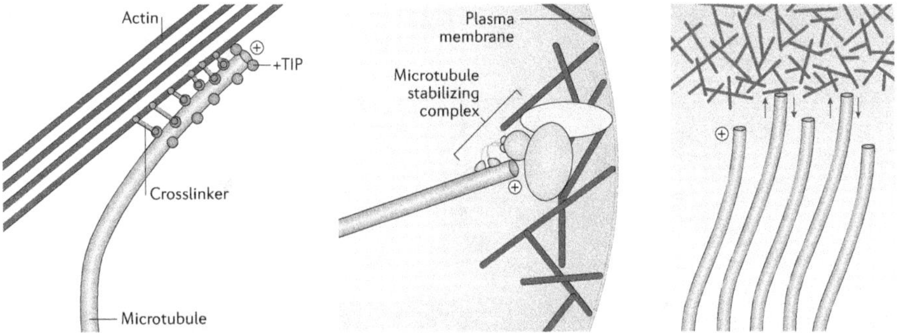

Fig. 5.7 Interaction between actin filaments and microtubules. *Left*: Microtubule guidance. *Center*: Microtubule anchoring. *Right*: Actin barrier (Dogterom and Koenderink, 2019)

back when their work is done to use the parts for making other proteins. Internal signaling and energy exchange play the role of a typesetter. The structure of the cytoskeleton is the subject of thorough investigation; besides a plethora of papers, there are quite a few monographs, e.g., Howard (2001), Lee (2013).

5.4 Filaments *in Vitro*

Away from complexities of the crowded environment of a living cell, biophysicists study the dynamics of filaments in *motility assays* containing purified mixtures of filaments and motors, complete with a supply of ATP sufficient to keep them moving. Here we step back from *live* to *active* matter, where dazzling patterns can be observed, no longer hidden and unconstrained by realities.

The first *in vitro* studies of this kind were carried out in a constrained 2D geometry with a mixture of microtubules and kinesin motors. Nédélec et al (1997) observed patterns consisting of dynamic asters at a higher, and vortex arrays at a lower motor

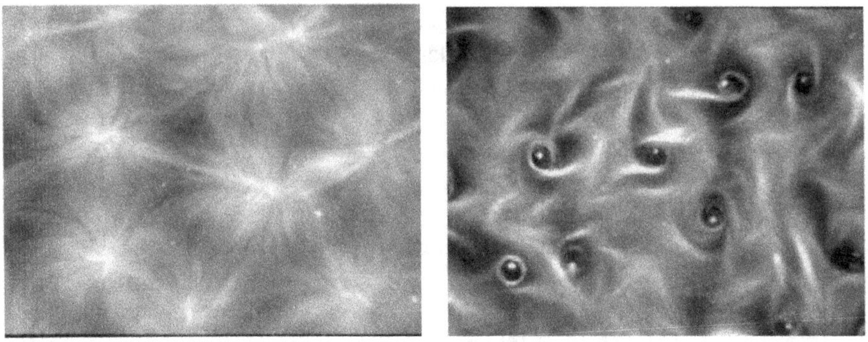

Fig. 5.8 Aster and vortex array patterns in a microtubule/kinesin mixture (Nédélec et al, 1997)

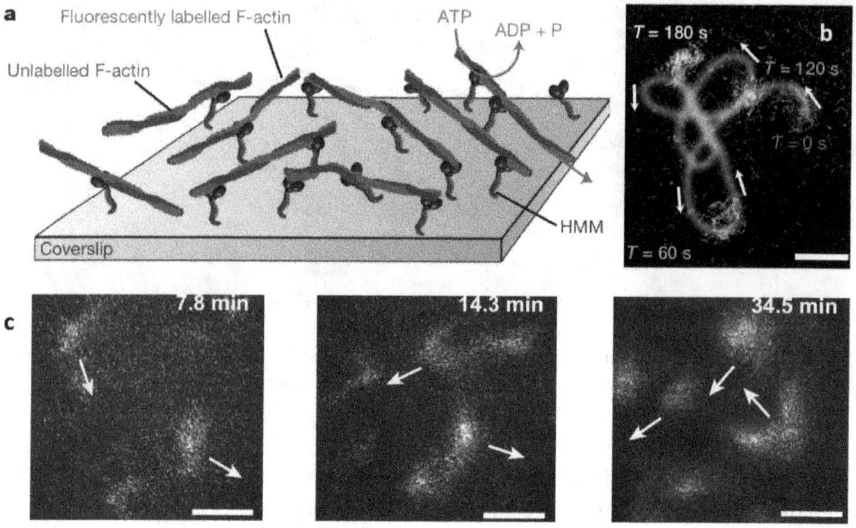

Fig. 5.9 (**a**) Scheme of the motility array. (**b**) Color-coded time overlay of snapshots showing the trajectory of a single cluster (Schaller et al, 2010). (**c**) Time trace of cluster formation (Suzuki and Bausch, 2017). Scale bars 50 μm

concentration (Fig. 5.8). Before that, asters were observed in this system in a less refined setting (Urrutia et al, 1991).

Bausch and coworkers have carried out a series of experiments with a mixture of actin filaments and myosin motors powered by ATP in a still more constrained setting of motility assays, schematized in Fig. 5.9a, where a particular kind of molecular motor (heavy meromyosin HMM) was immobilized on a coverslip surface. The motors, unable to move themselves, displace the filaments by stepping with their heads at a fixed velocity set by the ATP concentration. At low density, the filaments move randomly, but above a critical density, they self-organize to form coherently moving planar structures with persistent density modulations, forming long-lived clusters, swirls, and interconnected bands that can span length scales orders of magnitude longer than their constituents. The trajectory of a single cluster, moving as indicated by the arrows, is shown in Fig. 5.9b. Clusters develop from "seeds" containing several filaments. Most seeds last only for a short while without reaching any sort of a stable structure, but a few develop into persistent clusters (Fig. 5.9c).

At still higher densities, the filaments form a swirl encompassing the entire motility assay (Fig. 5.10a). Since all filaments move with the same velocity, there are large angular velocity gradients in the radial direction, leading to an inherent instability of the pattern, and the core of the swirl, where vorticity is maximal (Fig. 5.10b), is displaced (Fig. 5.10c), leading eventually to non-uniform swirling motion (Fig. 5.10d).

The immobilized myosin molecules used in these experiments could not exercise their most important function – stressing the actin network. When Vogel et al (2013) tied active filaments by *processive* myosin filaments in a different 2D setting which

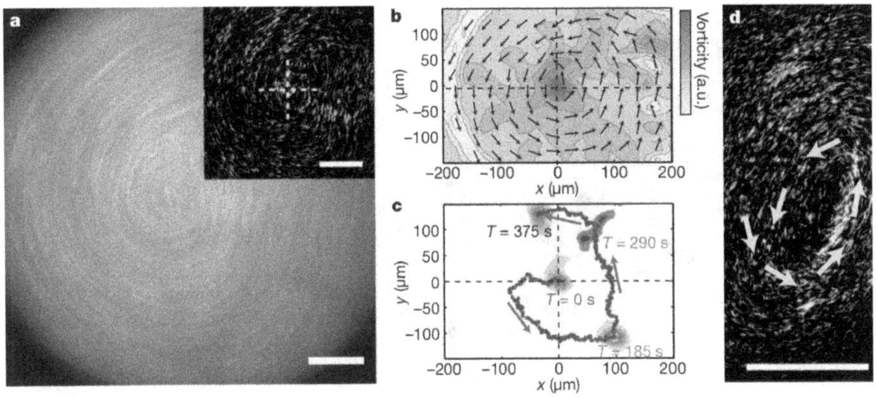

Fig. 5.10 (**a**) Swirling motion visualized in a time overlay of ten consecutive images, starting from the image depicted in the inset. (**b**) Map of the magnitude of vorticity. (**c**) Trajectory of the displaced core of the destabilized swirl. (**d**) Non-uniform swirling motion. Scale bars 50 μm. (Schaller et al, 2010)

they called a "minimal actin cortex", the result was completely different: actin filaments were broken and compressed, and coalesced under compressive stress over the course of a few minutes (Fig. 5.11).

Ennomani et al (2016) generated actin networks with a well-defined organization using a surface micropatterning technique. Based both on experiment and simulations, they related the contractility of actin networks with their *connectivity*, defined as the average number of connectors per actin filament. All relations, dependent on the type of network architecture and plotted in Fig. 5.12, reach a maximum centered on an optimal connectivity within the range between 2 and 4. This value corresponds to the percolation threshold – a critical point above which all filaments are connected together in a single cluster (Alvarado et al, 2013). The prevailing mechanism of contraction changes with changing connectivity and network architecture.

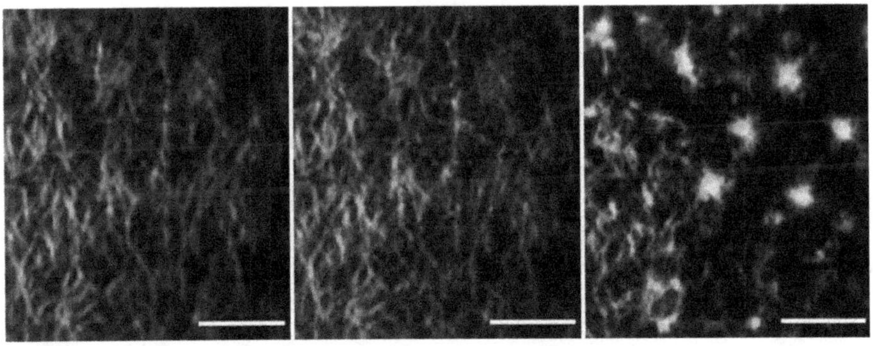

Fig. 5.11 Collapse of a a "minimal actin cortex". Actin filaments are labeled *green* and myosin filaments, *red*. Scale bars 10 μm (Vogel et al, 2013)

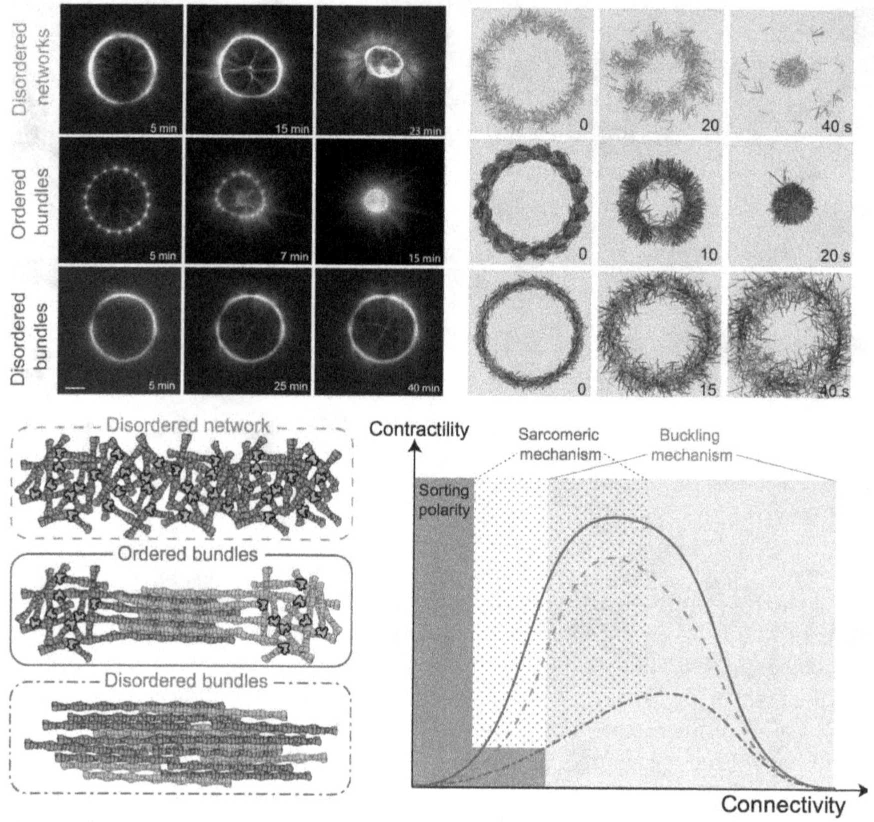

Fig. 5.12 Contractile response of different architectures of the actin networks. *Top*: Observed (*left*) and simulated (*right*) contraction sequences of actin rings. *Bottom*: Contractility of actin networks as a function of connectivity (Ennomani et al, 2016)

At low connectivity, it is weak, as sliding of actin filaments by the action of myosin motors leads to polarity sorting rather than contraction. As connectivity increases, sarcomeric-like and buckling mechanisms are switched on.

In vitro experiments by Ideses et al (2008) came closer to the cell's natural environment: they also included, besides actin and myosin, the bundling protein *fascin* (from the same Latin word for *bundle* as fascism) and ARP2/3 complexes (shown in Fig. 5.4b) facilitating branching. The sequence of transitions from asters to thickening stars to an entangled network is shown in Fig. 5.13.

In later experiments originating from the same group (Ideses et al, 2018), a 2D analogue of the cell cortex was fabricated in a flat chamber by polymerizing actin in the presence of the fascin cross-linker and clusters of myosin motors. The gel sheet, initially homogeneous, contracted, starting from the boundaries and spreading into the bulk, and eventually buckled, as shown in Fig. 5.14.

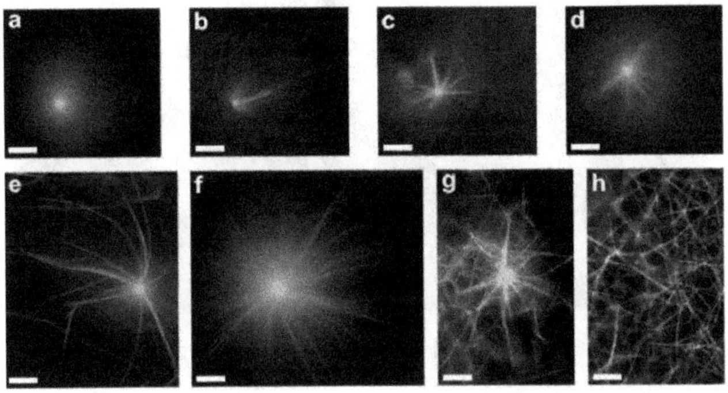

Fig. 5.13 (**a**) Asters at very low fascin concentrations. (**b**)–(**f**) Star-like structures form, with the density and length of the bundles emanating from the star core increasing with fascin concentration. (**g**) and (**h**) Formation of an entangled network of actin/fascin bundles. Scale bars 10 μm (Ideses et al, 2008)

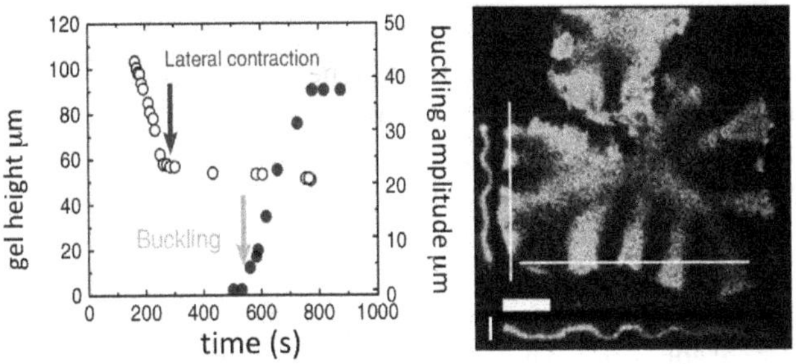

Fig. 5.14 Buckling of an acto-myosin sheet. *Left*: Evolution of the gel height and buckling amplitude with time. *Right*: Top view and side views along the white lines. Scale bars: horizontal 200 μm, vertical 80 μm (Ideses et al, 2018)

5.5 Adhesion

The 3D branched cytoskeleton structure of real cells neither collapses nor buckles under myosin-induced compressive stress, because it is not just suspended in a motility assay but fastened by a dense cortex envelop attached to a substrate or an extracellular matrix. Actin filaments are often bundled in stress fibers characterized by a striated morphology with alternating actin and myosin bands. They are organized across the cell on the ventral and dorsal sides[1] and as transverse arcs (Fig. 5.15a).

Stress fibers couple the actomyosin network to the substrate via *focal adhesions*, and serve as the strongest contractile elements in the network. They exert a traction force on the substrate, as shown in Fig. 5.15b and c. Ventral fibers anchored at each end to focal adhesions are predominant in elongated cells. Such cells may be approximated as force dipoles. By stressing the substrate, they interact with other cells, favoring their nematic ordering (Schwarz and Safran, 2013). It is a combination of adhesion force and myosin contractility that is "forcing cells into shape" (Murrel et al, 2015).

A focal adhesion is a sophisticated piece of hardware (Fig. 5.16) connected to the exterior of the cell by *integrin* molecules protruding through the plasma membrane,

[1] "Ventris" is Latin for stomach, the lower part of the cell, the opposite to "dorsal", relating to the upper side or back.

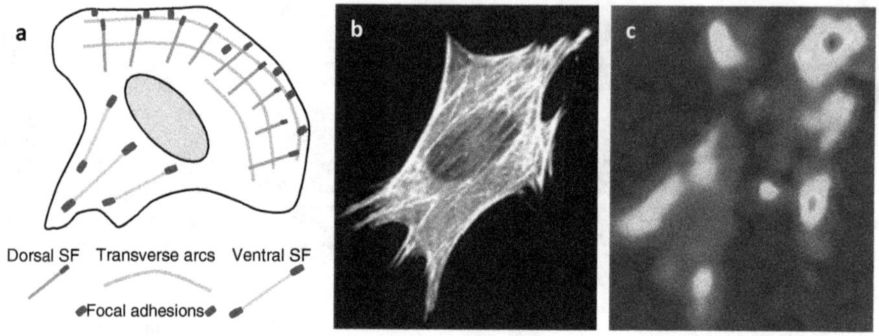

Fig. 5.15 (**a**) Stress fibers (Echarri et al, 2019). (**b**) and (**c**) Strength of the traction force exerted by the cell on a substrate, increasing from dark blue to red (Murrel et al, 2015)

and engaging the cytoskeletal filaments via a multicomponent protein clutch shown by multicolored blobs in the left panel. There is a complicated network of signaling proteins which conveys the mechanical state of cell–substrate interactions felt by integrins to the cytoskeleton and back to the focal adhesions. It involves not only actin filaments, but also microtubules, appearing in the right panel. Coupling of microtubules to integrin adhesions via KANK protein complexes limits the growth of focal adhesions, and its release activates the assembly of myosin filaments and binding stress fibers with focal adhesions. Through this sophisticated and not yet fully understood feedback network, the cell feels the state of the substrate and reorganizes its structure.

On softer substrates, cells contain smaller and more dynamic adhesions; the contractile network is stronger on stiffer substrates, as can be inferred by comparing (Fig. 5.17a and b). Cells prefer a harder substrate, and migrate accordingly if they feel

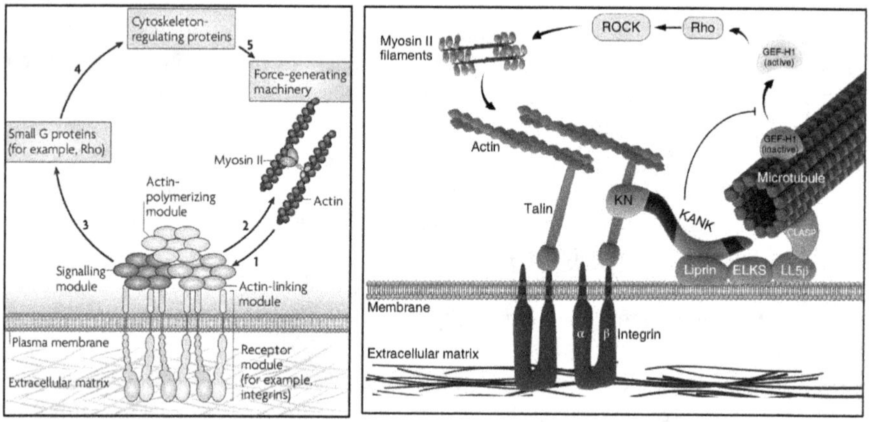

Fig. 5.16 Two views of the signaling network of a focal adhesion (*left*: Geiger et al, 2009; *right*: Rafiq et al, 2019)

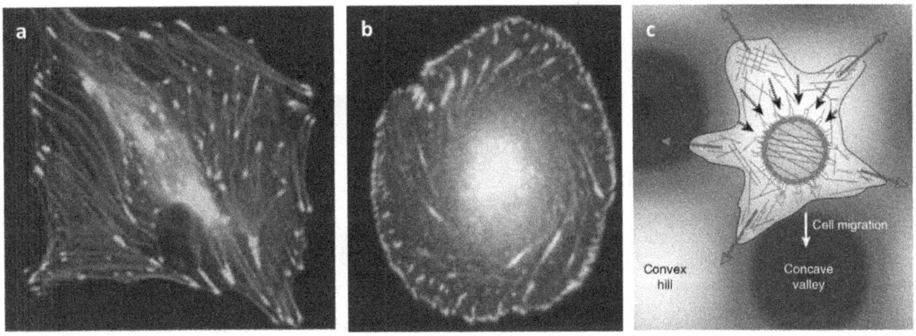

Fig. 5.17 A cell on a hard (**a**) and on a ten times softer (**b**) substrate (Geiger et al, 2009). (**c**) Curvotaxis (Pieuchot et al, 2018)

a gradient of stiffness: this is *durotaxis* (Schwarz and Safran, 2013). When placed on a curved surface, they migrate towards a concave depression where they have a stronger contact with the substrate: this is *curvotaxis* (Fig. 5.17c).

A cell placed on a round support fitting its size has nowhere to go, but its cytoskeleton keeps rearranging, acquiring a variety of structures, depending on the proteins that enable branching and bundling of actin filaments, on actomyosin contractility, and on the arrangement of microtubules. Among them, there are remarkable chiral structures which may be directed in different ways, depending on the type of actin

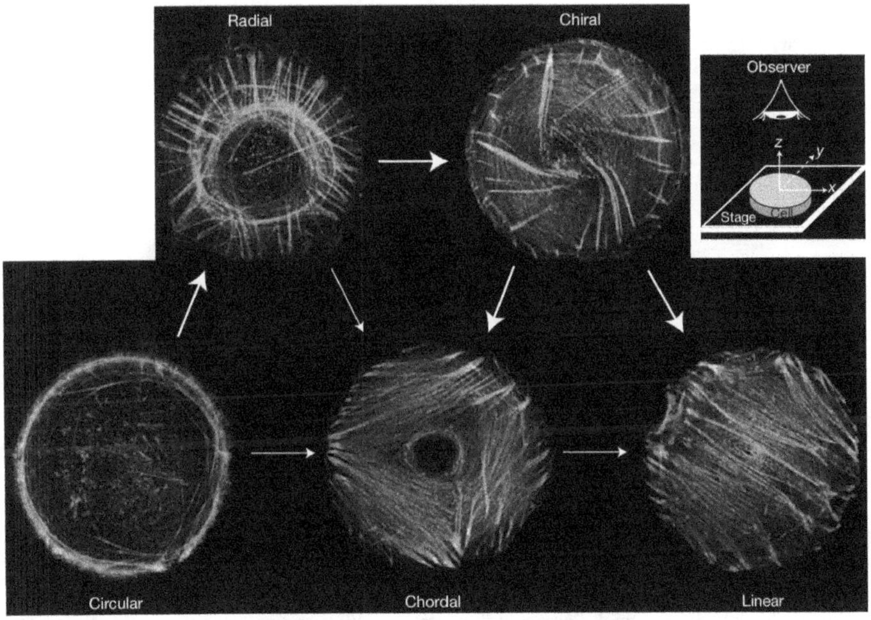

Fig. 5.18 Transitions among alternative structures of a cell on a round support (Tee et al, 2015)

monomers. The observed transformations among alternative structures are shown in Fig. 5.18, with prevalent transitions indicated by thick arrows.

5.6 Adhesive Crawling

Unicellular organisms need to move, and isolated cells move toward preferred environments. This is not an easy task when traveling on a solid substrate and having no flagella or other suitable appendages. Focal adhesions cannot step like feet; they have to be released at the hind end and created anew ahead, and motion requires realigning the entire structure of the cytoskeleton. As traction forces move the cell forwards, the actin network depolymerizes and focal adhesions disassemble at the rear of the cell, whereupon actin monomers are transported to the front to polymerize there. This is a slow process, limiting the speed to about a cell length per minute.

There are different manners of crawling. The leading edge is often an almost two-dimensional protrusion of the actin mesh, or *lamellipodium* (Latin for "thin-sheet foot"). It may inch ahead by tentatively protruding *filopodia* – again referring to a

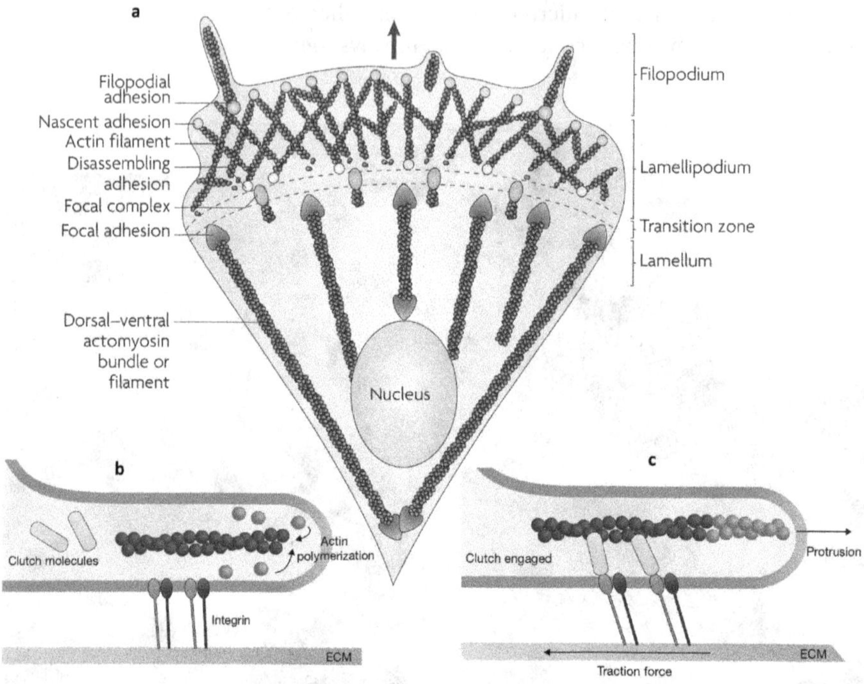

Fig. 5.19 (**a**) Crawling cell (Parsons et al, 2010). (**b**) and (**c**) Engagement of the molecular clutch (*yellow*). The pre-existing actin filament is shown in *dark blue*, and polymerizing actin monomers, in *light blue* (Case and Waterman, 2015)

foot, although those are not feet at all but aids to explore the way ahead: they can be either withdrawn or strengthened by actin filaments to pull the leading edge ahead (Fig. 5.19a). Nascent adhesions form in the lamellipodium, immediately behind the leading edge. Their assembly determines the protrusion rate of the leading edge and requires actin polymerization (see Fig. 5.19b).

Not all nascent adhesions persist, but the persistent ones evolve to larger focal complexes slightly behind the leading edge, and further into mature focal adhesions that reside at the ends of large actin bundles or stress fibers that extend to the centre or the rear of the cell. Maturation involves engagement of the molecular clutch (Fig. 5.19c), and the resulting traction force on the substrate or extracellular matrix (ECM) leads to a net edge protrusion.

However, the leading edge is a line rather than a single point, and protrusions have to be supported by the formation of an actin arc along the leading edge to ensure sustained motion. Burnette et al (2011), based on their observations, formulated the intermittent model of advance of the leading edge, going through the stages shown in Fig. 5.20a. Starting from the base of a previous retraction terminating at a newly created actin arc coupled to a focal adhesion, the new lamellipodial protrusion pushes the arc ahead, driving the membrane forward. During this protrusion stage, actin filaments polymerize behind the plasma membrane and depolymerize a few micrometres away from the edge. Actin filaments treadmill through the lamellipodium during protrusion, and a nascent adhesion forms. At the peak of the protrusion, myosin filaments form in the lamellipodium, and the local network contracts, driving the actin arc formation and edge retraction. The new actin arc slows down at the maturing focal adhesion, so that this cycle leads to a net advance. This mechanism is

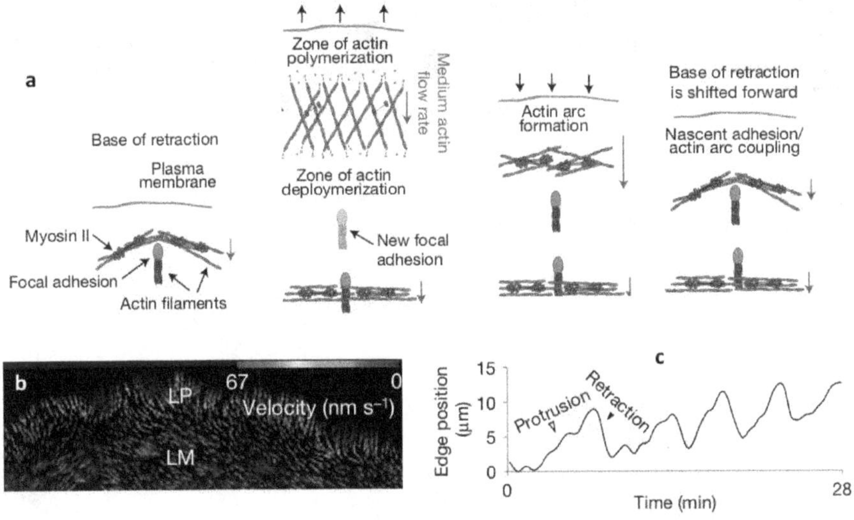

Fig. 5.20 (**a**) The protrusion/retraction stages of the leading edge. *Blue arrows* scale with the intensity of the actin retrograde flow. (**b**) Velocity of the leading edge (color coded). LM, lamella; LP, lamellipodium. (**c**) Change of the edge position with time (Burnette et al, 2011)

supported by the observed fluctuations in the leading edge advance velocity in space (Fig. 5.20b) and time (Fig. 5.20c).

Protrusions may be resisted by tension in the plasma membrane, keeping the leading edge smooth and the shape elongated normally to the direction of motion. This is characteristic of *keratocytes* taken from fish skin, favored by experimentalists for their relatively fast and persistently directed motion. Barnhart et al (2011) revealed a strong effect of adhesive properties of the underlying substrate on the morphology of crawling cells, caused by tight coupling of adhesion dynamics and actin polymerization. The different shapes and actin flow patterns shown in Fig. 5.21 demonstrate the effect of molecular clutches localized at adhesions coupling the actin network to the substrate. Most common fan-like shapes are observed when the cell–substrate adhesion strength is neither too high nor too low, but otherwise keratocytes migrate more slowly and acquire round or asymmetric shapes.

Stronger adhesion increases both friction and the traction transmitted to the surface, and slows down retrograde flow of the actin network, but, on the other hand, strong force transmission through the molecular clutches reduces their average life span. Thus, propulsion is most efficient under intermediate "Goldilocks" conditions. The shape of the cell is determined by the balance of the polymerization and retrograde flow rates, both color coded in Fig. 5.21. The former is greater at the cell front, while the relation is opposite at the cell end, and both rates are balanced

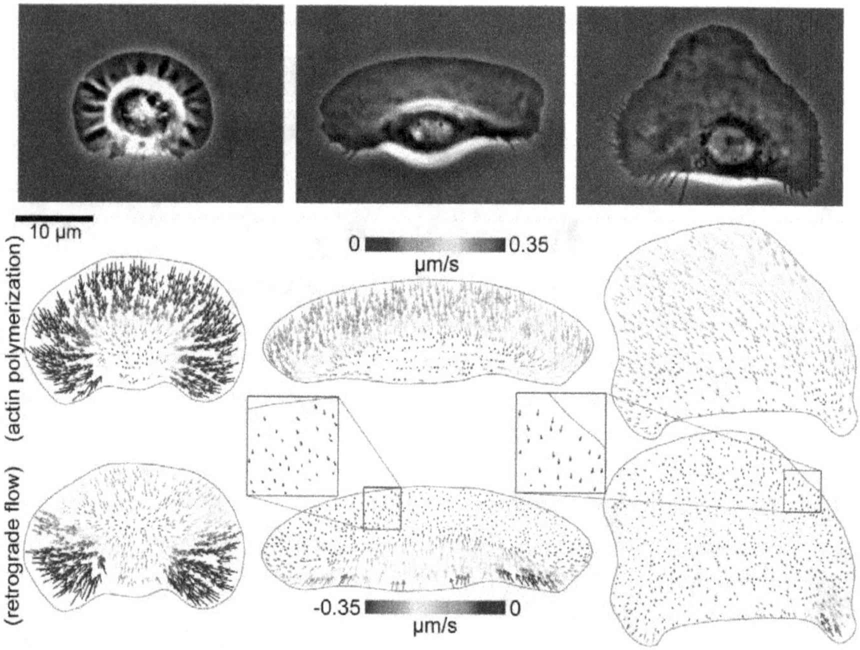

Fig. 5.21 Keratocyte cells crawling at low (*left*), intermediate (*center*), and high (*right*) adhesion strengths. *Top*: Phase contrast images. *Middle row*: Actin polarization rate. *Below*: Retrograde actin flow (Barnhart et al, 2011)

in the cell corners. In addition to these mechanical effects, adhesion strength and myosin contraction affect the organization of the actin network through various signal transduction pathways (recall Fig. 5.16).

5.7 Non-Adhesive Motility

There are different manners of motion, independent of adhesion. The alga *Euglena gracilis*, which can also swim by flagellar propulsion, crawls by gracefully reshaping its whole body (in what is known as *metaboly*), as shown in Fig. 5.22.

Motion not relying on adhesion and involving frequent shape changes is described as *amoeboid*. It is more characteristic of motion in a 3D tissue environment. Its advantage is the freedom to set routes dictated by the structure of a substrate, and the ability to rapidly maneuver in a 3D environment following environmental, e.g., chemotactic, cues. Leukocytes, scattered throughout the body and able to infiltrate any type of tissue, follow this strategy, and are able to migrate up to 100 times faster than adhesive epithelial cells. While protrusion, adhesion, and contraction are tightly coupled in adhesive cells, amoeboid cells protrude without anterior pulling forces, while the trailing edge displays an irregularly alternating pattern (Lämmermann et al, 2008). This is illustrated in Fig. 5.23 by successive shapes and myosin distribution patterns in a kind of a leukocyte, "dendritic cell", moving along a chemotactic gradient.

Amoeboid propulsion commonly involves *blebbing*, extending spherical protrusions (Fig. 5.24a). Blebs form by detachment of the membrane from the underlying

Fig. 5.22 Crawling *Euglena gracilis* (Agostinelli et al, 2019)

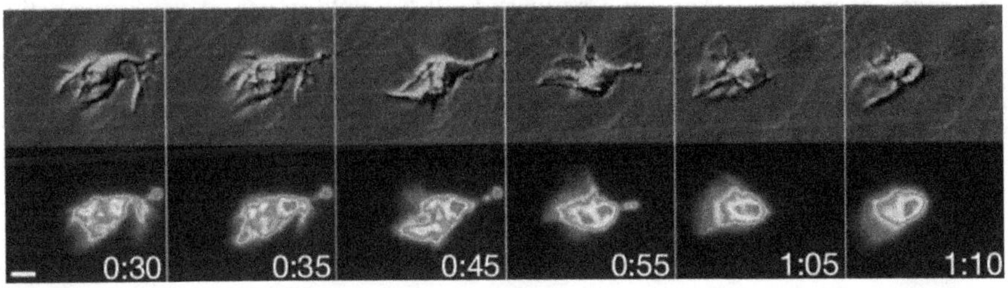

Fig. 5.23 Shapes of a moving dendritic cell (*top*) and myosin distribution patterns (*bottom*). *Red coloring* shows the highest myosin level; time in minutes; scale bar 5 μm (Lämmermann et al, 2008)

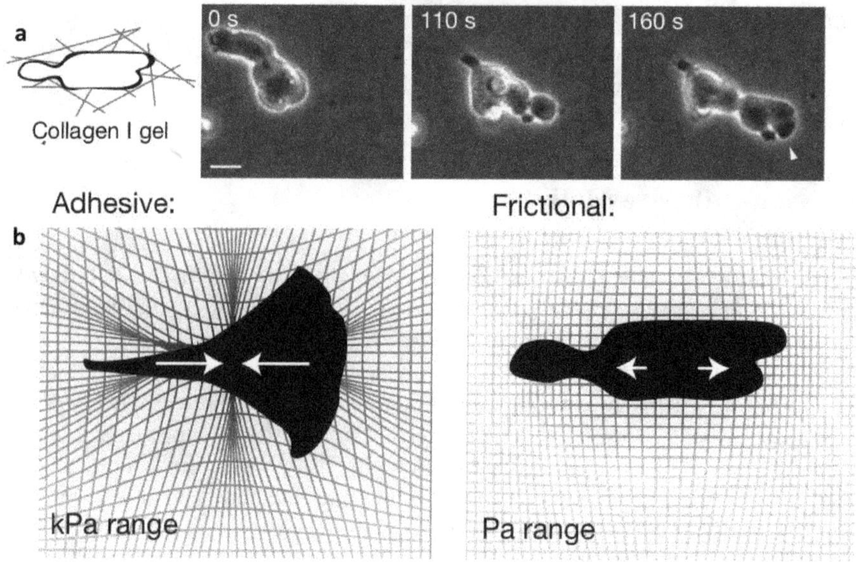

Fig. 5.24 (**a**) Blebbing cell moving through a gel. Scale bar 10 μm. (**b**) Comparison of the contractile stress exerted by an adherent cell and a much weaker tensile stress exerted by a blebbing cell on a medium with a thousand times weaker elasticity (Bergert et al, 2015)

cortex, triggered by their reduced mutual adhesion or by rupture of the cortex; they do not appear if the fluid pressure within the cell is insufficient (Charras and Paluch, 2008). For blebbing to result in movement, blebs need to be created only at the leading edge, and to exert a force on the substrate to translocate the cell's body. The forces propelling the cell are several orders of magnitude lower than during adhesion-based motility, and the force distribution is inverted (Bergert et al, 2015). This is shown in Fig. 5.24b, comparing the map of the stress of the substrate in adhesion-dependent motion, where it is compressive and highly concentrated, with a weak and evenly distributed tensile stress associated with blebbing.

A peculiar manner of motion is used by the bacterium *Listeria* to propel itself through the cytoplasm of an infected cell by constructing a "comet tail" behind it, composed of a cross-linked actin network built by making use of the host's actin machinery. The tail can grow to more than 100 μm long when depolymerization is slow. *Listeria* contributes only the transmembrane protein required to trigger actin polymerization and thus to induce the motion. Other bacterial and viral pathogens that exploit this machinery are known, and even a bead can be made to move in the same way when an appropriate nucleator is supplied.

By the Brownian ratchet model (Mogilner and Oster, 1996), the barbed end of an actin filament momentarily detaches from the surface due to elastic fluctuations, allowing a new actin monomer unit to squeeze in and push the propelled body ahead, in the same way as the plasma membrane is pushed when filaments grow within a cell (recall Fig. 5.5b). An alternative mechanism (Dickinson and Purich,

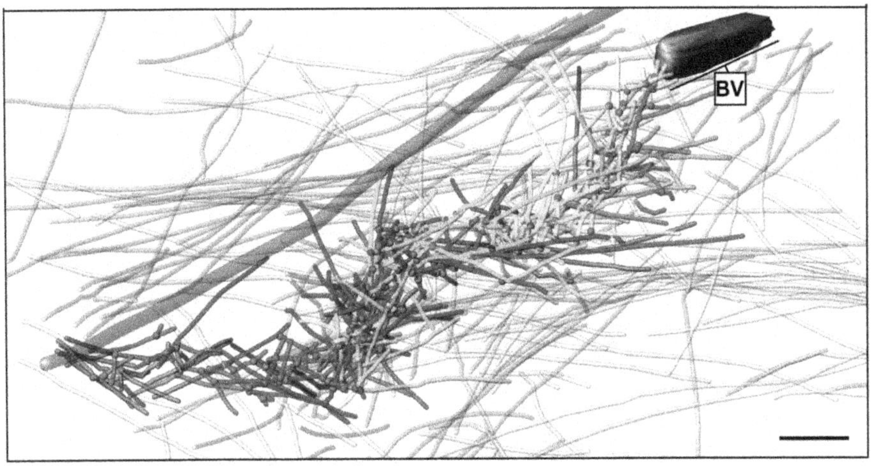

Fig. 5.25 Model of the "comet tail" of baculovirus (BV) based on an electron tomography image. The actin filaments of the host cytoskeleton (*translucent*) and a microtubule (*grey tube*) are also shown. Scale bar 100 nm (Mueller et al, 2014)

2002) assumes that the filament's terminal unit always remains tethered to the surface by some molecular clump that can shift between two energy wells, one of which leaves enough space to accommodate a new monomer unit. The propelling force is retained because the far end of the "comet tail" is anchored to the host's cross-linked network.

Mueller et al (2014) compared the statistics of deviations of the "comet tail" from the straight line generated by simulations of different models with the electron tomography images they recorded, and concluded that both fit only when the "comet tail" always remains tethered. However, this does not invalidate the Brownian ratchet model, since the tail can be attached at several points, which never detach simultaneously. Nucleation of filaments on the propelled surface is required not only to initiate movement, but also to correct sharp turns and accommodate fluctuations in the pushing strength. In this way, a branched structure is created, like the one shown in Fig. 5.25, similar to the structure of a cellular cortex network.

5.8 Cytoplasmic Flow

Motor-mediated active motion during cargo transport, remodeling of the cytoskeletal network, or cell migration also excite flow in the fluid component of the cytoplasm, the cytosol (Goldstein and van de Meent, 2015; Goldstein, 2016). In 2D simulations by Trong et al (2012), cytosol streaming followed the spatial organization of the motor velocity field, but velocities are reduced thousandfold in their magnitude (Fig. 5.26a). A system of localized vortices, as in Fig. 5.26b, may arise only in the unrealistic case of a perfectly aligned cytoskeletal network. Both the range and

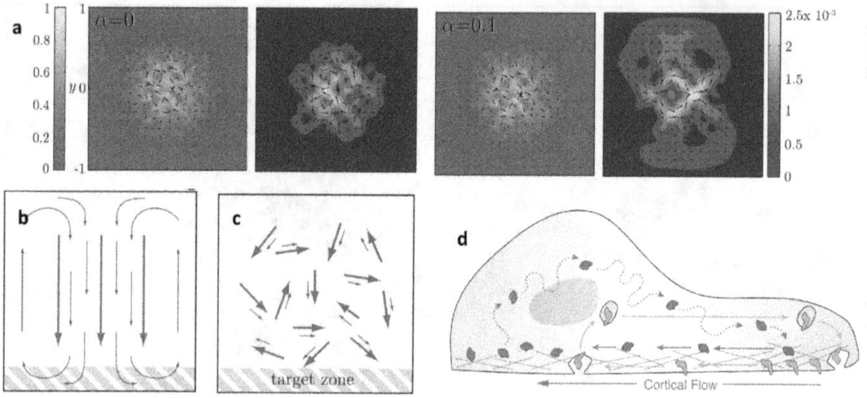

Fig. 5.26 (**a**) Motor (*red panels*) and cytosol (*blue panels*) flow fields, with scale bars of the same colors. A perfectly aligned motor velocity field ($\alpha = 0$) causes recirculatory fluid flow. The flow is anisotropically biased at $0 < \alpha < 1$. (**b**) A perfectly aligned cytoskeletal network (*red arrows*) causes recirculatory fluid flow (*blue arrows*) out of the target zone (*green dashed area*). (**c**) A disordered cytoskeletal network suppresses the range and magnitude of fluid flows (Trong et al, 2012). (**d**) Cytoplasmic flow pattern in a crawling cell (Illukkumbura et al, 2020)

magnitude of fluid flows are suppressed when the network is disordered, as it usually is (Fig. 5.26c). A better organized forcing on the scale of the entire cell comes from cytoskeleton restructuring, in particular, cell migration, which involves retrograde cortical flow and back-flow bringing actin monomers to polymerize at the leading edge (Fig. 5.26d).

Since cytoplasmic flow is hindered in a dense cytoskeletal network crowded by proteins and organelles, it cannot compete with efficient and well-addressed cargo transport by kinesin molecular motors on a microtubular railway, and concedes even to diffusion on the 10 μm scale of typical somatic cells. However, it becomes important in large oocytes during the first morphogenetic stages (see Sect. 8.2). Various aspects of cytoplasmic flow in the oocyte are reviewed by Quinlan (2016). The sketch in Fig. 5.27a shows cytoplasmic circulation induced near the cytoskeletal cortex adjacent to the egg chamber, but observed and simulated flow patterns (Fig. 5.27b–d) are far more complex.

Microtubules are nucleated or anchored in the oocyte at the cortex and grow into the bulk to form a mostly disordered mesh. Delivery of cargo by kinesin motors along this network is most efficient and fast, but it is estimated that a large part of the traffic is driven by motors indirectly, through cytoplasmic streaming excited by their motion. Trong et al (2015) initiated their simulations by seeding points for microtubule nucleation at random positions along the oocyte cortex, with their density accounting for the observed inhomogeneities of the cortex microtubule density, to make the computed meshwork grown from these seeds as close as possible to observations. These thorough preparations enabled a close resemblance of computed cytoplasmic flow fields to *in vivo* flows, as evidenced by comparing Fig. 5.27c and d.

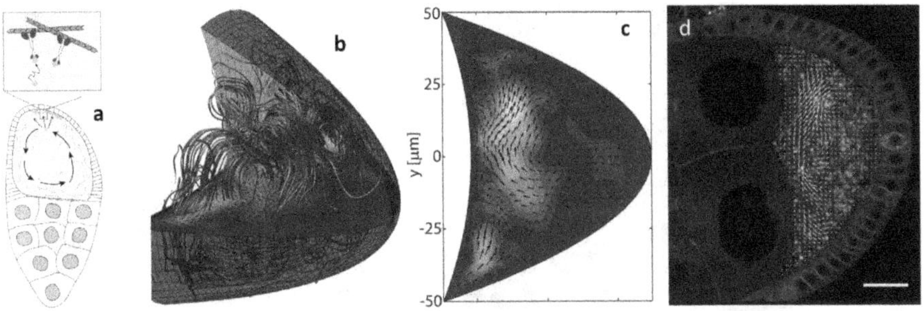

Fig. 5.27 (**a**) Sketch of circulation induced by the actomyosin cortex (Illukkumbura et al, 2020). (**b**) Streamlines (*light blue lines*) of the 3D cytoplasmic flow field. The horizontal plane shows its 2D cross-section. (**c**) Cross-section through the 3D field shown in panel (**b**), with *arrows* indicating flow directions and colors coding flow speeds. (**d**) Confocal image of a live stage 9 oocyte. *Arrows* show the flow field computed from particle image velocimetry in the *Drosophila* oocyte (Trong et al, 2015)

This flow mechanism prevails when the egg chamber is stiff, as it is, besides *Drosophila*, in other animals with a *bilateral* body plan. However, it is different in the soft oocytes of starfish, an animal with *polar* symmetry. Klughammer et al (2018) modeled flow driven by a band of increased surface tension moving over the oocyte cortex and compared the results with experimental flow and deformation patterns. A good fit could be taken as a proof that the latter are fully determined by surface contraction waves.

This kind of intracellular flow is characteristic of blebbing motion (Fig. 5.28), where cytoplasm passively fills amoeboid protrusions, which are formed due to cortex contractions. It is an efficient means of transportation, akin to peristaltic streaming. *Physarum polycephalum* plasmodial slime mold (Fig. 5.29a), a very special unicellular organism joining many nuclei into a huge cell spreading out up to a meter long, builds up a true vascular network (Fig. 5.29b), where cytoplasm is driven peristaltically. This network develops at the leading edge of the spreading organism and thereby shapes its morphology. It is reputed to be a clever planner: if put in a maze, the network reshapes to connect two exits by the shortest path. If pieces of food are put at certain locations, slime mold connects them by the optimal road network taking into account density of traffic between nodes. This planning ability

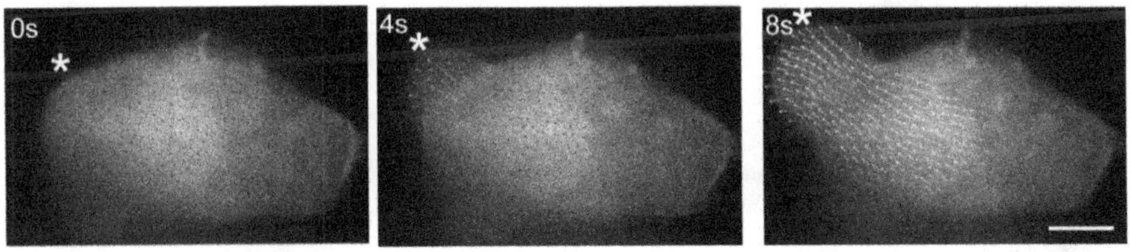

Fig. 5.28 Cytoplasmic flow during the creation of a bleb. Scale bar 5 μm (Goudarzi et al, 2019)

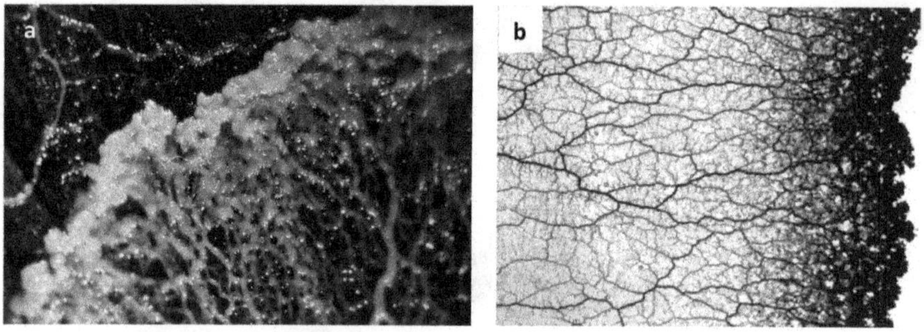

Fig. 5.29 (**a**) *Physarum polycephalum* (CC). (**b**) Network of tubular veins in the *Physarum polycephalum* strand. The cell extends and propagates to the right. The dense apical front (*at the right*) develops a vast network of tubular strands, where protoplasm is transported by peristalsis. Dimensions: 6.14 × 4.60 cm (Haupt and Hauser, 2020)

was tested in a hard task: designing a railroad network in the Tokyo metropolitan area (Tero et al, 2010) .

Chapter 6
Active Gels

6.1 Cytoskeleton as a Continuum

What is a rational approach to the mind-boggling complexity we touched upon in the preceding chapter, even restricting to mechanical aspects and forgetting for a while the chemistry of proteins and nucleotides, which really run the entire show? A tentative answer is the *active gel* theory. The term, first appearing in the paper by Kruse et al (2004), predates the notion of active matter; earlier usage of this term was mostly medical. However, mechanical modeling of a viscoelastic cytogel, though lacking both the appellation "active" and a reference to orientational order, goes twenty more years back (Oster and Odell, 1984; Dembo and Harlow, 1986).

In a later review, Prost et al (2015) summarized the way to construct an active gel theory:

> One has first to identify slow variables relevant for a macroscopic or mesoscopic description. [...] Clearly, overall mass, solvent mass, energy and momentum are conserved and, on timescales short compared to production and degradation rates, we can consider the number of actin monomers and motors to be conserved. In the following we consider isothermal systems; thus we can ignore energy conservation.

Clearly, energy is *not* conserved in living systems, nor in any non-equilibrium system interacting with environment, and the authors realize it:

> [...] these systems lack time reversal symmetry, because energy is constantly transduced.

Thus, we can be satisfied that energy conservation is ignored, although isothermicity is not a sufficient reason. The authors' outlook is far more optimistic than the opening sentence of this chapter:

> Fortunately, after a century of skilled experimental work, a certain simplicity has begun to emerge.

Indeed, the physicist should not care for interconnections of the cytoskeletal maze and an unceasing dance of actin monomers and molecular motors any more than for the molecular chaos in common fluids:

L. Pismen, *Active Matter Within and Around Us*, The Frontiers Collection,
https://doi.org/10.1007/978-3-030-68421-1_6

> By writing down generic equations in the hydrodynamic limit, one has replaced a hundred thousand variables by only a few in a field theory. From our experience in soft condensed matter, one can infer that the space of solutions is still large enough to describe most experimental situations within a unified framework.

As we shall see, the problem with this space of solutions might be the opposite: within its range of applicability, it may be so large that the same experimental situation can be described in different ways.

The unified framework of the active gel theory extends the theory of orientationally ordered active fluids (Sect. 2.5), retaining polarization equations derived, as before, by varying the appropriate energy functional, and replacing the Stokes equation of a viscous fluid by viscoelastic transport equations. Numerous coefficients of the theory, including mechanical and orientational elasticities, viscosity, and activity, have to be obtained from experimental data. The effective mechanical elasticity of living tissues is feeble; hence the term *gel*. The elasticity of a network including stiff polymers may be highly anisotropic.

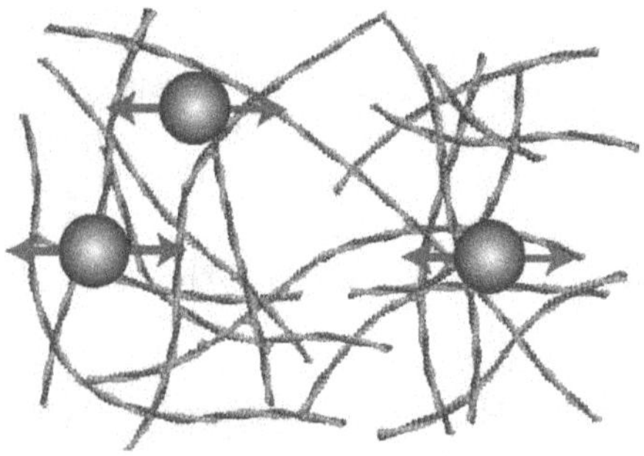

Fig. 6.1 Magnetic beads embedded in an actin network. Crosslinker proteins are shown by *red dots* (Bausch and Kroy, 2006)

The actin network behaves as an elastic solid on short timescales, and as a liquid on longer time intervals; the characteristic relaxation time may vary in different cells from tenths to tens of seconds (Mogilner and Manhart, 2018). Viscous and elastic response can be distinguished, for example, by the frequency dependence of the displacement of imbedded magnetic particles in microrheological measurements (Fig. 6.1). A fine example is the extraction of the physical parameters of the actomyosin cortical layer *in vivo* from laser ablation experiments with embryonic cells (Saha et al, 2016). However, different methods and procedures, including whole-cell deformation, bead-based measurements, atomic force microscopy, and more, as reviewed by Wu et al (2018), lead to different results, which may be more or less relevant in a particular situation.

Since the medium is anisotropic, both mechanical and orientational elastic coefficients, as well as viscosity, are, generally, expressed by fourth-rank tensors, and their full characterization is practically unavailable, as already mentioned in Sect. 2.5. The momentum conservation equations include activity, which generally depends on the underlying chemistry. The mechanical balance equations necessarily include pressure, although, as mentioned in Sect. 1.3 and illustrated by Fig. 1.7, pressure is not a well-defined variable in active systems (Solon, Fily, et al, 2015).

While keeping in mind these drawbacks, we have to be aware that no feasible alternative exists, save for following all microscopic motions, which would not be

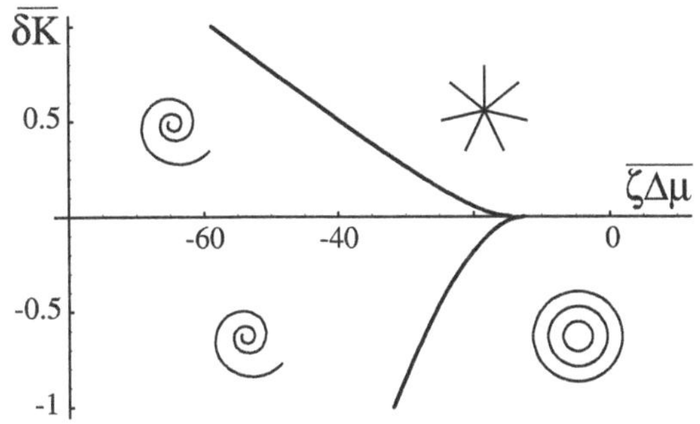

Fig. 6.2 Diagram of stability regions of spirals, asters, and vortices, as indicated by their outlines. *Ordinate*: Scaled difference between the bend and splay orientational elastic moduli. *Abscissa*: Scaled activity (Kruse et al, 2005)

sufficient either if chemistry remains behind the scene. Of course, Prost et al are well aware of this:

> Some aspects of biological systems do escape this generic description. The active gel theory has to be complemented by what is commonly called signaling. Note that signaling pathways can depend in subtle ways on mechanical stresses, which brings a further twist to the nonlinear physics aspect of active gel behavior.

The debut application of the active gel theory (Kruse et al, 2004, 2005) was the description of common structures observed most clearly in *in vitro* assays (recall Fig. 5.8): asters, vortices, and spirals. All these are topologically equivalent forms of unit-charge defects of either sign, characteristic to polar media. Activity causes spontaneous pairwise generation of defects of opposite sign, leading to a persistent turbulent pattern. We have already discussed this in Sects. 2.5–2.7 for the case of active fluids; topology is insensitive to physics and is the same in viscous and elastic media. However, the question of which particular structure of a defect is preferred depends on quantitative details. The diagram in Fig. 6.2 shows that spirals are stable when activity is contractile and the absolute value of the difference between splay and bend orientational elasticity is not too great, whereas asters and vortices are preferred in the case of tensile or not too great contractile activity, with asters being stable when the bend elastic modulus is larger, and vortices, the other way around. The location of curved boundaries in Fig. 6.2 depends on other parameters of the problem, and would also change if the anisotropy of the viscoelastic coefficients was taken into account.

Some caution is required here. We can speak about the structure of the inner core of a defect only when its size is much smaller than the "macroscopic" scale of the system but much larger than the "microscopic" scale. For the cytoskeleton, the former is the cell size measured in tens of microns. However, the latter is not a molecular scale of actin monomers but a characteristic distance between nodes of the network, which may be in the range of tens of nanometers or more. There may not be much leeway here, but it is more important that the alignment of the filaments is constrained by the structure of the network: recall that even angles at branching points are fixed (Sect. 5.3). In *in vitro* assays, there are no such constraints, and the scale contrast is better: the macroscopic scale is that of the laboratory device, but the microscopic scale is not molecular either, as it is set by the persistence length of rather stiff filaments, which is in the range of 10 μm. Therefore the pattern in Fig. 5.8 rather than the cytoskeleton could be a testing ground for the particular prediction in Fig. 6.2.

6.2 Crawling Cells

The testing ground for the active gel theory is the description of cell motion. Various studies have approached this problem using different variants of a continuous theory and at a different level of detail. In a review encompassing both continuous models and those based on direct description of growing actin filaments, Holmes and Edelstein-Keshet (2012) graded prior work by the level of biological detail and computational complexity, with the latter sometimes going down to 1D, hardly a physically relevant setting, and topped by 3D two-phase computations (Herant and Dembo, 2010).

Motility comes about naturally in a contractile active polar viscous or gel-like droplet. Simha and Ramaswamy (2002) proposed the mechanism of spontaneous symmetry breaking illustrated in the left panel of Fig. 6.3. When the orientation is undistorted, contractile forces, shown as blue arrows, are balanced (top). A splay causes an imbalance, which induces downward motion (middle), and this augments splay (bottom). The instability is counteracted by orientational elasticity, and motion arises at a critical level of contractile activity ζ_c as a result of a pitchfork bifurcation (Fig. 6.3 right), or second-order phase transition, using the term preferred by physicists. This mechanism is viable but in no way universal, and other patterns of distortions accompanying the mobility transition are feasible.

A difficult part of any problem involving a mobile reshaping body is to delineate its boundary. A radical way to solve it is to eliminate it in the way Alexander cut the

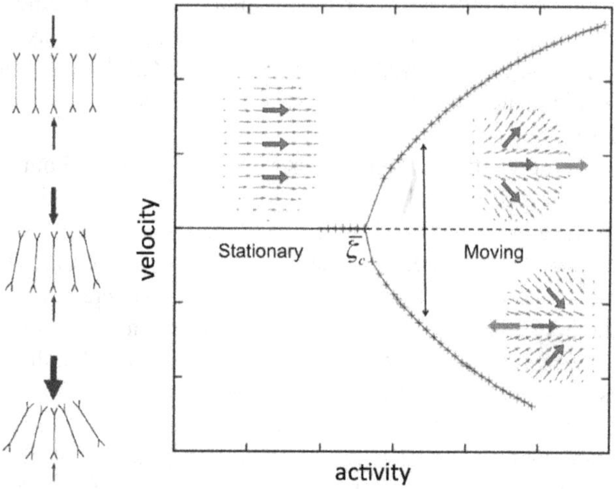

Fig. 6.3 *Left*: Mechanism of splay instability in a contractile active medium. The *size of the arrows* matches the strength of the contractile force. *Right*: Bifurcation to a motile state (Marenduzzo, 2016)

Gordian knot. A formal variable ϕ, called the *phase field*, is introduced, governed by a diffusion equation with a cubic nonlinearity that has two stable uniform solutions, say, $\phi = 1$ and $\phi = 0$ which, if they coexist, are separated by a narrow border region. The two values are associated first with the interior, and second with the exterior of the cell; thus, the boundary, rather than being sharp, becomes *diffuse*. This formal variable can be associated with the thickness or mass per unit area of the cell.

The phase field should be coupled to physical variables. In simulations by Shao et al (2012), it was part of a full-fledged model based on the viscous flow equation of the actin network and including contributions from the membrane surface tension and friction due to adhesion to the substrate. The hydrodynamic part was complemented by convection–diffusion equations for actin and myosin, taking into account the actin polymerization rate. The phase variable enters these equations in such a way that all terms in the transport equations, as well as the velocity, vanish at $\phi = 0$, and the flow feeds back by advecting the phase variable. A plethora of parameters in this system should in principle be sufficient to describe all experimental situations, and the publication contains many pictures of the various shapes of migrating cells.

Conversely, Ziebert and Aranson (2013) adopted a radical approach. The only physical field in their model is the averaged vector field $\mathbf{p}$ representing the actin filament network, which defines both its magnitude and its polar orientation. Its scalar product with the gradient of the phase field imitates the advection of the cell's interface due to polymerization or breakdown of actin filaments; in a later publication (Ziebert and Aranson, 2014), it is also made dependent on the density of adhesions. In its turn, the phase field affects the physical vector variable in such a way that its magnitude can be finite only when ϕ approaches unity and its orientation is correlated with the gradient of the phase field in a way encouraging polarization of the cell; moreover, the strength of this correlation is associated with the adhesion strength, and the model is extended to include substrate compliance. A global constraint involving both variables keeps both the cell area and the magnitude of $\mathbf{p}$ (imitating actin density) finite. This is a kind of stripped-down active gel model,

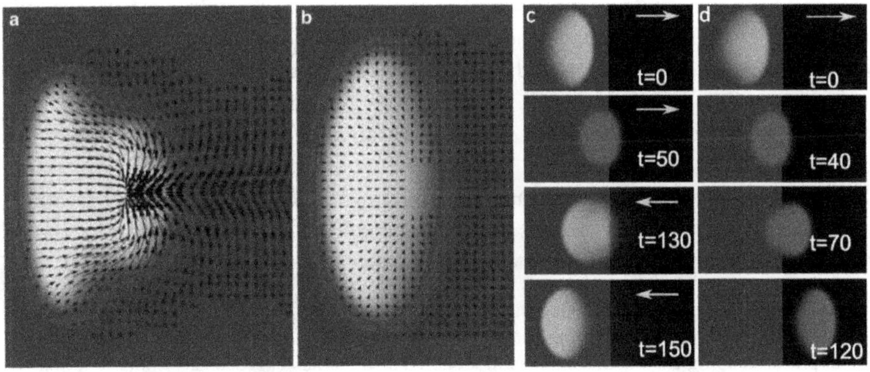

Fig. 6.4 (**a**), (**b**) Shapes of crawling cells and traction patterns (Ziebert et al, 2016). (**c**), (**d**) A cell reacts upon encountering a softer substrate, colored *black* (Ziebert and Aranson, 2013)

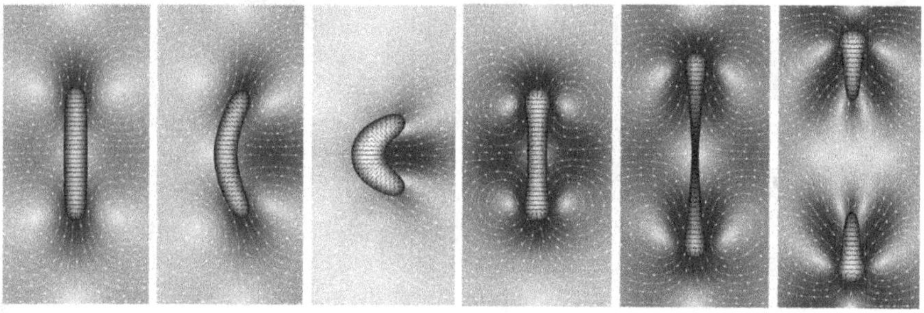

Fig. 6.5 Two sequences leading to motion (*left*) and division (*right*) of an active nematic droplet. The velocity of the surrounding fluid increases from lighter to darker regions (Giomi and DeSimone, 2014)

with orientational and mechanical elasticities merged in the same way that the orientation and propagation directions are merged in the Vicsek model (Sect. 1.1), making the momentum conservation equation superfluous.

Two examples of cell shapes, complete with a map of the substrate traction patterns effected by the cell, are shown in Fig. 6.4a and b. The cell feels the stiffness of the substrate and changes its shape when encountering a softer area (Fig. 6.4d), and can even reverse its direction of motion, as it prefers a stiffer substrate (Fig. 6.4c).

On this level of description, the only difference between the motion of a viscoelastic gel and a viscous droplet is in adhesive action. Giomi and DeSimone (2014) used the same phase field approach to model a droplet filled by an active nematic fluid, but as in Sect. 2.5, used separate equations for nematic orientation and viscous flow, encompassing also an isotropic passive fluid outside the drop. Two droplet deformation sequences are shown in Fig. 6.5. Remarkably, the droplet may break up. Since the interface is diffuse, the problem of a singularity at the pinching point mentioned in Sect. 3.6 does not arise here.

The phenomenology of the basic model by Ziebert and Aranson (2013) appears to be as rich as that of the far more complicated model by Shao et al, which tops other 2D models on the rating scale by Holmes and Edelstein-Keshet, but still does not include all elements of the active gel theory, not speaking of the biological realities. Nevertheless, Ziebert and Aranson (2014), anticipating possible objections, suggested a "modular approach", so that more "modules" fitting particular aims could be added on demand. One of them has already been mentioned above; the others include taking into account bending energy of the cell membrane and cortex, responsible for the effective surface tension that restricts the boundary curvature, and incorporating various specific propulsion mechanisms. Quoting the editorial of the Discussion and Debates issue where this paper was published (Pismen, 2014),

> The qualitative discussion of modeling of the various motility mechanisms in the concluding part of the paper opens interesting perspectives that might lead eventually to the convergence of phenomenological and mechanistic models. This "modular" approach is reminiscent of a tale about a soldier promising a peasant woman to make a soup out of an ax. Starting with this base, he gradually asks the woman to add more and more nutritious ingredients, so that in the end the soup comes out tasty! The tale can indeed be applied to all generic models.

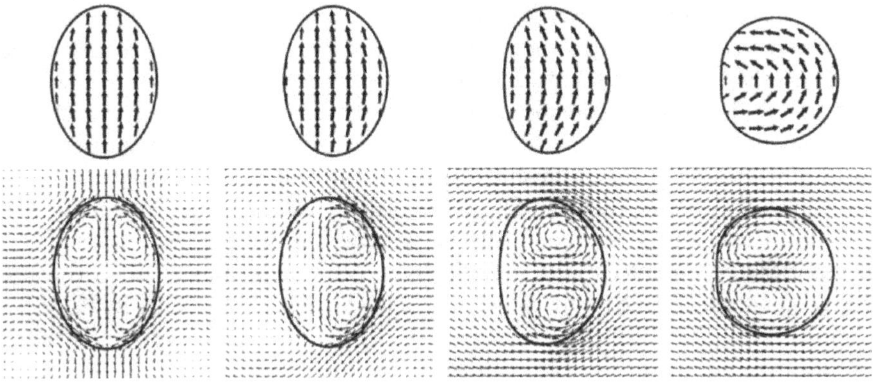

Fig. 6.6 Simulated active contractile polar cell moving to the right. *Top*: Orientation field. *Bottom*: Velocity field in the comoving frame (Marth et al, 2015)

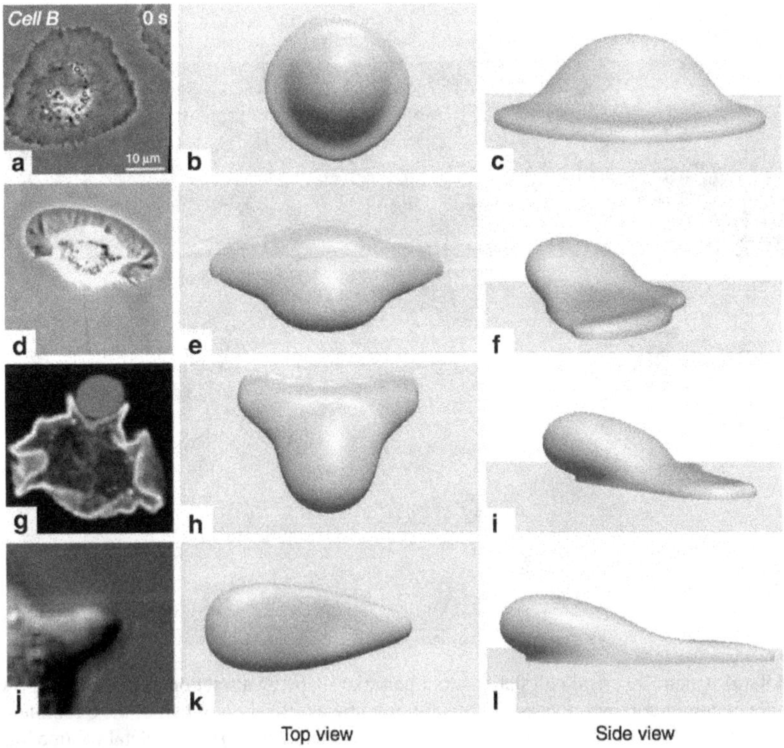

Fig. 6.7 Experimentally observed shapes of crawling cells (*left column*) and 3D shapes simulated by Tjhung et al (2015) in top and side view

The equations of a viscous polar gel with boundaries imitated by the phase field have become the prevailing choice in the single-phase continuous model. Marth et al (2015) also included the surface tension of a cell membrane, admitting, however, that it was of less relevance for the motility. One of the simulated sequences is shown in Fig. 6.6. We can discern in the upper row a splay instability accompanying the onset of motion, but it is in a direction perpendicular to the one in Fig. 6.3.

Tjhung et al (2015) extended to 3D this, as they call it, "minimal" model that includes all components of the active gel theory except mechanical elasticity, with the boundaries still delineated by the phase field. Several shapes observed in experiments and approximated by the fitting parameters of the model are shown in Fig. 6.7.

6.3 Two-Phase Models

While concentrating on the cytoskeleton responsible for the cell's mechanical strength in the preceding chapter, we have so far passed over its liquid component – *cytosol*. Two-phase models including viscous motion of both cytoskeleton

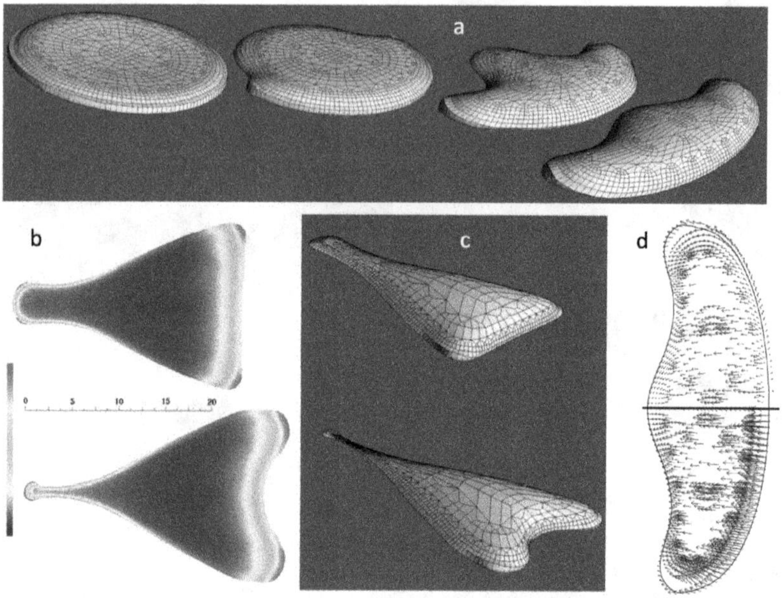

Fig. 6.8 (**a**) A hemispherical cell flattens to a pancake shape by attraction to the substrate, while the entire circumference is activated. This is followed by deactivation of half of the circumference, gradually leading to a stable gliding shape resembling a keratocyte. (**b**) Cytoskeletal volume fraction (scale from 0 to 2‰) and (**c**) the corresponding shapes; *surface lines* show the computational mesh. (**d**) Dorsal (*top half*) and midheight (*bottom half*) cytoskeletal velocity field in the cell frame (Herant and Dembo, 2010)

and cytosol are not uncommon, as reviewed by Cogan and Guy (2010) and Mogilner and Manhart (2018). The top place in both biological detail and computational complexity was awarded by Holmes and Edelstein-Keshet (2012) to the 3D two-phase simulations by Herant and Dembo (2010). Their model viewed cytosol as an incompressible fluid moving within the cytoskeleton as within a porous body. The cytoskeleton was also viewed as viscous rather than elastic, and neither was its polarization included, but its density was subject to change due to polymerization as well as transport. The cell was kept in shape by attraction to the substrate and specific interactions with the membrane. The latter, most ingenious part of the model included both network–membrane repulsive stress and generation at activated portions of the three-phase contact line of a polymerization promoter diffusing to the bulk. In this formulation, there was no need for the artificial phase field that later came into fashion. Notwithstanding all differences between this model and the later 3D model by Tjhung et al (2015), the shapes it produced, some of which are shown in Fig. 6.8, are rather similar. This is quite understandable. Both models contain a number of parameters which can be adjusted to make the result look like experimentally observed crawling cells. Recall the similar convergence of 2D single-phase models in the preceding section.

The cytoskeleton was assumed to behave on the time scale of the cell motion as a viscous fluid in this and all the above-mentioned models, so not all ingredients of the active gel model have been tested here. Joanny et al (2007) extended this model to include both a viscoelastic gel and viscous fluid cytosol, as well as diffusion equations for whatever chemical species should be essential for a particular problem.

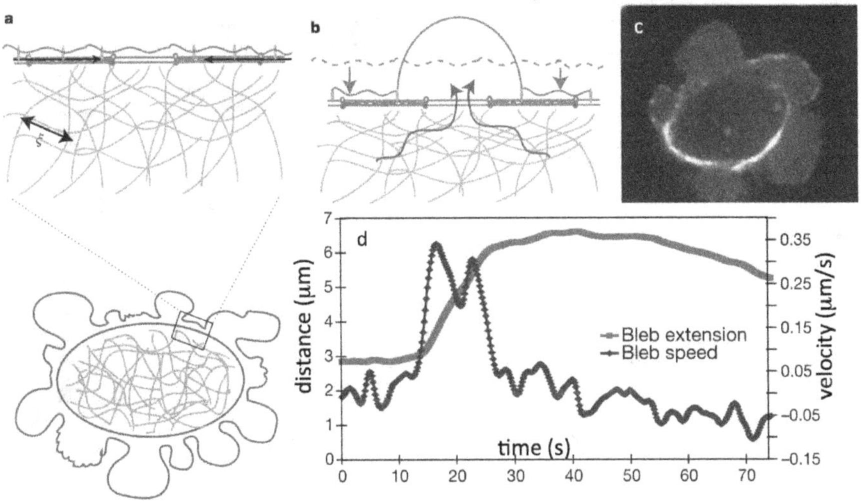

Fig. 6.9 (**a**), (**b**) The mechanism of blebbing initiated by cortex contraction (see the text for explanation). The actin cortex is drawn in *red*, the membrane in *magenta*, and the cytoskeletal network in *green*. (**c**) Image of a blebbing cell. Actin-enriched areas are rendered *lighter*. (**d**) Change in the bleb extension and velocity in time (Charras et al, 2005)

No attempt to solve this formidable system in a realistic setting has been undertaken in this paper, which was apparently meant as an inspiration for its followers. The antecedent of this approach lies far from biophysics. This was the *poroelastic* theory originating in the studies of fluid-saturated porous soils (Biot, 1941; Coussy, 2004). It was not mentioned by Joanny et al, but cited and applied to the model motion of blebbing cells in the earlier paper by Charras et al (2005).

The theoretical arguments were only qualitative in this paper, but they emphasized the elastic effect not accounted for in other models: compression of the cytosol by the contractile cytoskeletal network. This mechanism is illustrated in Fig. 6.9a, where part of a blebbing cell adjacent to the membrane is enlarged in the upper panel. A local myosin-driven contraction in the cortex neighborhood (black arrows) leads to shortening of the cortical periphery and therefore to compression of the porous cytoskeletal network that fills the cell. The compression creates a hydrostatic pressure, which drives the cytosol (blue arrows in Fig. 6.9b). If there is a local defect in the attachment of the membrane to the cytoskeleton (the dashed mauve line

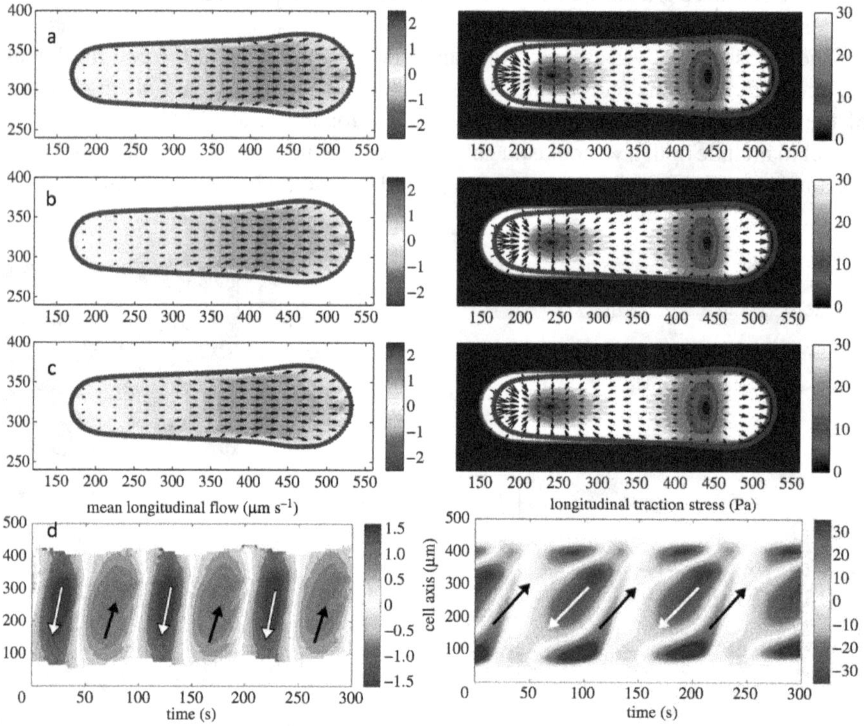

Fig. 6.10 Peristaltic propulsion in the poroelastic model of *Physarum polycephalum*. (**a**)–(**c**) Sequences of flow (*left*) and traction (*right*) fields. *Arrows* indicate the direction; the *colour map* indicates the projection of the flow velocity (μm/s) onto the cell axis and the magnitude of the stress field (Pa), respectively. (**d**) Simulated evolution of mean longitudinal flow (*left*) and traction (*right*) in time. *Black* and *white arrows* indicate forward and backward flow, respectively (Lewis et al, 2014)

indicates its original position), a bleb is extruded. Bleb expansion is opposed by two forces: extracellular osmotic pressure and membrane tension. The typical dynamics of a protrusion is shown in Fig. 6.9d.

Apparently, the earliest simulations of a crawling cell based on the poroelastic model were carried out by Taber et al (2011), where the same ubiquitous kerocyte-like shape, seen on many pictures in this and the preceding sections, were produced. Lewis et al (2014) applied this model to the dynamics of *Physarum polycephalum* plasmodial slime mold (Sect. 5.8). The centerpiece of poroelastic models, the active contractile force that drives deformation of the cell and the flow of cytosol, was assumed in this work to be generated by a traveling wave of isotropic contractile stress that corresponds to the peristaltic propagation mechanism, with waves of contraction and flow traveling from posterior to anterior along the long axis of the cell, as observed in the experiment (Matsumoto et al, 2008). The simulated intracellular 2D flow and traction fields are shown in Fig. 6.10. The cell reshapes and advances as the flow and traction oscillate.

Polymerization waves, mentioned among promising additional "modules" by Ziebert and Aranson (2014), were later discussed in detail by Dreher et al (2014) and Kruse (2016). They related waves of this kind to the dynamics of nucleators of actin filaments, illustrated in Fig. 6.11a–c. The sequence of images shown in Fig. 6.11d has been obtained by solving the dynamical equations of actin and nucleators together with the equation of the phase, velocity, and polarization fields. The cell advances here in an oscillatory fashion, similar to simulations by Lewis et al (2014), but this is not the only way it can move. Under different conditions, simulations also show oscillatory and erratic motion induced by different internal wavy patterns of polarization and activity.

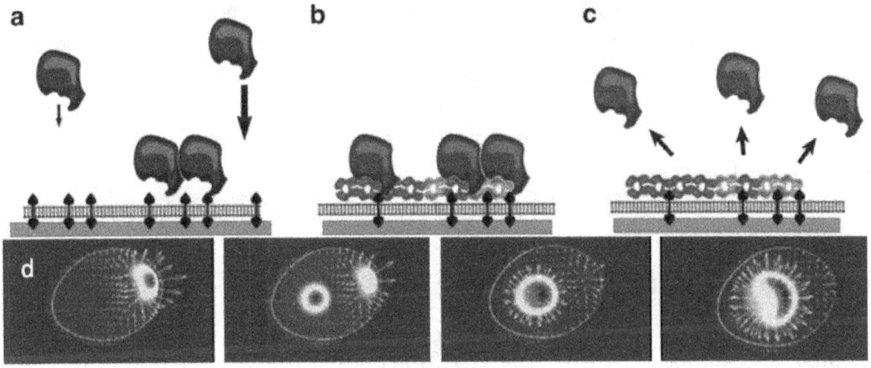

Fig. 6.11 Illustration of nucleator dynamics. (**a**) Nucleators (*blue*) exist in an active state or are connected to the substrate (*grey*) by adhesion molecules (*black*). Inactive nucleators bind cooperatively to the membrane and thus become activated. (**b**) Active nucleators generate new actin filaments (*red*). (**c**) Actin filaments feed back on the nucleators and inactivate them. (**d**) Subsequent snapshots of a moving cell, showing the polarization, indicated by *arrows*, and color-coded actin density increasing from blue to red (Kruse, 2016)

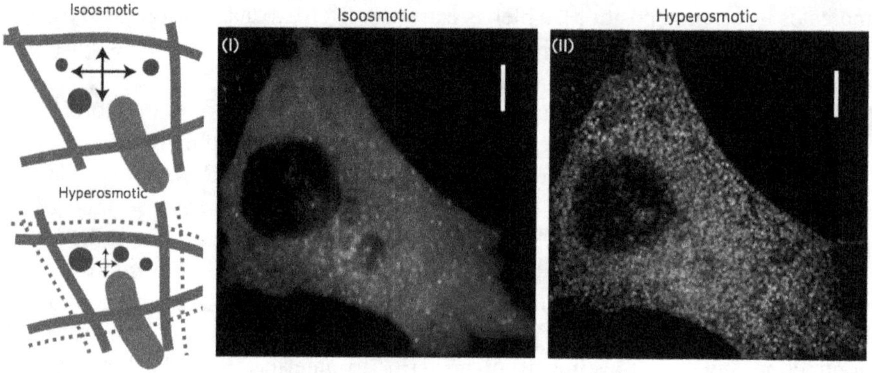

Fig. 6.12 *Left*: Change in the cytoplasmic pore size due to contraction under hyperosmotic conditions. A cellular organelle is colored *red*, and a protein molecule, *black*. *Center*: Blurred images of freely diffusing quantum dots. *Right*: The images become distinct when the pores shrink, restricting motion. Scale bars 10 μm (Moeendarbary et al, 2013)

In all the above models, the cytosol volume was assumed fixed, although water can easily seep in and out of the cell through the membrane. Moeendarbary et al (2013) looked for reasons why this effect has not been considered previously. Their explanation was that most techniques for studying the mechanical properties of cells (Wu et al, 2018) take into account only volume-conserving (isochoric) deformations, and do not apply to situations when changes in osmotic pressure induce water flux into or out of the cell. The cell can be viewed as a porous body where interstices shrink when water is driven out under hyperosmotic conditions, as sketched in the left panel of Fig. 6.12). This slows down diffusion of macromolecules, including protein enzymes and circulating actin monomers, and thereby affects both cytoskeletal mechanics and biochemical reactions. Suppression of diffusion was demonstrated by time-projection of images of light-emitting quantum dots. Under isoosmotic conditions (panel I), they are freely moving, and the images are blurred, but individual quantum dots are seen clearly when pores contract, thereby restricting their motion (panel II).

6.4 Chemo-Elastic Instabilities

Complaints by Moeendarbary et al (2013) about the lack of attention to osmotic effects were not quite justified. Salbreux et al (2007) accounted for water transport through the membrane in their theory of oscillations involving calcium ions. It is known (Alberts et al, 2002) that active tension in the cytoskeleton is regulated by the calcium concentration. Its local increase causes the actin cortex to contract; at the same time, it stretches on the opposite side of the cell, triggering there calcium influx through stretched activated channels in the membrane and enhanced actin

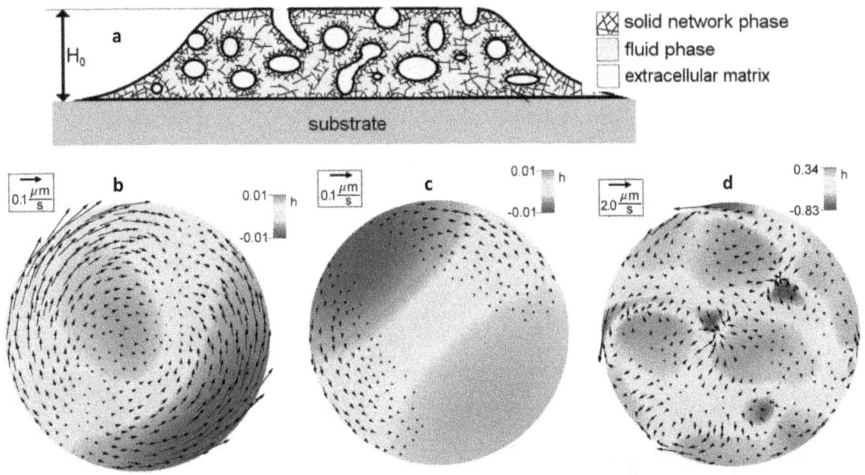

Fig. 6.13 (**a**) Drawing of a *Physarum* microdroplet in side view, showing the invaginations filled with extracellular matrix (*light blue*), the fluid phase of the cytoplasm (*blue*), and the solid filamentous phase (*black*). (**b**)–(**d**) Wave patterns in the model *Physarum* droplet: rotating spiral (**b**), standing wave (**c**), and irregular pattern (**d**). The relative height deviation is color-coded, and the protoplasmic flow field is shown by *arrows* with length proportional to the velocity magnitude (Radszuweit et al, 2014)

polymerization. At the same time, with increasing calcium concentration, the cytosol becomes hyperosmotic, causing a water influx that thins out (*solates*) the gel. In this way, both actin tension and membrane protrusions oscillate around the circumference of the cell.

Radszuweit et al (2014) took into account calcium dynamics in the poroelastic model of the oscillations in *Physarum microplasmodia* cells. They worked with the droplet model sketched in Fig. 6.13a. Their model was also built around the observation that calcium ions both solate actomyosin gels and stimulate its active contraction, and incorporated water and calcium dynamics in the full two-phase viscoelastic description. The downside is that the round perimeter of the cell was fixed, and only its height was allowed to change. The simulations produced various wave patterns; some examples are shown in Fig. 6.13.

Interactions between elastic deformation, polarization, and concentration fields are a rich source of instabilities. Köpf and Pismen (2013a) studied them in an abstract setting, with periodic boundary conditions commonly used in the theory of non-equilibrium patterns. The minimal scheme of interactions, shown in Fig. 6.14a, includes production of a signaling species c induced by deformation, polarization $\mathbf{p}$ caused by the gradient of this species, and deformation caused by polarization. The bifurcation diagram in Fig. 6.14b is obtained by linear stability analysis of the uniform quiescent solution with no polarization and signaling species. The diagram is spanned by the parameters quantifying the strength of interactions. The activity parameter q is the proportionality coefficient between the polarization vector and the active force

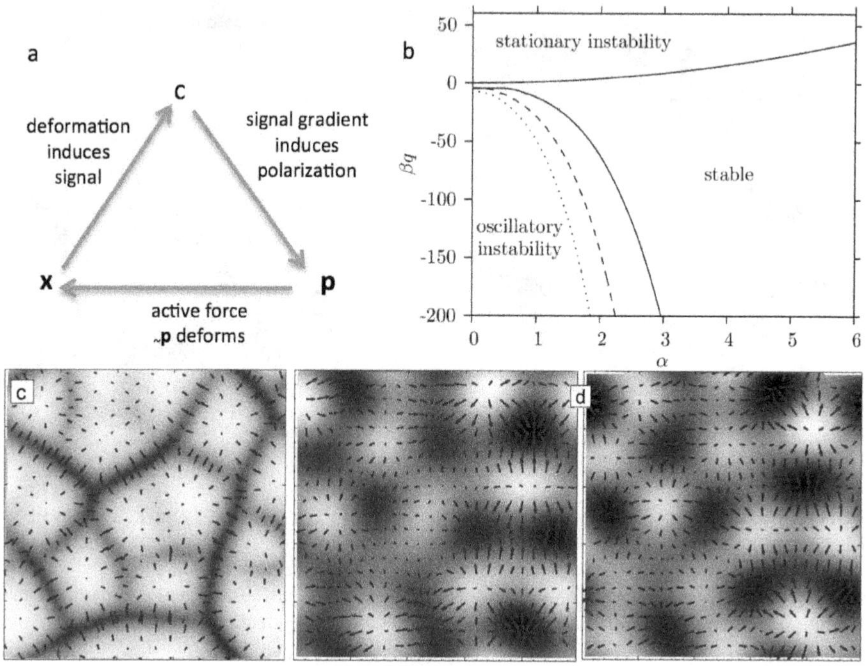

Fig. 6.14 Spontaneous emergence of deformation, polarization, and a signaling species in a minimal interaction model. (**a**) Scheme of interactions. (**b**) Bifurcation diagram (see the text for explanations). (**c**) Stationary pattern midway through a coarsening sequence (**d**). Two antiphase snapshots of the oscillatory pattern (Köpf and Pismen, 2013a)

that causes deformation of the elastic medium. It appears in combination with the parameter β that quantifies the production of signaling molecules due to deformation. The parameter α defines the polarization strength. An additional dynamic parameter, the ratio of characteristic times of the deformation and polarization dynamics, affects only the domain of oscillatory states, which shrinks as it increases, as the oscillatory instability locus retreats from the solid to dashed to dotted lines in the diagram. The oscillatory instability sets in at a certain wavelength, increasing with α, but the preferred wavelength of the stationary instability is infinite, so the pattern, starting from an initial perturbation, coarsened with time.

6.5 Modeling Tissues

A tissue can also be viewed as an oriented active gel in the framework of a continuous approach, abstracting from its cellular structure, upon which we shall concentrate in the next chapter. This approach is most suitable in the application to epithelial layers, the most commonly encountered 2D dense cellular assembly, sketched in Fig. 6.15.

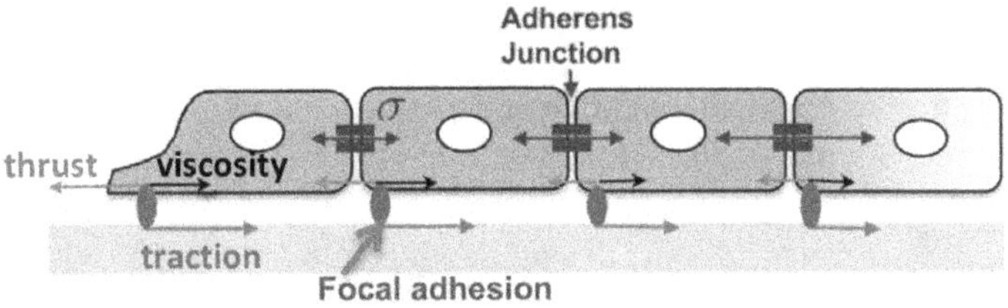

Fig. 6.15 Sketch of a flat epithelial layer. Stress σ (*blue arrows*) is transmitted along the layer though adherens junctions; the thrust (*green arrows*) is balanced by substrate traction (*red arrows*) and the viscous friction force (*black arrows*). Adapted from Notbohm et al, 2016

It consists of cells tightly connected with their neighbors by force-transmitting *adherens junctions*, while individual cells are connected by focal adhesions to the substrate or extracellular matrix. Modeling flat epithelium as a viscous polar or nematic active medium in a confined domain does not differ much from the approach of Sects. 2.5–2.7. Notbohm et al (2016) solved, in a circular domain, the active gel model amended by the convection–diffusion equation of a chemical, produced in proportion to the cellular stretching and promoting polarization of the medium along its gradient, according to the scheme in Fig. 6.14a.

The model produced chaotic wave patterns similar to those observed in their experiment. Figure 6.16 shows color-coded maps of the dependence of the radial velocity, stress, and radial traction, all of them averaged over the azimuthal angle, on the radial position and time. The velocity field, as well as stress and traction, alternate between inward and outward motion, in accordance with experimental observations. Recall the diagram in Fig. 6.14b, indicating that oscillations are expected in a model coupling polarization, deformation, and chemical signaling. The produced patterns certainly contained intermittent production and annihilation of topological defects, but this, of course, does not affect regular oscillations seen upon angular averaging.

The most interesting application of continuous tissue models is the description of an *advance* of a cellular layer into an unoccupied area, relevant, in particular, for *wound healing* (more on this in Sect. 7.2). In this setting, we again encounter a difficult moving boundary problem that was dealt with by introducing a nonphysical

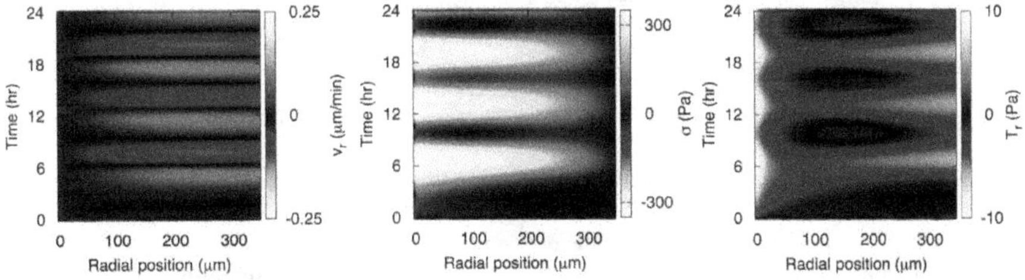

Fig. 6.16 Color-coded maps of the dependence on the radial position and time of the radial velocity v_r (*left*), the magnitude of the stress σ (*center*), and the radial traction T_r (*right*) averaged over the azimuthal angle (Notbohm et al, 2016)

phase field when the motion of a single crawling cell was considered in Sect. 6.2. Here, as in the crawling cell problem, different approaches were attempted, reflecting physical reality with variable veracity and tinkering efforts. The phase field is no longer needed because, unlike volume-conserving cells, the layers *grow* due to cell division, and therefore tend to advance into available space, in what is called *kenotaxis*, from the Greek κενός, meaning "void". The tendency to advance may be strengthened by the force deriving from the tendency of cells to form focal adhesions. This kind of motion is called *haptotaxis*, from the Greek ἅπτω, meaning "touch". It is similar to *wetting* by common fluids, which is likewise driven by molecular fluid–substrate interactions.

Arciero et al (2011) managed to fit their observations of the closing gap in a tissue with the help of a model based on the Darcy law, which itself originated in the theory of flow in porous media, wherein the velocity is proportional to a pressure gradient. The pressure was connected to the cell density by the ideal gas law (!), and the layer was referred to as "a compressible inviscid fluid", although, of course, the Darcy law is based on viscous motion. Neither polarity nor activity nor cell division were included, but the layer nevertheless moved under the action of a force applied to the free boundary, and the model had enough parameters to produce convincing pictures and acquire a fair number of citations.

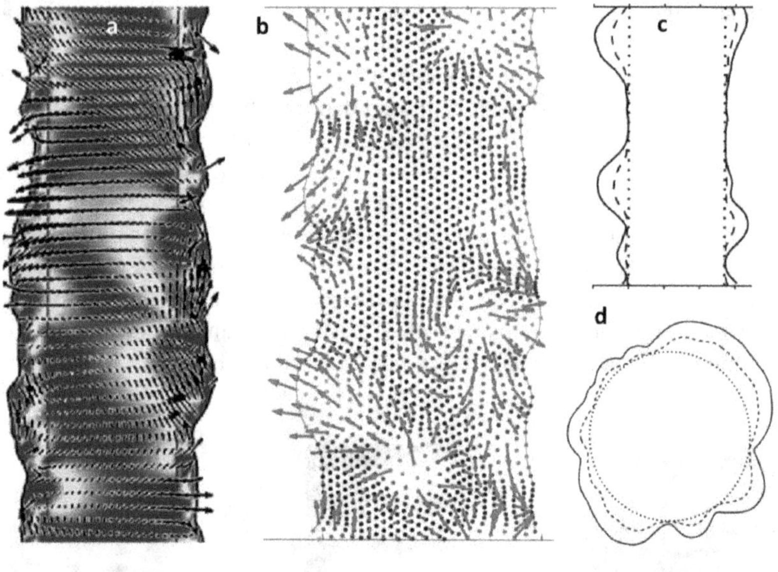

Fig. 6.17 (**a**) Advance of a rectangular layer in the model by Lee and Wolgemuth (2011). *Arrows* show the local velocity; the traction force exerted on the substrate is color-coded, increasing from blue to red. (**b**)–(**d**) The model by Köpf and Pismen (2013b). (**b**) Swirling motion initiated by noisy initial conditions. *Arrows* indicate the polarization amplitude and direction. *Points* may be interpreted as locations of cell centers, so that cells are larger in the dilute areas. (**c**), (**d**) Advance of rectangular or circular layers from the original (*dotted*) to dashed and solid boundaries

A coeval model by Lee and Wolgemuth (2011) did include the Stokes equation of a viscous fluid, nematic elasticity, and the tensile active force along the nematic director, but assumed the cell density to be constant. The free boundary of the layer advanced to close a circular wound or to distort a rectangular layer, as shown in Fig. 6.17a. Save for the meaning of the arrows and coloration, this picture looks very much like Fig. 6.17b, originating from a totally different model.

Köpf and Pismen (2013b) considered an elastic rather than viscous layer, and as in the scheme in Fig. 6.14a, included interactions among three fields: vector polarization, deformation, and a chemical signal. However, this triad interacted in a different way: advance along the polarization direction was proportional to the concentration of a chemical, which also enhanced polarization. This feature was grounded in experiments by Nikolić et al (2006) showing that injury-induced activation of a signaling species is essential for collective cell migration after wounding. On the other hand, Poujade et al (2007) observed expansion of a layer into unoccupied space with no injury, although this did not mean that chemical signaling was absent.

The model faithfully reproduced the essential features of unconstrained spreading observed by Poujade et al, including fingering, swirling motion, as in Fig. 6.17b, and a faster advance near the leading edge. Fingering is a ubiquitous feature of spreading layers that can be attributed just to the enlargement of the contour of the advancing free boundary in protruding regions. This basic factor operates in the model by Lee and Wolgemuth (2011), while in the model currently discussed it is enhanced by higher chemical activity in expanded areas. The computation used a Lagrangian algorithm, whereby points on the original grid advanced with the local velocity. If each point in the Lagrangian grid corresponds to a cell, this means that cells in more strongly polarized and faster moving areas grow larger, as seen in Fig. 6.17b, and can be identified with enlarged *leader cells* observed in protruding fingers. Later experiments, also by Silberzan's group (Reffay et al, 2014), identified enhanced chemical activity in the leader cells, in particular in the finger area (Fig. 6.18a), and a stronger traction force near the leading edge (Fig. 6.18b and c), bringing together chemical and mechanical cues.

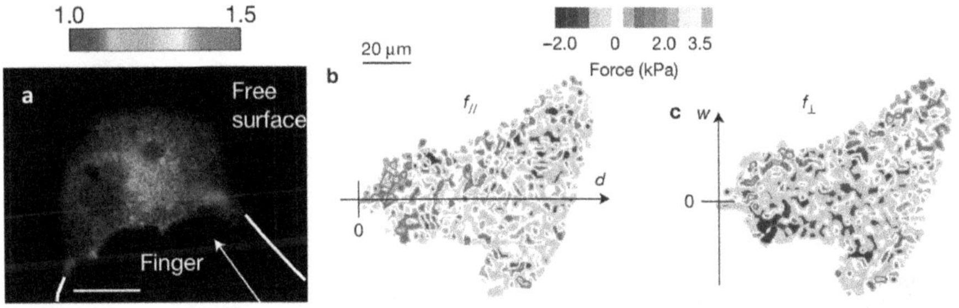

Fig. 6.18 (**a**) Mapping of chemical activity. Scale bar 10 μm. (**b**), (**c**) Mapping of the traction forces obtained by the micro-pillar technique, showing the components in the longitudinal (**b**) and transverse (**c**) directions (Reffay et al, 2014)

6.6 Network Restructuring

The ultimate challenge to the active gel theory is a dynamic nonlinear dependence of its parameters on applied or emergent forces. Cytoskeleton is a *live* permanently restructuring material. Therefore, it reacts to applied force not just by extending or contracting as an iron rod would do, and not just in a more sophisticated fashion as a rubber string or jelly would react, but by adjusting its structure in a way dependent on the strength of the force and its temporal changes. In a supplement to their paper, Trepat et al (2007) briefly review the early appreciation of uncommon mechanical properties of living matter:

> The hallmarks of "soft glassy matter" including shear fluidization (thixotropy), crowding, and trapping of particles by their neighbors, had already been identified in the living cell – although not without substantial controversy – as long ago as the late part of the 19th and early part of the 20th centuries. But with the subsequent discovery of the polymeric cytoskeleton, the emergence of polymer physics, and modern theories of semiflexible networks that stiffen with strain instead of being fluidized, these early seminal observations were all but forgotten.

Stiffening mechanisms were emphasized in the late 20th century. The guiding principle was *mechanotransduction* (Ingber, 1997), whereby mechanical signals modify intracellular biochemistry, e.g., through influx of signaling molecules through stretch-sensitive ion channels, and promoting synthesis of molecular enhancers of filament polarization and branching. In this way, stretching a tissue sample leads to the assembly of cytoskeletal filaments, and relaxation of tension promotes their disassembly. There is also a purely mechanical reason for the stiffening of a wormlike polymer chain before it breaks at a critical tension. This is *entropic* elasticity due to a decreasing number of possible configurations of a stretched chain. The theory based on this principle was extended to networks of semi-flexible filaments (MacKintosh et al, 1995), explaining significant strain hardening at modest strains, as well as a stiff power-law dependence of the elastic modulus on actin density due to increased entanglement. A stress-stiffening material offers little resistance to small deformations, allowing it to be easily remodeled locally, but strengthens at larger strains to ensure cell and tissue integrity.

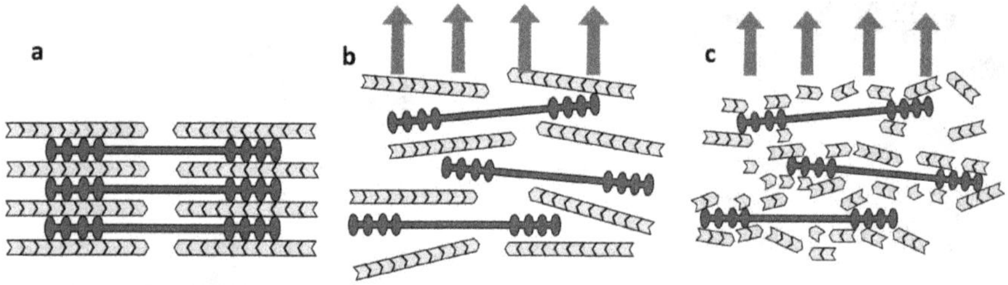

Fig. 6.19 Sketch of the break-up of an actomyosin bundle under transverse forcing. (**a**) Initial quasisarcomeric structure of actin (*yellow*) and myosin (*blue*) filaments. (**b**) External transverse forces (*red arrows*) disrupt actin–myosin connectivity, which may lead (**c**) to the break-up of actin filaments (Morozov and Pismen, 2011)

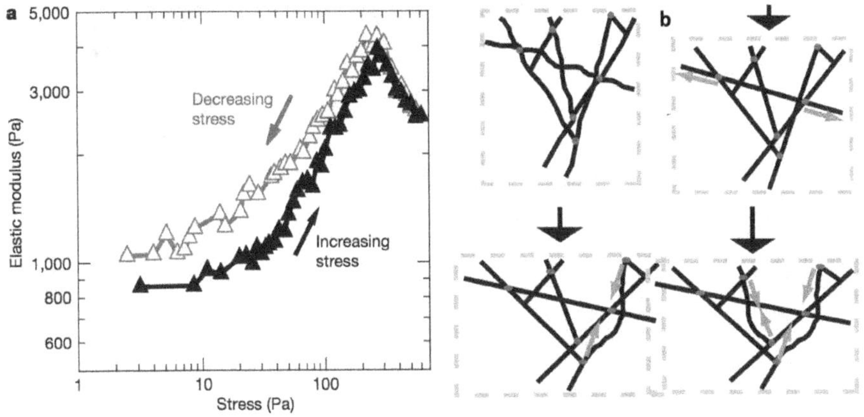

Fig. 6.20 (**a**) Dependence of the elastic modulus on the applied stress in the stress-increasing (*black*) and stress-reducing (*red*) sequences. (**b**) Scheme of consecutive stress-stiffening and stress-softening regimes of an entangled network. Starting from an unstressed network, compressive stress is applied, as indicated by the *black arrows* with proportionally increasing length. The material expands laterally, and some filaments and crosslinkers are stretched, as indicated by *green arrows*, leading to a stress-stiffening regime. When the stress is increased above a critical level, some filaments resisting compression (*green arrows*) buckle and no longer contribute to the elasticity, but network connections prevent them from collapsing. As the stress is further increased, more filaments buckle and the elasticity of the network drops further, leading to the stress-softening regime (Chaudhuri et al, 2007)

Stiffening and fluidization are not exclusive but depend on the direction of the applied force relative to particular cytoskeletal structures. For example, an actomyosin bundle may fluidize under the action of a transverse force that severs links between myosin and actin filaments, triggering the latter's break-up, as sketched in Fig. 6.19. Stiffening and softening under stress can arise in branched networks owing

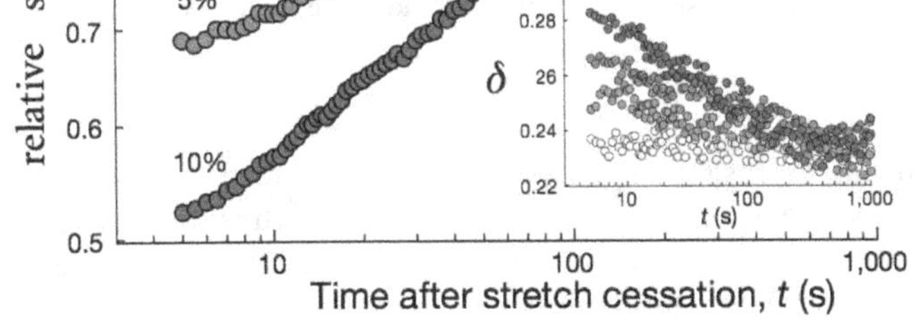

Fig. 6.21 (**a**) Evolution of stiffness (relative to the pre-stretch value) for no prestretch (*open circles*) and after a single transient stretch of 2.5% (*green*), 5% (*blue*), and 10% (*red*). (**b**) Evolution of the phase angle δ after the same stretch (Trepat et al, 2007)

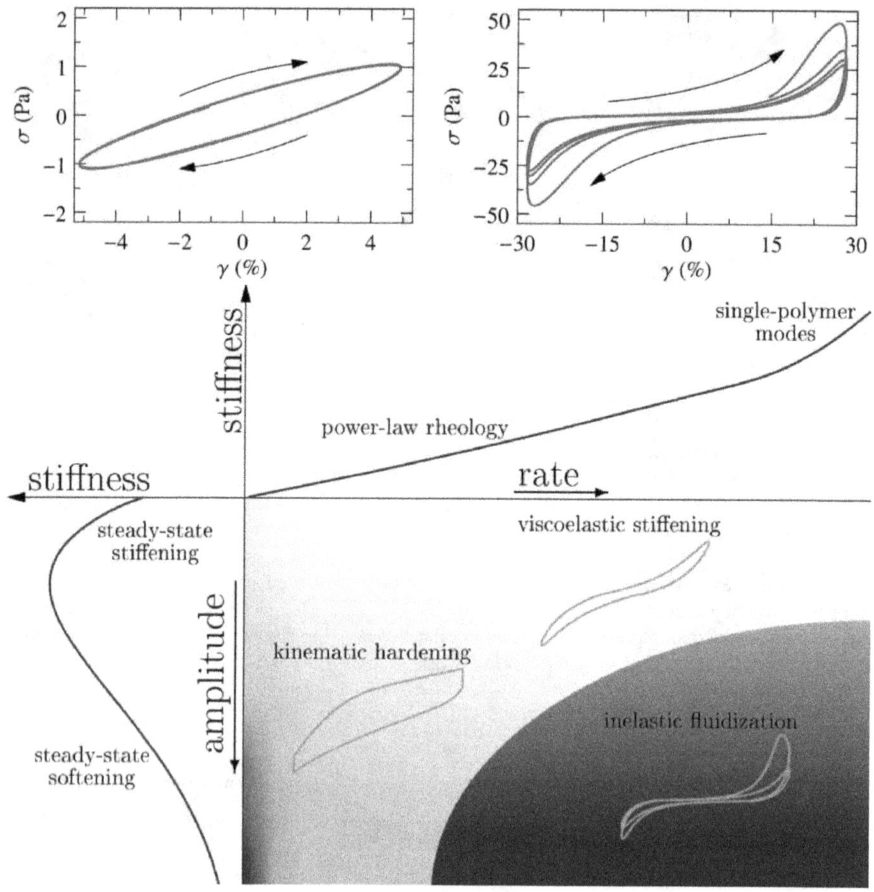

Fig. 6.22 *Top*: Experimental response of the instantaneous stress σ on small (*left*) and large (*right*) strains γ. Upward bending of the ellipses signals stiffening, while their concave regions near the maximum strain imply softening. *Bottom*: Constitutive diagram of stress–strain relations in the course of oscillatory forcing obtained in model simulations. The main *colored quadrant* gives a qualitative graphical summary of the mechanical response predicted by the model as a function of the amplitude and characteristic rate of imposed deformations. At high rates and high amplitudes, a steep initial stiffening with subsequent inelastic fluidization and slow recovery governs the response leading to complicated dynamics. Viscoelastic stiffening prevails at high oscillation frequencies and kinetic hardening at high amplitudes. The limiting behaviors at vanishing rate and vanishing amplitude, i.e., near the coordinate axes, are characterized in the *side diagrams*. The *upper quadrant* presents the power-law frequency dependence of the shear modulus on a log-log scale at low amplitudes. The *left quadrant* shows the nonlinear steady-state shear modulus showing stiffening and softening (Wolff et al, 2012)

to some filaments resisting extension and buckling of filaments resisting compression. Diverse ways a crosslinked network reacts to compressive stress are sketched in Fig. 6.20b. Stiffening under growing stress shown by the black curve in Fig. 6.20a turns into stress-softening when the stress exceeds a critical level. This process may be reversed as buckled filaments unbuckle once the stress is reduced, as shown by the red curve.

Since the network is dynamic, its response to an external force depends not only on its instantaneous strength, but on the *history* of forcing. In experiments by Trepat et al (2007), a single transient stretch drove the stiffness down, whereupon it recovered on a scale of minutes, as shown in Fig. 6.21a. The viscoelastic phase angle δ, equal to 0 in a purely elastic medium and $\pi/2$ in a purely viscous medium, also increases under stretch, indicating fluidization of the cytoskeleton, and gradually recovers (Fig. 6.21b).

Wolff et al (2012), working with an *in vitro* model cytoskeleton, subjected it to periodic stretching that caused the instantaneous stiffness to oscillate, as shown in the upper panels of Fig. 6.22. Oscillations at a small strain amplitude γ (left) are regular, but, as the amplitude increases (right), they become highly nonlinear and do not repeat the same orbit in a chaotic fashion. In their simulations using a phenomenological extension of the wormlike chain model, they took into account a slowdown of the long-wavelength bending undulations of the polymer backbone represented by a stretching of the relaxation spectrum. Beyond a characteristic interaction wavelength of the order of the entanglement length, the relaxation spectrum gives rise to a dramatic slowdown of the dynamics at long times or small frequencies, producing the power-law strain/stress dependence that is typical for cells. At low rates, in a quasistatic regime, it exhibits stiffening at low amplitudes, where entropic stiffening of the polymer backbone dominates, while at high amplitudes the stiffening eventually gives way to softening, which accounts for the distinction between oscillatory changes in the stress–strain relations. The modeling results are summarized in the diagram of the stress–strain relations presented in the lower panel of Fig. 6.22.

Tissues exhibit a similar response. In experiments by Walker et al (2020), microtissues strain-softened to maintain their mean tension but did not fluidize, and regained their initial mechanical properties upon loading cessation. This is evidenced by depolymerization of actin filaments in the course of 20-minute stretching and its restoration during a recovery period of the same duration, as shown in Fig. 6.23.

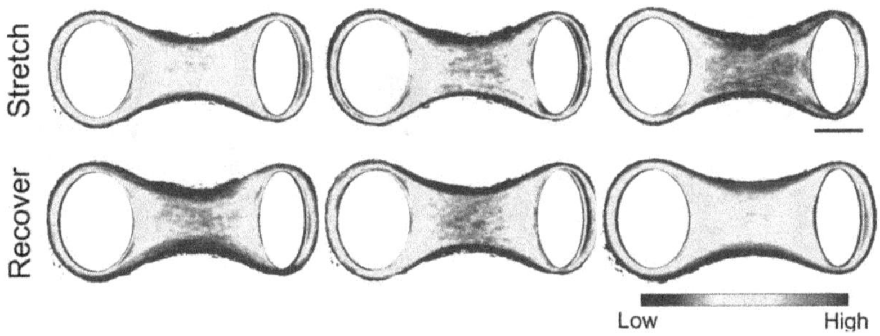

Fig. 6.23 Evolution of the filamentous actin concentration (color coded) in the microtissue during stretch and recovery. Scale bar 100 μm (Walker et al, 2020)

6.7 Geometric Activity

Many (but not all) complications inherent in living tissues may be avoided in *synthetic* active mater. A biomorphic material should be soft, compliant, and ductile; at the same time, it has to possess its own intrinsic properties responsive to external signals. Soft materials may reshape in various ways following *plastic* deformations due to volume change or anisotropic expansion/contraction. This sort of response, not driven by internal energy sources of the kind molecular motors would supply, can be characterized as *geometric* activity. This behavior is neither exotic nor sophisticated. Complex buckling patterns arise due to plastic deformations at the edge of a torn sheet (Sharon et al, 2007). They can be presented as a superposition of waves of different wavelengths varying with the speed of the crack propagation and the distance from the torn edge (Fig. 6.24a), and they reveal a set of transitions, each of which adds a new oscillatory mode.

Various shape-changing materials with alternative actuation mechanisms can be characterized as geometrically active. Polymers change their volume when they crystallize or undergo a glass transition. *Hydrogels*, networks of hydrophilic polymer chains imbibed with aqueous solutions, mimic humidity-induced swelling and shrinking of plant cells and are also responsive to other environmental factors, like temperature, illumination, and ionic strength. Some examples of reshaping under nonuniform illumination are shown in Fig. 6.24b (Kuksenok and Balazs, 2013). A variety of shapes can be created in layered or anisotropic flexible materials. All these capabilities have found a variety of applications (Ionov, 2013).

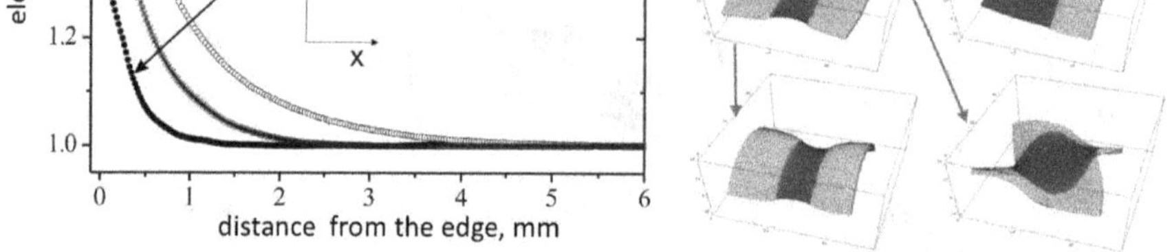

Fig. 6.24 (**a**) Elongation in the x direction (parallel to the torn edge) as a function of the distance from the edge of a polyethylene sheet, and buckling patterns at the point shown by *arrows*. The crack velocity is 0.5 cm/s for the *top curve* and 5 cm/s for the others; the *two upper curves* correspond to the steady state, and the *lower curve*, to the initial crack propagation (Sharon et al, 2007). (**b**) Deformation of a light-sensitive gel under non-uniform illumination. Illuminated regions, which are shrunk, are shown in *dark blue* (Kuksenok and Balazs, 2013)

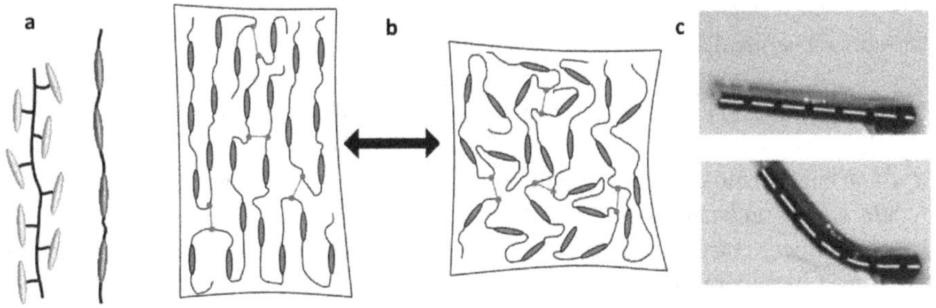

Fig. 6.25 (**a**) Schematic structure of side-chain and main-chain nematic elastomers. (**b**) Working principle of an artificial muscle: reshaping due to a phase transition between the isotropic and nematic states (Ohm et al, 2012). (**c**) Bending of a nematic elastomer plates due to one-sided actuation (Camacho-Lopez et al, 2004)

The most versatile responsive material, imagined by Pierre-Gille de Gennes (1975) in the role of an artificial muscle but able to do a finer work, is a *nematic elastomer* reshaping in response to changes in molecular orientation. The idea was realized by Finkelmann et al (1978), and a cholesteric elastomer was also briefly synthesized by the same group. Molecules of a *mesogenic* compound displaying liquid-crystalline properties, elongated as in the cartoon of Fig. 2.2, can be either connected into a polymeric chain or attached to a polymeric skeleton; both varieties, main- and side-chain elastomers, are sketched in Fig. 6.25a. Physical and mechanical properties of both nematic and cholesteric elastomers are reviewed by Warner and Terentjev (2003).

When this material is *actuated*, prompting a phase transition from the nematic to the isotropic state, it shrinks along the director and accordingly elongates in the normal directions to preserve its volume; it goes the opposite way when the transition is reversed, as shown in Fig. 6.25b. Shrinkage or elongation is greater in main-chain elastomers. In this way, a liquid crystal elastomer can really work as a muscle, lifting a considerable weight. If the phase transition is enforced on one side only, as in a layered material or when only one side of a flat strip is heated or illuminated, the strip bends, as shown in Fig. 6.25c (Camacho-Lopez et al, 2004). Much more can be accomplished through reversible phase transitions under the action of light, temperature, or chemical agents: strings, sheets, or shells of liquid crystal elastomers can be induced to walk and swim under repeated nonuniform dynamic actuation.

Deformed sheets of various elaborate shapes can be obtained by actuation of flat sheets with a prearranged nematic texture. This is commonly done by preparing the texture in a thin layer of nematic liquid on an appropriately patterned substrate and polymerizing it to fix the pattern. Computing the shape to be obtained upon actuation is a straightforward problem, although it can solved analytically only in exceptionally simple cases. The *reverse* problem of finding a texture producing a given shape is far more difficult, and its solution may not be unique or not exist at all (Griniasty et al, 2019).

The closest analogy to living tissues becomes apparent when deformations are caused by chemical interactions. The nematic–isotropic transition may be shifted by changing the concentration of the mesogenic component. This can be done by adding a non-mesogenic dopant or through isomerization induced by illumination or chemical agents. A more subtle effect is interaction of *gradients* of the concentration of a non-mesogenic component and nematic orientation that can be accounted for by adding the relevant term to the energy density (2.1) or, more appropriately, to the more precise expression based on the nematic tensor (2.2). The energy may be minimized by separating the nematic and isotropic phases. Depending on the sign of the gradient interaction term, either parallel or normal nematic orientation can be favored on the interphase boundary (Köpf and Pismen, 2013c).

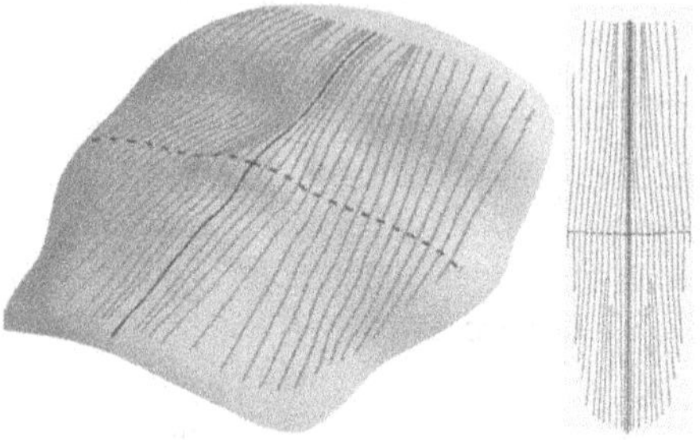

Fig. 6.26 A flat sheet deformed to a semblance of a human face (Griniasty et al, 2019)

Some dopant concentration patterns developing in a uniformly orientated material are shown in Fig. 6.27. The orientation of the stripes relative to the nematic director depends on the sign of the gradient interaction parameter. These patterns are transient: they coarsen with time to minimize the length of the boundary between the dopant-rich isotropic and dopant-poor nematic cases. The change in the nematic order parameter between the regions with different dopant concentration has to cause deformations, which are not taken into account in Fig. 6.27. An example of a deformed state of a rectangular sheet with separated isotropic and nematic phases is shown in Fig. 6.28a. In this computation, the higher solvent concentration plays the role of a dopant that causes a transition to the isotropic state, so that the isotropic domain is swollen.

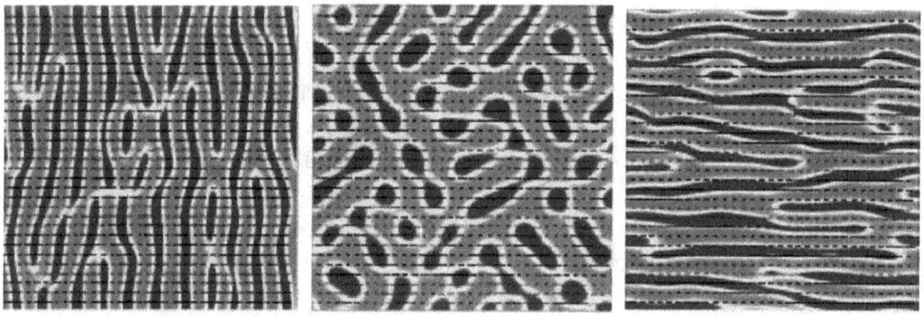

Fig. 6.27 Dopant concentration distribution (color coded) and nematic director orientation (*dashed lines*) at different values of the gradient interaction parameter: negative (*left*), zero (*center*), and positive (*right*). Snapshots of an intermediate stage of the coarsening sequence are shown (Köpf and Pismen, 2013c)

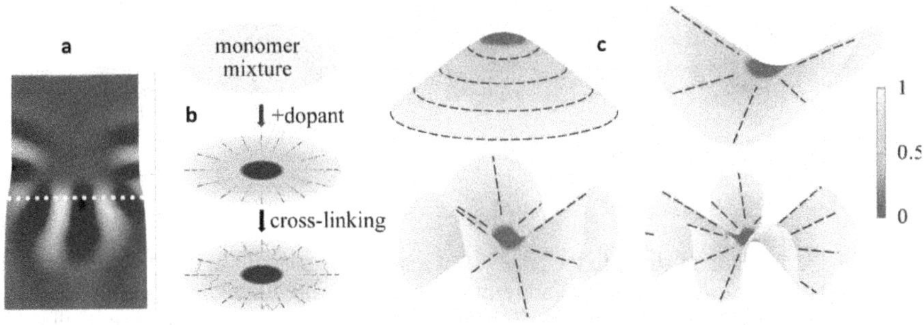

Fig. 6.28 (**a**) Deformation of a flat sheet near the boundary (shown by the *white dashed line*) between the isotropic (*top*) and nematic (*bottom*) phases. Depressions and bulges are shown by *dark* and *light colors*, respectively (Zakharov and Pismen, 2017a). (**b**) Scheme showing the establishment of a nematic texture in the presence of a non-mesogenic dopant. (**c**) Deformation of a circular patch. The *color scale* shows the nematic order parameter (Zakharov and Pismen, 2017b)

Patterns and shapes of chemically modified nematic elastomers can also be built up starting from a monomer mixture. No prepatterning is needed then; instead, a dopant is added at a chosen location, as sketched in Fig. 6.28b, and the nematic pattern is established spontaneously in accordance with boundary conditions. If equilibrium separation of isotropic and nematic phases is allowed to be attained, the dopant concentrates in a compact isotropic domain with a boundary of minimal length. Irrespective of the sign of the gradient interaction parameter, the nematic director should rotate by 2π around a closed boundary of the isotropic domain. By symmetry, we expect the isotropic domain to be placed centrally; then, in spite of the circulation of the director, no defects arise. In the absence of other constraints, the director will align either radially or circumferentially, depending on this parameter's sign. In the latter case, transition into the isotropic state causes the radius of the flat disk to extend and its circumference to shrink, whence its nematic domain will bend into a cone upon actuation. More interesting shapes arise in the opposite case of radial alignment. A shrinking radius and extending circumference generate one of these alternative forms in Fig. 6.28c, where the number of "petals" grows with the extension ratio.

Similar actuation of curved surfaces affects the number of defects. The nematic alignment field on a cylinder would commonly be smooth, but a point dopant source at the boundary induces circulation of the nematic director by π around the isotropic domain. The total circulation caused by two such sources at both ends is compensated by two defects with the charge $-1/2$. The defects are placed axially on the side opposite to the sources in a long cylinder (Fig. 6.29a), but are shifted to a middle location with circumferential separation when the cylinder is squat, as in Fig. 6.29b. The sign reversal of the gradient interaction coefficient between the pictures on the left and on the right just causes the entire distribution to rotate by $\pi/2$, but the shapes are, of course, very different, with radial bulging and shrinking interchanged. Local deformation at defects is not resolved at the scale of the picture and would be suppressed if the layer thickness exceeded the size of the defect core. Textures with

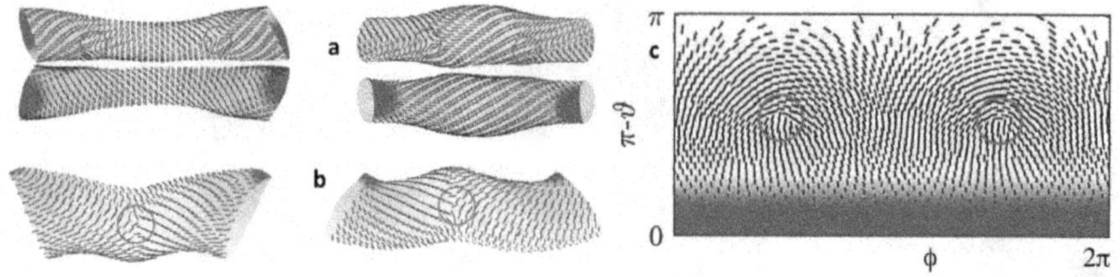

Fig. 6.29 Deformation of shells in the presence of a non-mesogenic dopant. Nematic alignment is shown by *black dashes* and defects are enclosed by *red circles*. The *color scale* showing the nematic order parameter is the same as in Fig. 6.28. (**a**) Elongated cylindrical surface; both sides are shown. (**b**) Squat cylinder. The textures *on the left* and *on the right* in (**a**) and (**b**) correspond to the positive and negative gradient interaction coefficients, respectively. (**c**) Spherical surface in the Mercator projection. The isotropic domain is at the south pole (Zakharov and Pismen, 2017b)

defects are specific to a cylinder, and defects disappear if the cylinder is cut into a square sheet, lacking the periodicity along its circumference.

With no dopant, the nematic alignment field on a sphere features four defects with the charge $+1/2$ placed symmetrically at vertices of a tetrahedron. Rotation by π around the isotropic domain reduces the required number of $+1/2$ defects to just two, as shown in Fig. 6.29c, where the Mercator projection is used to make the location of defects more evident.

6.8 Biomorphic Motion

Repeated restructuring may cause chunks of shape-changing polymers to move. Most experimental studies use *illumination* to actuate geometrically active materials dynamically. The effect of light commonly reduces just to heating the material, thereby forcing a volume change or a transition from an ordered to the isotropic state, but light can also cause structural changes, e.g., altering the gel's hydrophilicity,

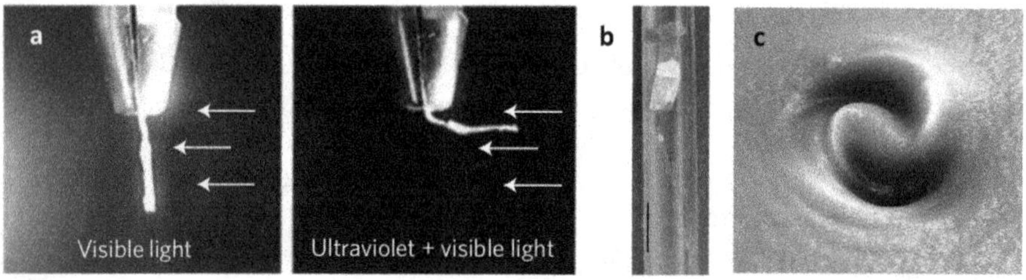

Fig. 6.30 (**a**) Artificial cilia bending due to isomerization caused by ultraviolet light (van Oosten et al, 2009). (**b**) Artificial flagellum driven by the same mechanism. Scale bar 5mm (Huang et al, 2015). (**c**) Spiral-shaped relief patterns caused by isomerization in the polymer film illuminated by a focused beam with helical wavefront. The side of the square is 6 μm. The elevation is indicated by the shade changing from dark at −80 nm to light at 50 nm (Ambrosio et al, 2012)

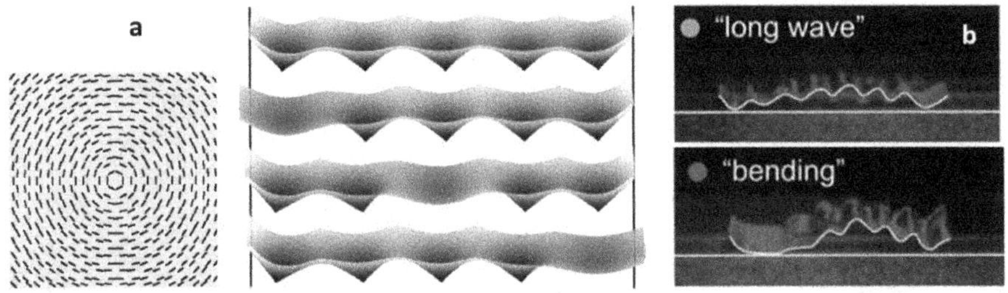

Fig. 6.31 (**a**) Simulated walker made of five sequentially actuated squares with the nematic pattern shown *on the left*, which turns into conical "legs" when heated (Zakharov and Pismen, 2016). (**b**) Crawling "caterpillar" (Rogóż et al, 2016)

as in experiments causing a strip of gels to advance by repeatedly passing a light beam (Kuksenok and Balazs, 2013). Van Oosten et al (2009) constructed artificial cilia in which the driving deformation is caused by isomerization under ultraviolet light, as shown in Fig. 6.30a. The same mechanism works to drive artificial flagellar swimmers (Fig. 6.30b). Polarized beams can carve complex forms into the films on a microscopic scale, as exemplified in Fig. 6.30c.

Zakharov and Pismen (2016) modeled a walker built of units with circular nematic alignment and a unit-charge defect in the center, as suggested by Ware et al (2015). Each unit forms a conical leg when heated and moves by lifting them sequentially one by one (Fig. 6.31a). Independently, Rogóż et al (2016) built a "caterpillar", stepping in a similar way when scanned along its body by the laser beam (Fig. 6.31b).

Palagi et al (2016) modeled and fashioned a millimeter-sized nemato-elastic "worm" crawling peristaltically when actuated by a dynamic light field that enforces a wave of transition from the isotropic to the ordered state and back (Fig. 6.32a). This device can swim as well. A versatile simulated swimmer shown in Fig. 6.32b is a twisted yarn comprising interwoven actuated and passive strands. When the nematically ordered strain shortens, a straight-line yarn turns into a helix; it can acquire a variety of shapes when only a fraction of its length is actuated, as in the

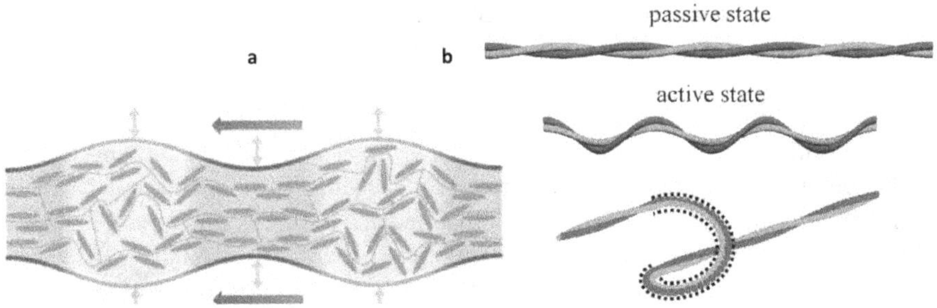

Fig. 6.32 (**a**) Peristaltic locomotion in the direction shown by the *arrow* of a nemato-elastic "worm" (Palagi et al, 2016). (**b**) Reshaping of a Janus filament. The illuminated segment in the *lower panel* is marked by *dots* (Zakharov et al, 2016).

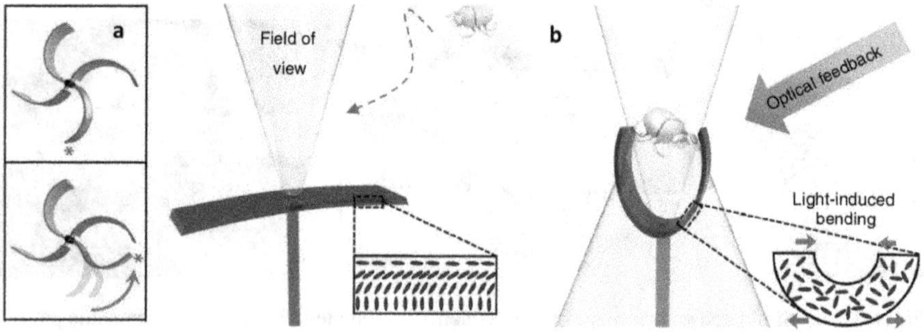

Fig. 6.33 (**a**) Light-driven "mill" rotating counterclockwise. The light source is positioned on the *top right* in the *upper panel*, and the focus point of the light is represented by a *yellow spot* on the right blade. *Light-gray images* in the *lower panel* show the initial and intermediate positions of the actuated blade (Vantomme et al, 2017). (**b**) Artificial flytrap. *Insets* show the nematic structure of the layer, flat when illuminated and bent when darkened (Wani et al, 2017)

lower panel. Zakharov et al (2016) computed various trajectories of this yarn in a viscous fluid under an actuating beam. "Janus filaments" merging nematic and passive polymers can be woven with passive strands to generate a great variety of structures (Zakharov and Pismen, 2019).

Another light-driven device is a "mill" (Vantomme et al, 2017) with flexible blades bending under one-sided illumination (Fig. 6.33a). As a blade rotates and moves out of the light spot, it relaxes; at the same time, rotation brings a new blade to be exposed to light. Successive bending and unbending of the blades gives rise to continuous rotation. A particularly inventive contraption is the imitated Venus flytrap (Wani et al, 2017), which closes when an object enters its field of view and causes optical feedback (Fig. 6.33b).

More devices based on actuation of nematic elastomers and other shape-changing materials are reviewed by Shang et al (2019), Pilz da Cunha et al (2020), and others. The review by Oscurato et al (2018) emphasizes applications based on light-induced changes in chemical structure, able to operate down to a nanoscopic range. Although reshaping of nematic elastomers is most versatile, they are still practically inconvenient, as they become sticky when heated and are apt to deteriorate with time. Practical applications are largely centered on hydrogels, which are similar in their hydraulic action to plant tissues (Sect. 7.8). Success depends here on creating stiffer and faster materials.

In a truly biomorphic material, internal chemical reactions would produce a species changing the gel volume or modifying the nematic alignment, and the resulting deformations would affect the rate of its production, as commonly happens in animal cells and tissues. Rhythmic oscillations of a gel undergoing the Belousov–Zhabotinsky reaction (Kuksenok et al, 2008) most closely approach internally driven motion, but they lack a feedback effect of the mechanics on the kinetics of the chemical reactions.

Chapter 7
Live Tissues

7.1 Cellular Models

Tissues consist of discrete entities, cells. Why then model them by continuous equations, which are discretized anyway? A discretized description of cell shapes in a 2D layer goes back to Graner and Glazier (1992), who applied an extended Potts model, originally devised in a 1951 Ph.D. thesis to describe interacting spins on a crystalline lattice. The addition or removal of a lattice site associated with a given cell captures protrusion and retraction of the cell boundary. The standard Cartesian grid used by Graner and Glazier is hardly compatible with living geometry, and a large number of squares per cell is needed to approximate a natural shape. A triangular grid, which can be readily subdivided when it is necessitated by large gradients, is now generally preferred in all kinds of computations, and Potts models on a triangular or hexagonal grid are also encountered, but not in biophysical applications.

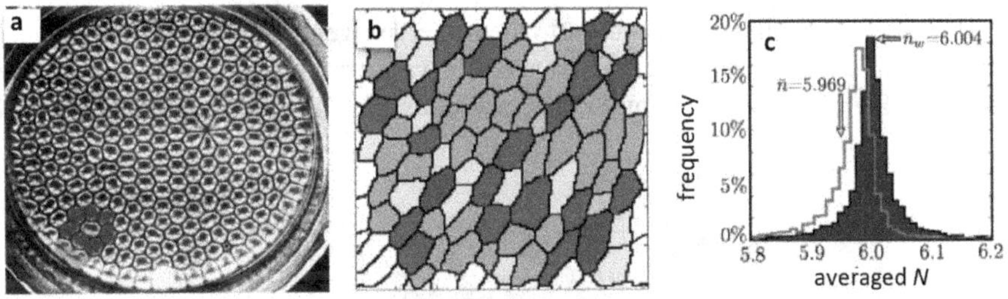

Fig. 7.1 (**a**) Photo of Bénard's original experiment with a penta-hepta defect accentuated by colors (public domain). Note also the rosette defect. (**b**) Example of an epithelial layer with cells color-coded by the number of neighbors: green (4), yellow (5), gray (6), blue (7), purple (8). (**c**) Probability distribution of the mean number of neighbors in 5000 random rectangular subsamples of a base sample of cells from 1200 *Drosophila* wing discs (*red*), compared with the theoretical distribution in the same sample according to Euler's theorem (*blue*), Miklius and Hilgenfeldt (2011)

L. Pismen, *Active Matter Within and Around Us*, The Frontiers Collection,
https://doi.org/10.1007/978-3-030-68421-1_7

A natural and economical way of simulating rearrangements of structured layers is the *vertex model*, widely used in the description of all kinds of structures composed of nearly-uniform domains of polygonal or polyhedral shape (Stavans, 1993). Natural planar tilings are *hexagonal*. Hexagonal patterns arise in symmetry-breaking bifurcations due to the resonance interaction of a triplet of modes. This causes, for example, common patterns of Bénard convection, driven either by gravity or surface tension gradients, shown in Fig. 7.1a. But a still deeper reason lies in topology. The generic number of edges joining a vertex in a random tiling is three. Even in such an artificial tiling as the map of the USA there is a single "four-corners" point. If each of N cells in an infinite tiling has Q edges, the number of vertices is $V = \frac{1}{3}QN$ and the number of edges is $E = \frac{1}{2}QN$. These numbers are related by the classical Euler theorem: $N + V - E = 1$. Neglecting unity in an infinite tiling, this yields $Q = 6$, i.e., a hexagonal pattern.

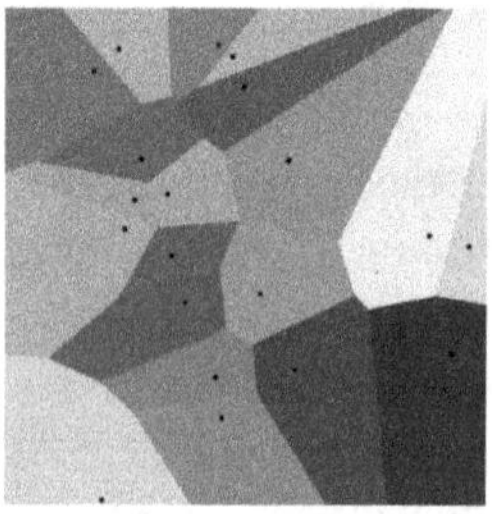

Fig. 7.2 Example of a Voronoi tessellation (CC)

Cellular patterns in epithelial layers, like the one in Fig. 7.1b, are not as neat, but the average number of neighbors comes very close to six anyway. Miklius and Hilgenfeldt (2011) compiled thorough statistics of different samples and detected a minuscular shift, easily attributed to the inability to resolve some edges separating neighboring cells, leading to apparent four-fold junctions between cells, and the finite size of the experimental samples.

Even rather regular patterns, like the one in Fig. 7.1a, contain defects. The generic topological defect in a hexagonal tiling is the *penta-hepta* defect, a pair of cells with five and seven neighbors, accentuated by colors in Fig. 7.1a. A non-generic defect, also present in this picture, is a *rosette*, which is easily resolved by inserting a small cell in its center.

A simulation may be initiated by a tiling generated by *Voronoi tessellation*. A random set of generating points is chosen, and cell borders are drawn normally to the lines connecting nearby points as far as intersections with other lines. Formally, each tile should contain points closer to the respective generating point than to other points from the generating set. This creates a very irregular pattern, like the one shown in Fig. 7.2. In some simulations, positions of generating points, interpreted

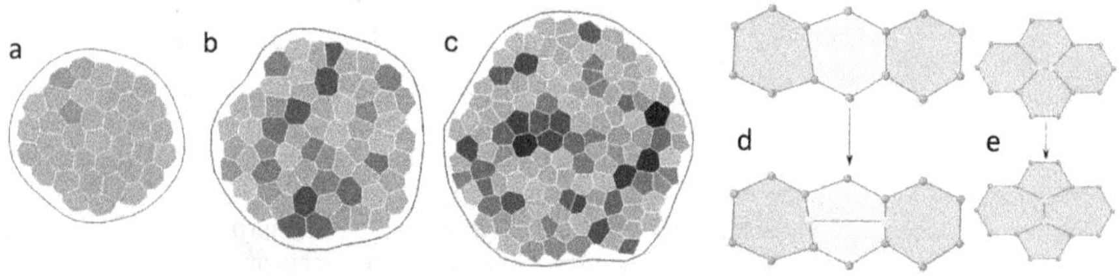

Fig. 7.3 (**a**)–(**c**) Sequence of tissue growth and randomization due to cell division. Cells color-coded by the number of neighbors, growing from dark blue to dark purple (Barton et al, 2017). (**d**) Cell division creates a pair of penta-hepta defects. (**e**) Intercalation (Salm and Pismen, 2012)

as cell centers, evolve, which necessitates recomputing the tiling at each step. Even if a simulation starts from a regular hexagonal tessellation, defects naturally emerge as a result of cell division, as sketched in Fig. 7.3d. Repeated divisions in the sequence Fig. 7.3a–c lead to growing randomization of the tiling. Another generic rearrangement event, also adding to disorder, is *intercalation* (Fig. 7.3e) proceeding via an evanescent four-fold junction.

The vertex model was originally applied to soap foams evolving to minimize the total length of bonds between vertices (Nagai et al, 1988). This, naturally, led to a coarsening sequence, with the size of cells increasing as vertices collide and bonds annihilate. When Nagai and Honda (2001) applied this method to cellular patterns, cell mergers had to be prevented. They proposed, therefore, to evolve a vertex model in a direction minimizing the *energy*, defined as the sum over the tiling of the terms characterizing both length and area elements. The former, as in the case of foams, is the sum of the energies of all bonds, which are now better referred to as edges, defined as their lengths multiplied by some coefficients quantifying the respective line tensions depending on the properties of adjacent cells. The second group of terms are square deviations of the cell area from a preferred standard value, which may also be different for cells of different kinds, multiplied by appropriate coefficients. The essential area and boundary terms are also present in formulations of the cellular Potts model.

The mechanical force acting on a node, or vertex, is defined as the derivative of the energy with respect to its position. Rather than solving a huge system of differential equations, evolution can be followed by testing the change of energy due to displacement of individual nodes, but, since all of them are tied in a common interaction network, the results of a simulation may be sensitive to a sequencing algorithm. The parameters entering each term can be assigned according to specific features of a particular problem at hand, and the evolution of the tiling can be coupled with other processes, as in the examples to follow. In the simulations by Nagai and Honda, all coefficients were identical and only the relative strength of area and line contributions could be adjusted. Farhadifar et al (2007), often cited as the originators of the method, added to the energy expression the sum of the squared cell perimeters multiplied by a coefficient that could reflect, for example, contractility of the actin–myosin cortex. Their simulations took into account cell proliferation alongside their reshaping.

A simple example of implementation of the vertex model is *cell sorting* – the same problem that had much earlier inspired the first biological application of the Potts model (Graner and Glazier, 1992). Coarsening of domains occupied by cells that adhere better to cells of the same kind than to those of the alternative kind, seen in Fig. 7.4, naturally follows from

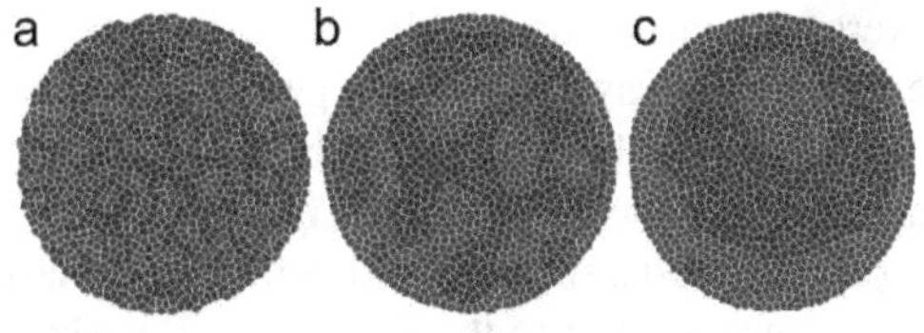

Fig. 7.4 Coarsening sequence in cell sorting (Barton et al, 2017)

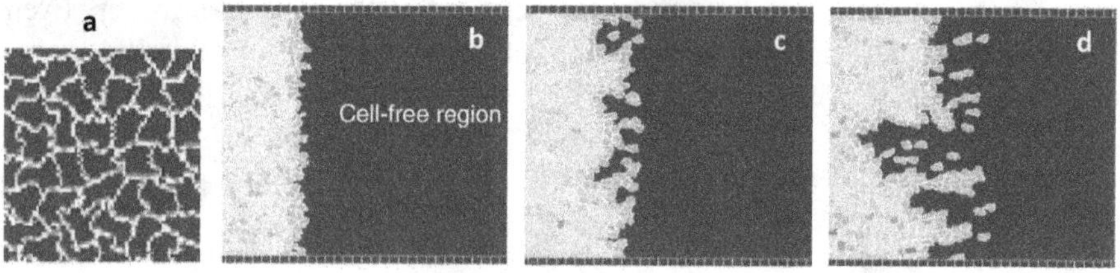

Fig. 7.5 (**a**) Snapshot of a simulation implementing the cellular Potts model. (**b**) Sequence of advance of a cellular layer in a Potts model simulation (Khataee et al, 2020)

assigning a higher energy to edges between cells of different kinds. Other applications will be encountered further in this chapter. Barton et al (2017) published a pedagogical introduction with many colorful pictures. The program TissueMiner (Etournay et al, 2016) contains a panoply of useful tools for evolution, visualization, and analysis of cellular patterns, including both vertex models and triangulation of the network.

The vertex model provides a coarse-grained representation of the cell shape that does not capture finer details, such as curvature of boundaries. Another restriction is the inability to change the layer's topology: to split it or punch a hole, in contrast to the Potts model, which is free of this constraint, as seen, for example, in the picture of an advancing layer in Fig. 7.5b–d. Nevertheless, the vertex model works well in numerous simulations, and generates more natural cell patterns than the Potts model with its staircase-like cell borders, as in Fig. 7.5a.

The evolution of a cellular layer is commonly influenced by enzyme action and chemical signaling. Active motion is controlled by polarization, and all factors are united by an interaction scheme, as in the basic example drawn in Fig. 6.14a. Salm and Pismen (2012) applied the vertex model to the wound-healing problem discussed in Sect. 6.5. They used a simple energy expression, including only the weighted sum of area and perimeter terms. Specifying energies of individual bonds is not needed when all cells are identical, but could be employed to express a dependence of the edge tension on its orientation relative to the cell polarization; however, there is insufficient information available for its proper definition.

The mechanical force was derived by varying the energy as stated before, but it was complemented by an active propulsion force acting in the direction of the average polarization of adjacent cells and proportional to the concentration of an enzyme that was proved in experiments by Nikolić et al (2006) to be essential for spreading. The force acting on border nodes also included a wetting force promoting advance into the empty area. The nodes moved only when the absolute value of the force exceeded a set threshold, with the velocity equal to the force times the mobility coefficient, similar to the Darcy law. Polarization evolved to adjust to the direction set by superposition of mechanical and wetting forces, biased by random noise. The model contained a more complex chemical scheme than the continuous simulations by Höpf and Pismen (2013), including, besides the enzyme that activates propulsion, a signaling species penetrating from the leading edge, which triggers production of

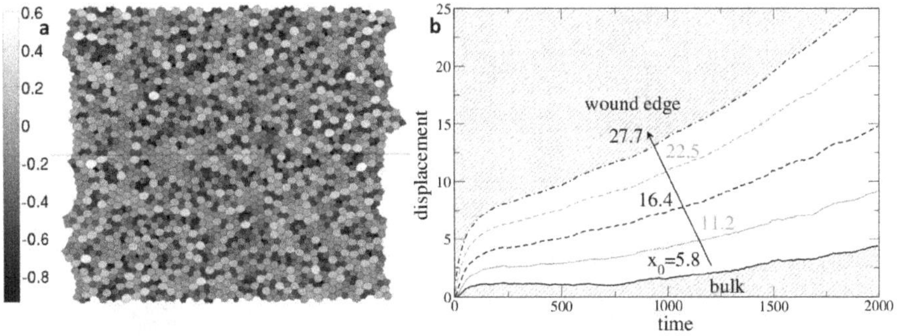

Fig. 7.6 (**a**) Snapshot of a simulation run. The *gray scale* indicates the change in cell area relative to the standard value. The upper and lower boundaries are tied by the periodic boundary condition, so that only side boundaries are free. (**b**) Average displacement of nodes as a function of the initial position indicated by the respective curves (Salm and Pismen, 2012)

this enzyme. Both chemicals exchanged between cells at a rate proportional to the length of their common edge.

The model also included cell division at a rate increasing with the number of edges. Intercalation was triggered when the length of an edge fell below a certain limit. A vertex model may also include elimination of a cell when its area drops below some threshold, but there is no need for this feature in an expanding layer. The results are more detailed than in a continuous simulations: the entire picture of a layer is generated, as in Fig. 7.6a, so that the statistics of sizes and numbers of edges can be collected. In agreement with experiment, the detected advance of cells decays with the distance from the free boundary (Fig. 7.6b), as the influence of both the wetting force and signaling attenuates. The interplay of forces triggered the entire spectrum of cell deformation and rearrangement, but fingering was less pronounced than in the continuous model, apparently, due to the restriction imposed on the cell size.

7.2 Forces in Migrating Layers

Tight junctions between epithelial cells transmit forces that accompany cell rearrangement and migration. Tambe et al (2011) developed the method of *stress microscopy*, based on recording cell-generated displacements of fluorescent markers embedded near the surface of a gel substrate[1]. The map of local deformations of the gel can be converted to a map of the traction forces exerted by the monolayer, which, in turn, is used by applying the 2D balance of forces to obtain the distribution of forces everywhere within the cell layer.

[1] In other works, e.g., Saez et al (2007), traction forces in spreading layers were measured using a dense array of micropillars.

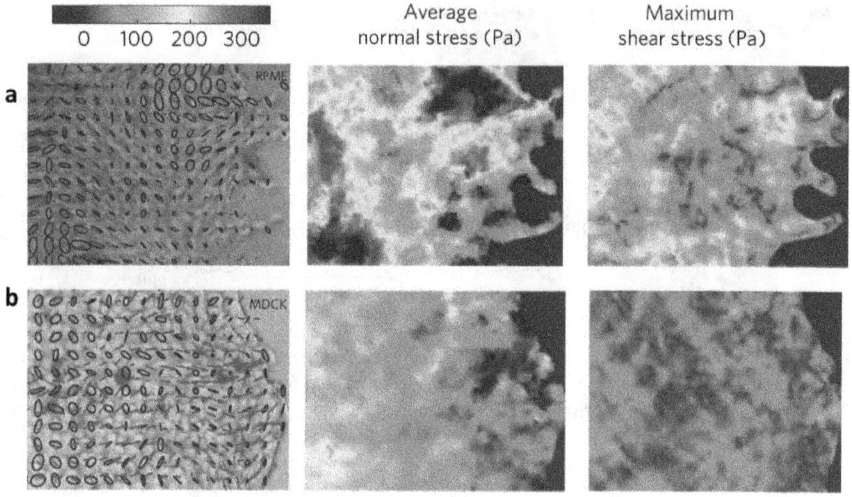

Fig. 7.7 Stress maps in elongated (**a**) and rounded (**b**) cells. The *left panels* show stress ellipses (*blue*) and velocities (*red arrows*). *Central panels* show the average normal stress (the sum of the diagonal components of the 2D stress tensor), and *right panels*, the maximum shear stress, defined as half of the difference between the stresses in the two principal directions (Tambe et al, 2011)

The results strongly depend on the type of cells. The stress maps in Fig. 7.7 show a sharp distinction between elongated (a) and rounded (b) cells. The former developed a rugged stress landscape with high, predominantly tensile, stresses alternating with regions of weakly compressive stresses, and the layer advanced to form a rough boundary with pronounced fingers. The stress contrast and roughness are much less pronounced for rounded cells, but in both cases local cell motions tended to follow local principal stress orientations, as can be seen in the left-hand panels.

Enlarged *leader* cells are often seen on the tips of promontories on the margins of spreading epithelial layers. Vishwakarma et al (2018) employed the same traction and stress microscopy method to understand the way these leaders are "elected". They found that a leader's emergence depends on the dynamics of its follower cells, which manifest enhanced stresses and traction well before these leaders display prominent phenotypic traits, such as a specific shape and prominent lamellipodial protrusions. This is demonstrated by the sequence in Fig. 7.8, which shows the evolution of the traction, the average normal stress, and the "shape index" measuring the ratio of the perimeter to the square root of the projected cross-sectional area. Alongside a rough traction and stress landscape, similar to that in Fig. 7.7, the cellular shape index also shows the same kind of spontaneously emerging heterogeneity.

Increased normal stress was already evident at the initial stage (left column), before a protrusion emerged and a large rounded leader cell became evident at the second stage (central column). At this stage, one can see a tail of followers exerting increased traction behind the leader. This is attributed to the integrity of cytoskeletal elements extending from the single-cell to a multicellular level. Such changes in traction precluded the emergence of another leader at a nearby location, shown in

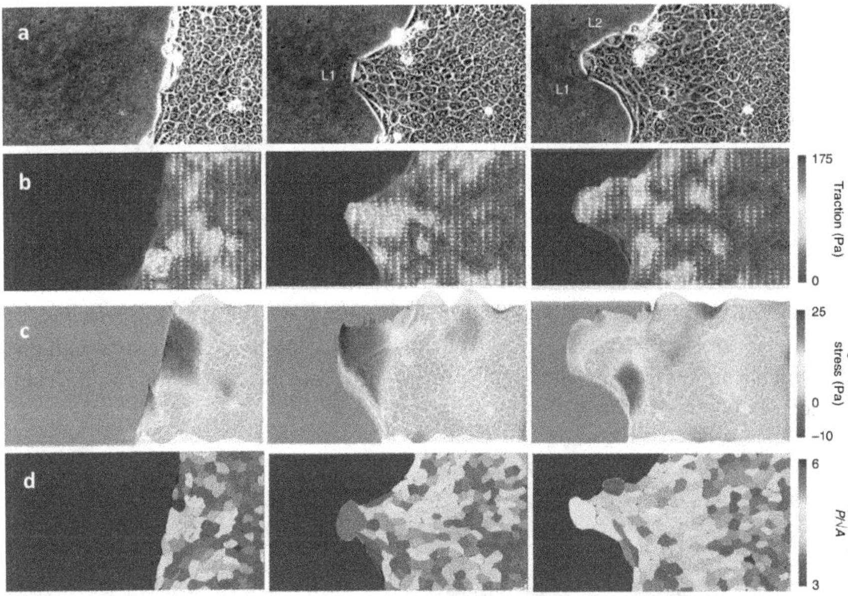

Fig. 7.8 Emergence of leader cells (L1, L2). Snapshot sequence showing the phase contrast image of the spreading layer (**a**) and maps of the traction (**b**), average normal stress (**c**), and cell shape index (**d**), Vishwakarma et al (2018)

the right column. The separation between adjacent leader cells at the layer's margin indicates the average number of cells that could collectively integrate their forces through their junctions, and its distribution is in correspondence with the distribution of the force correlation length.

The authors emphasize that, whereas the emergence of leader cells is an interfacial phenomenon, it is regulated by processes in the bulk of the spreading layer, originating from the dynamics of follower cells. However, it is still unclear what is

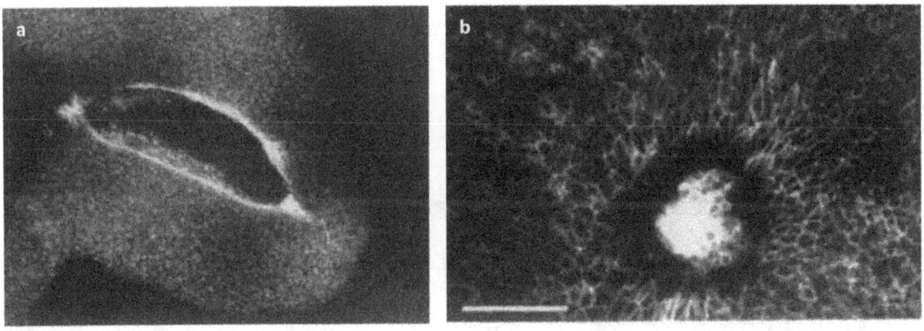

Fig. 7.9 (**a**) Supracellular actomyosin cable (*white*) surrounding a wound. (**b**) After the wound has closed and the cable has disappeared, the cells in the remaining mound still contain a high concentration of filamentous actin in a disorganized arrangement (Martin and Lewis, 1992)

the cause and what is the effect. As mentioned in Sect. 6.5, fingering of propagating interfaces is a generic phenomenon. In continuous computations, it is hard to decide whether fingers are pushed by internal dynamics, also defined as *plithotaxis*, from the Greek πλήθος, denoting a crowd or a swarm, or pulled by kenotactic or haptotactic action of the unoccupied area: both forces are present, and either would suffice. The leaders may be selected and strengthened in their advance by the pushing crowd or rise due to favorable conditions at the interface and empower followers by their pull.

Propagation of the epithelial layer is the essential part of wound healing, which, unlike the above experiments and the modeling in Sect. 6.5 and Sect. 7.1, usually involves closing a damaged area within the layer. It is commonly attributed to a combination of cell crawling and "purse-string" contraction of a supracellular actomyosin cable (Fig. 7.9), although in simple models, such as Arciero et al (2011), wounds safely close without such an actin ring.

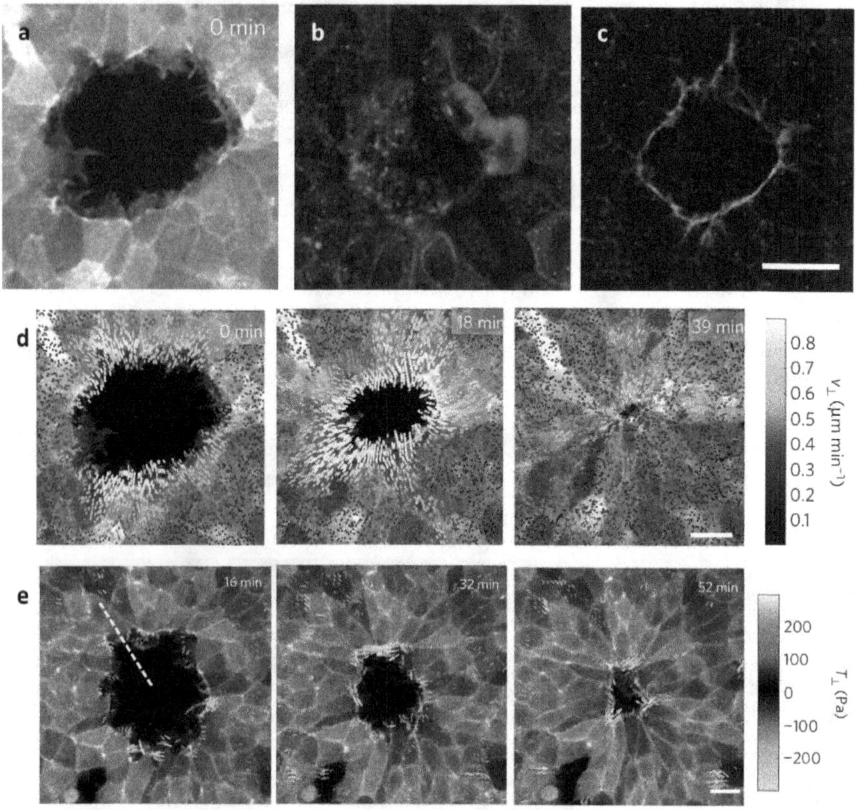

Fig. 7.10 (**a**) Wound generated by laser ablation. (**b**),(**c**) Distribution of actin (*red*) and myosin (*green*) at the apical (**b**) and basal (**c**) sides. (**d**) Evolution of cell velocities in the direction normal to the wound edge. (**e**) Evolution of traction forces in the direction normal to the wound edge. All scale bars are 20 μm (Brugués et al, 2014)

This picture was refined with the help of modern methods of traction-force microscopy and particle image velocimetry (Brugués et al, 2014). The wound shown in Fig. 7.10a was generated by laser ablation, and at the beginning everything proceeded along the classical scenario of cell protrusion followed by accumulation of actin and myosin at the wound edge, as seen in Fig. 7.10b and c. Cells adjacent to the wound became elongated, and their constricted front edges created a rosette-like geometry. However, the velocity of each cell row around the wound exhibited a non-monotonic time evolution: these are largest in the middle panel of the sequence shown in Fig. 7.10d. The experimenters were surprised to observe traction forces pointing towards the wound, colored green in Fig. 7.10e, at concave segments of the leading edge, which showed little protrusive activity. In contrast, convex segments showed pronounced protrusive activity and traction forces pointing away from the wound. The difference in the rate of advance reflects the usual instability of the propagating boundary, but the reversal of traction was an unexpected effect. The traction pattern also exhibited strong force components tangential to the wound. This pattern is attributed to non-uniform tensions caused by the heterogeneity of the actomyosin ring and transmitted to the underlying substrate through focal adhesions. This reveals the important role played by the substrate. The effect is enhanced on soft substrates, where displacements are higher, and may explain why isolated cells are slower on soft substrates whereas wound-closure rates are not.

The same kind of discrepancy between directions of motion and force was detected in experiments with a wound replaced by an island where cells could not adhere (Kim et al, 2013). It is clearly seen in the close-up views shown in Fig. 7.11. Local tension builds up, on average, from zero at the advancing edge towards the bulk of the monolayer, as shown by the color scale. It grows progressively as a result of a cellular "tug-of-war", with each cell pulling not only on the substrate but also on the cell behind. In the bulk, individual cells tend to migrate along the local orientation of the maximal principal stress, i.e., along the longer axis of the stress ellipse, but this tendency is frustrated at the edge of the unaccessible island, where local velocity vectors veer away from the orientations of both the principal stress and the local traction, which pulls almost perpendicular to that edge, as if trying but failing to extend the monolayer into the unfilled space.

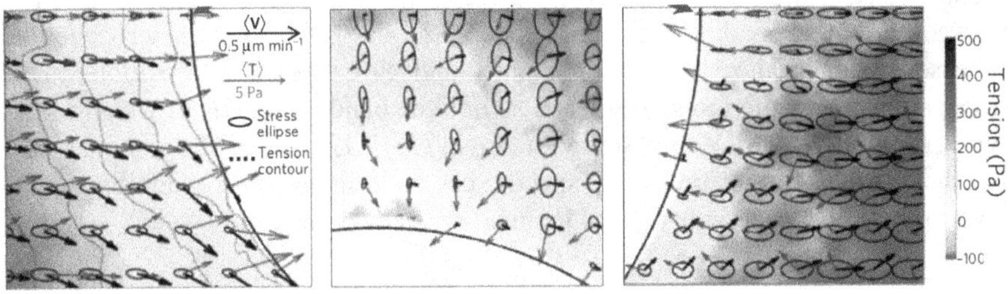

Fig. 7.11 Directions of velocity and traction force, stress ellipses, and color-coded tension levels near an island where cells could not adhere (Kim et al, 2013)

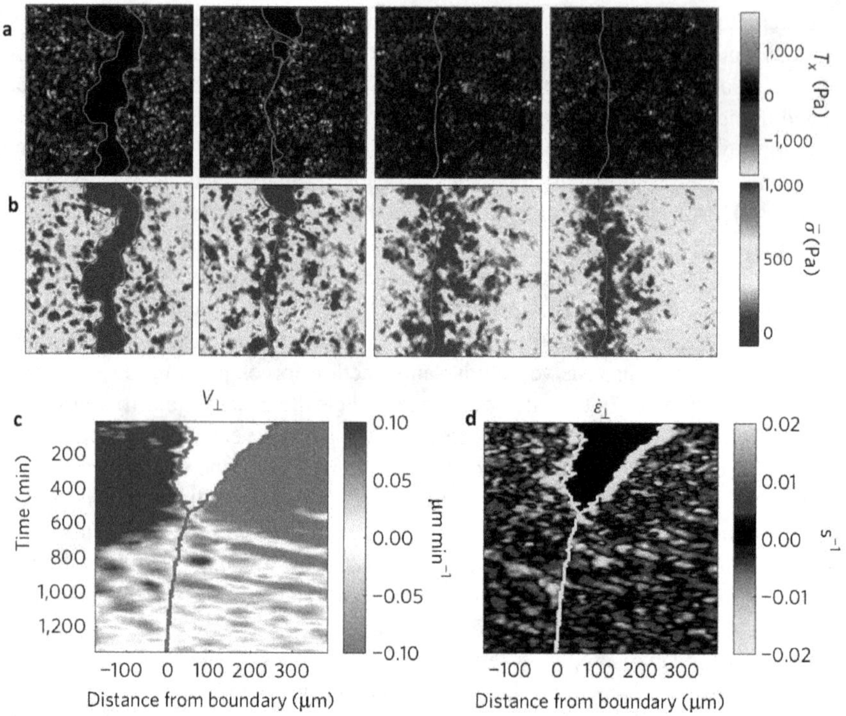

Fig. 7.12 (**a**), (**b**) Evolution of the traction force in the direction of confluence T_x (**a**) and of the average normal stress $\overline{\sigma}$ (**b**) before and after the collision of two tissues. (**c**), (**d**) Kymographs of velocity $V_\perp$ (**c**) and strain rate $\dot{\varepsilon}_\perp$ (**c**) in the direction normal to the boundary (Rodríguez-Franco et al, 2017)

The interplay of forces, elucidated by the same powerful method of stress microscopy, also underlies the dynamics of confluent tissues of different kinds, and this plays an important role in the formation of tissue boundaries during development (Dahmann et al, 2011). Rodríguez-Franco et al (2017) studied the collision of two tissues spreading until the advance was stopped by their repulsive interactions. The fluctuating pattern of the traction force in the direction of confluence T_x and of the average normal stress $\overline{\sigma}$ are shown in Fig. 7.12a and b. Dynamic heterogeneities increased with distance from the boundary and gave rise to deformation waves propagating across the monolayer, seen as oblique bands of alternating sign in the kymographs of velocity $V_\perp$ and strain rate $\dot{\varepsilon}_\perp$ in Fig. 7.12c and d. The authors assert that these waves are not caused by specific chemical interactions at the boundary, but appear to be a generic feature of jammed epithelial contacts.

7.3 Jamming and Liquefaction

We have seen in Sects. 1.6 and 1.7 that jamming is associated with crowding, but high density does not totally freeze motion. Cates et al (1998), considering concentrated colloidal suspension of hard particles or sandpiles, contended that these jammed systems are fundamentally different from an ordinary solid and can be attributed to a special class of materials: "fragile matter", as they can rearrange even in response to small changes in the applied stress. Liu and Nagel (1998) proposed a qualitative phase diagram for jamming shown in Fig. 7.13a. It can be extended to living matter, with load and temperature reinterpreted as averaged and fluctuating components of activity.

Motion never fully stops in dense bacterial swarms and colonies. *Mesenchymal* cells (Sect. 7.4) behave in a similar way. In the high density regime, cells compete for voids, moving slowly through the crowded environment, as shown by the trajectories of selected cells in Fig. 7.13b. The situation is different in epithelial layers, which are always dense and tied up by focal adhesions. However, they are also mobile, as witnessed by all examples in the preceding section. Even when no forces drive collective motion, it is possible to distinguish between the *glassy* state when all cells are "caged", just fluctuating around a fixed position, and a sort of liquid state allowing for large displacements.

Bi et al (2015) asserted that the onset of rigidity at constant density is governed by a single geometric parameter – the ratio p_0 of the target cell perimeter to the square root of the target cell area. They based this conclusion on simulations of a vertex model in an area confined by periodic boundary conditions, minimizing the energy in the form suggested by Farhadifar et al (2007) – recall Sect. 7.1 – but without distinguishing between energies of different tiles and their boundaries, since the properties of all cells were presumed identical. Cell division and apoptosis were excluded, and mobility could be attained only through intercalations. The

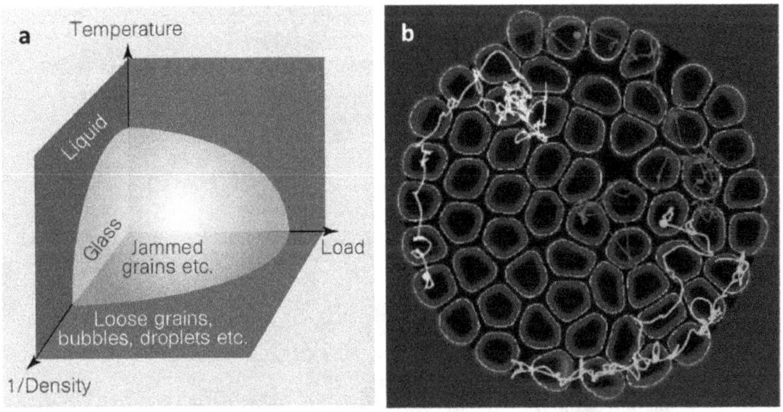

Fig. 7.13 (**a**) Possible phase diagram for jamming (Liu and Nagel, 1998). (**b**) Crowded assembly of mesenchymal cells with trajectories of some of them emphasized by colors (Löber et al, 2015)

computations produced the critical number $p_0^* = 3.81$. Targeting the parameters to produce more elongated cells with $p_0 > p_0^*$ led to a fluid pattern, while more squatty cells produced at $p_0 < p_0^*$ formed a glassy tessellation. Intuitively, intercalations should be easier for elongated cells containing shorter edges, and indeed, measured average energy barriers of intercalation decreased almost linearly with p_0. The critical value just slightly exceeds the corresponding value for the regular hexagonal pattern, $p_{\rm hex} \approx 3.72$, and is very close to the value for a regular pentagon. However, it is impossible to tile a plane with pentagons, and setting $p_0 > p_{\rm hex}$ always produces a disordered pattern, although it is still solid below p_0^*.

The prediction found its confirmation in experiments with human bronchial epithelial cells taken from asthmatic and non-asthmatic donors (Park et al, 2015). Both evolved from an immature fluid-like phase approaching the jamming transition at the predicted geometrical threshold, but in the case of asthmatic donors, jamming was delayed substantially or disrupted altogether. The bronchial epithelium is subject to repeated mechanical perturbations and is exposed to harmful environmental pollutants, allergens, etc., and healthy tissues self-repair by resettling to a quiescent glassy state. Thus, such a simple indicator as the prevailing cell shape appears to have a diagnostic value.

Bi et al (2016) refined the theory, assigning to each cell self-propelled motility with a constant velocity v_0 along a randomly fluctuating polarization vector. Since the identity of the cells became important in this formulation, it was rational to follow the displacement of cell centers rather than that of vertices, so the Voronoi tessellation method, mentioned in Sect. 7.1, was employed. Motility caused cells with $p_0 < p_0^*$ to fluidize, as shown in Fig. 7.14, where the cell trajectories in the right-hand panels demonstrate the contrast between the glassy and fluid states. In a later study by the same group (Czajkowski et al, 2019), cell division and death were

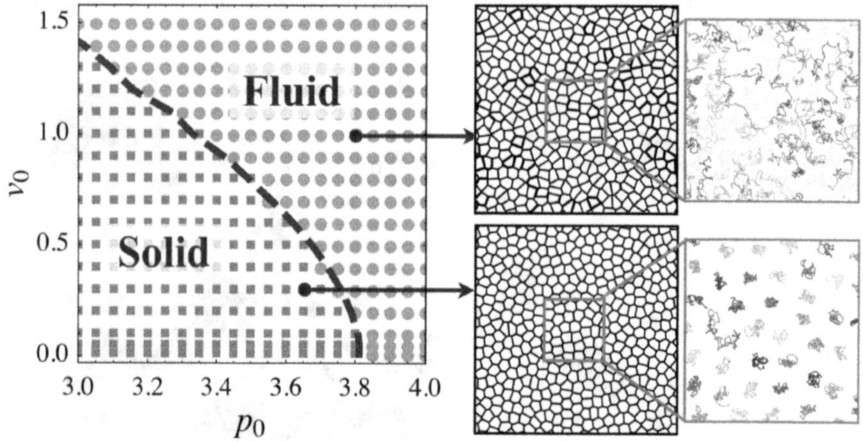

Fig. 7.14 Phase diagram of the jamming transition as a function of cell motility v_0 and target shape index p_0 at a constant noise level. Trajectories of the cells within the *red squares* are shown in the *right-hand panels* (Bi et al, 2016)

found, as expected, to shift the jamming transition in a similar way to smaller values of p_0. The model was further extended allowing for cell overlap, interpreted as a precursor for cell extrusion (Loewe et al, 2020). In this case, the melting transition became discontinuous, and an intermittent regime with alternating glassy and fluid states was observed close to the transition. Li et al (2019) found that heterogeneity of the target shape factor p_0 always enhanced the layer's rigidity.

However, the above results, united by a common approach, are still not the end of the story, as they do not elucidate the influence of many important factors, such as chemical signaling, substrate traction, and the internal structure of the cell. The target form factor p_0 is a formal parameter, and, although it certainly plays a crucial role in the jamming transition, it remains unclear which physical and biochemical factors govern the choice of its level. Complex biochemical cues related to asthma were at work in the study by Park et al (2015), but it remained unclear in what way they influenced the cell geometry.

Some answers may be found in the work by Malinverno et al (2017), who treated a jammed layer with a regulator protein which is known to promote tumor invasion. It works by enhancing membrane traffic, *endocytosis*, which brings substances into the cell. The treatment strengthened both cell–cell adhesion and focal adhesions exerting traction force on the substrate. This is made evident by comparing the left and right panels of Fig. 7.15, showing how cell junctions tighten and substrate traction strengthens under the action of the master regulator. Enhanced coordination of neighboring cells caused alignment of their polarities and hence also their velocities. Saraswathibhatla and Notbohm (2020) also detected changes in traction induced by another protein regulator, which preceded changes in the cell form factor. Based on this, they challenged the common notion that cell perimeter and hence also the form factor are controlled primarily by cortical tension and adhesion at each cell's

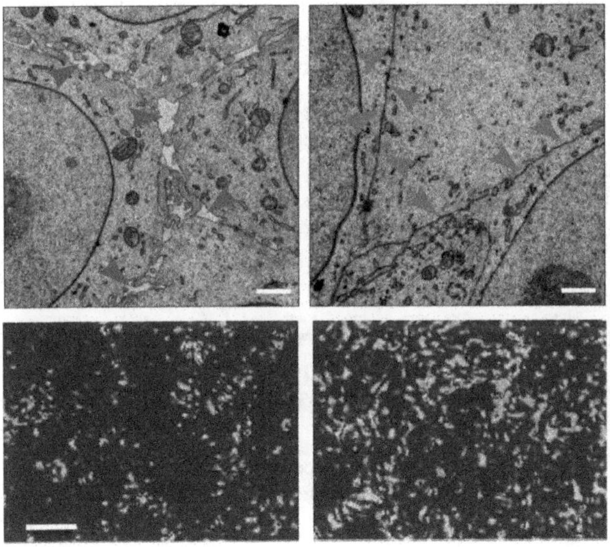

Fig. 7.15 Comparison of an untreated layer (*left*) and a layer with an altered level of the master regulator (*right*). *Top*: Electron microscopy images. Large spaces between cell–cell contacts, indicated by *blue arrows*, tighten, as shown by *red arrows*. Scale bars 1 μm. *Bottom*: Color-coded magnitude of the traction force, increasing from blue to red. Scale bar 25 μm (Malinverno et al, 2017)

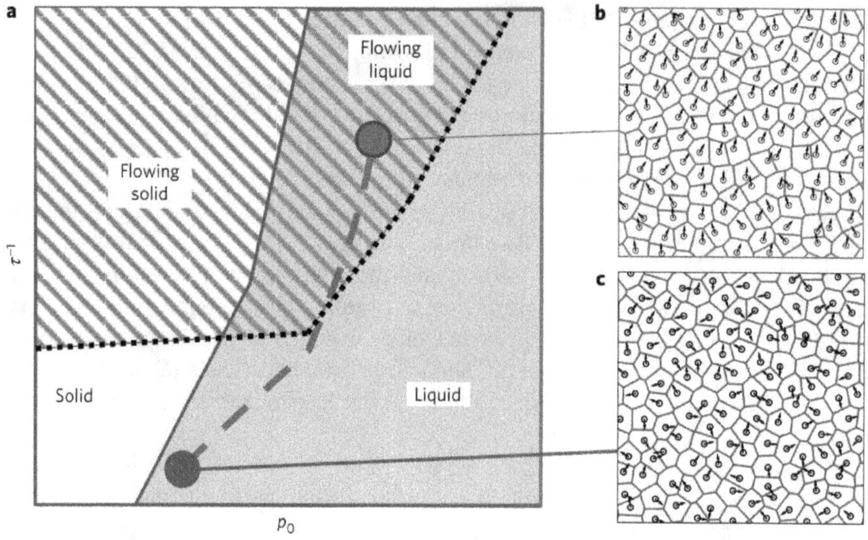

Fig. 7.16 (**a**) Qualitative phase diagram depending on the target shape index p_0 and the inverse reorientation time τ^{-1}. The *dashed red line* is a schematic representation of the trajectory in the phase space corresponding to the regulator-induced "reawakening". (**b**), (**c**) Velocities of the cells in an untreated layer and in a layer "reawakened" by altering levels of the master regulator (Malinverno et al, 2017)

periphery, and emphasized the role of stress fibers that produce tractions at the cell–substrate interface.

Malinverno et al (2017) compiled the phase diagram shown in Fig. 7.16a from simulations which differed from those by Bi et al (2016) in the treatment of activity. The direction of active propulsion, rather than fluctuating randomly, relaxed to the average orientation of neighboring cells with a characteristic response time τ. The diagram distinguishes not only between liquid and solid phases, but also between disorganized and coherent states. The coherent liquid state attained under the action of the master regulator flows in the common polarization direction, as shown in Fig. 7.16b, in contrast to the disorganized motion in Fig. 7.16c. The phase with correlated alignments but no local cell rearrangements is dubbed a "flowing solid" even though it cannot flow.

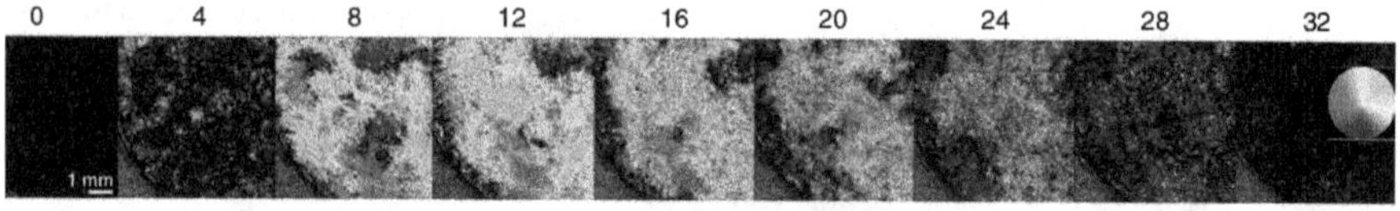

Fig. 7.17 Serum-stimulated collective migration in a quiescent cell sheet. *Colors* indicate direction of movement, as coded *on the right*, while *color intensity* indicates migration speed. *Numbers* show the time in hours following the stimulation (Lång et al, 2018)

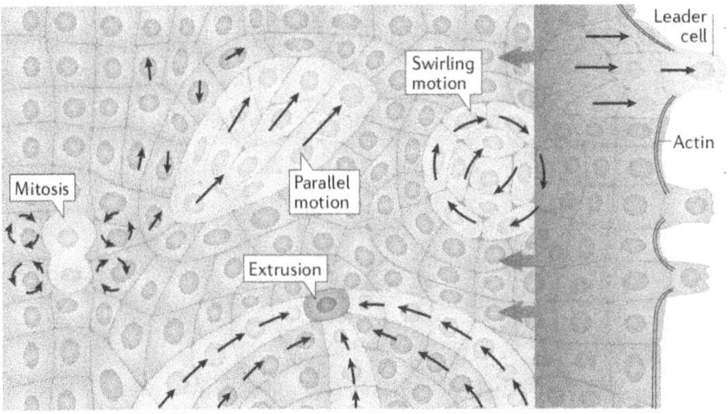

Fig. 7.18 Sources of collective motion. *Black arrows* show the direction of cell displacement and *brown arrows*, the strain wave induced by the leading edge (Ladoux and Mège, 2017)

The formation of a coherent liquid state signals transition to *collective motion*. In wound healing and similar assays (Sect. 7.5), collective motion is caused by a common force acting on the cells to excite any type of "taxis" mentioned there, but in the model by Malinverno et al (2017), it appears to arise spontaneously, with no particular direction imposed. This kind of spontaneous flow was induced in experiments by Lång et al (2018) in cultured cells by contact with blood serum that contains several essential wound-healing factors, after a period of serum starvation. Common orientation of motion is shown by patches of an intense common color in Fig. 7.17. The effect persisted for many hours, but eventually subsided. The flow was accompanied by spontaneous polarization, leading to asymmetric cell division with uneven inheritance of cellular components by front and rear daughter cells.

This is an example of coordinated motion which is not caused by the presence of a free edge of the tissue, as in Sect. 7.2, and does not involve leader cells. Motion in a cellular layer can also be stirred by local perturbations, due to cell division, apoptosis, or extrusion exciting an evanescent vorticity pattern over considerable distances away from their source. Several ways of triggering collective motion are brought together in Fig. 7.18. Different aspects of collective cell migration were reviewed by Danuser et al (2013), Hakim and Silberzan (2017), Ladoux and Mège (2017), and Banerjee and Marchetti (2019). Collective motion is particularly important in the rearrangement of cellular layers during development; more on this in Sect. 8.6.

7.4 Mesenchymal Cell Migration

Rearrangement of tissues strongly depends on the cohesion between cells. *Mesenchymal*, unlike epithelial, cells do not make mature cell–cell contacts. Nevertheless, they are capable to interact and migrate collectively, as their transient cell–cell contacts may be sufficient to polarize the cells and induce their motion in a correlated direction. This association of polarization with the direction of motion is reminiscent of the Vicsek model (Sect. 1.1), where the two are identified.

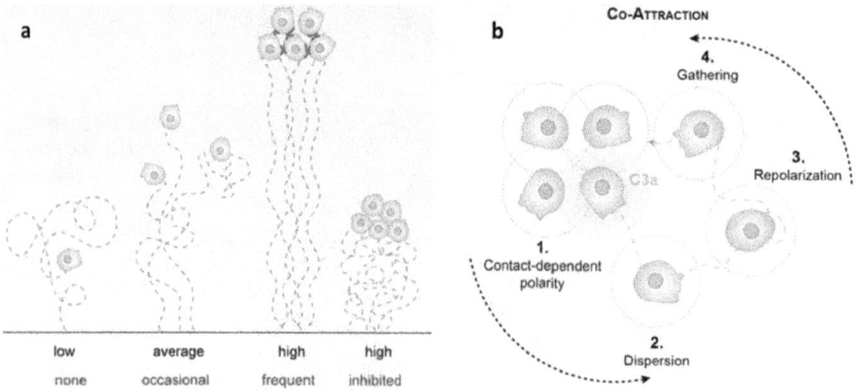

Fig. 7.19 (**a**) Migration in the attractant gradient (shown by *shades of green*) dependent on cell density (*black labels*) and frequency of cell-cell interactions (*red labels*). (**b**) Cycle of contact-dependent polarization, loss of polarity, and gathering induced by the attractant C3a, shown by *shades of blue* (Theveneau and Mayor, 2013)

Collision with other cells causes *contact inhibition* of locomotion: cells collapse their protrusions (lamellipodia, filopodia – recall Sect. 5.6), stop migrating, and polarize in the opposite direction (Abercrombie and Heaysman, 1953). Since the process is driven by collisions, it depends on density. The polarity of an isolated cell is unstable, and chemotactic motion is inefficient. At a higher density, cells acquire a common polarity in the propagation direction and advance as a coherent group (Fig. 7.19a). An additional factor leading to collective alignment is an attractant C3a secreted by cells that enhances their gathering after their polarity acquired at a collision is lost, as sketched in Fig. 7.19b. Camley et al (2016) showed, based on a model including both contact inhibition and attraction of cells, that clusters of cells may chemotax even when single cells do not.

Löber et al (2014) modeled the migration of mesenchymal cells without involvement of an attractant, but took into account interactions through substrate deformation, in addition to collisional interactions. They used the amended basic phase field model (Ziebert and Aranson, 2013) that was earlier applied to a single crawling cell (Sect. 6.2). Interactions between particles were accounted for by two nonlinear terms: an algebraic one, preventing overlap, and a term proportional to the scalar product of the phase field gradients of adjacent cells, which is large on their boundaries and regulates cell–cell adhesion. The model was complemented by the viscoelastic equation for the substrate that governs its deformation under the traction force exerted by the cells. It feedbacks on the cell motion through stepwise detachment of adhesive bonds when the substrate displacement exceeds a set threshold.

The sequences shown in Fig. 7.20 do not involve cell–cell adhesion. The mechanism of alignment as a result of an inelastic collision, illustrated in Fig. 7.20a, operates in a similar way to contact inhibition. The other two sequences of snapshots show an example of a few motile cells setting all other cells in motion (b) and the emergence of collective migration in an originally disordered group of cells (c).

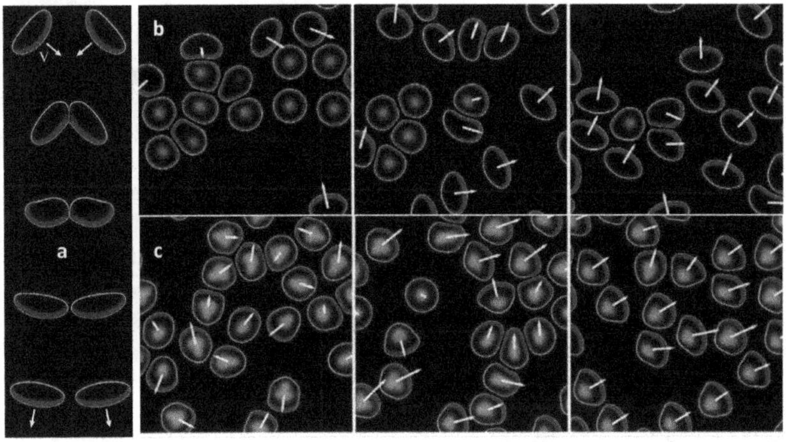

Fig. 7.20 (**a**) Cell alignment following an inelastic collision. (**b**) A few motile cells excite the motion of all cells. (**c**) Emergence of collective migration in a periodic domain without cell–cell adhesion (Löber et al, 2014)

Even immobile cells can be induced into directed motion by mobile cells. Theveneau et al (2013) studied interactions between highly mobile neural crest (NC) cells and placode cells, which engage in a "chase-and-run" behavior. NC cells chase placode cells by chemotaxis, and placodes run away when they are contacted by NCs, as illustrated by the classic picture of a donkey chasing a carrot (Fig. 7.21a). When left to themselves, NC cells move randomly, while placodes are immobile (Fig. 7.21b), but when they encounter one another due to chemotactic attraction, both populations switch to directional migration (Fig. 7.21c and d).

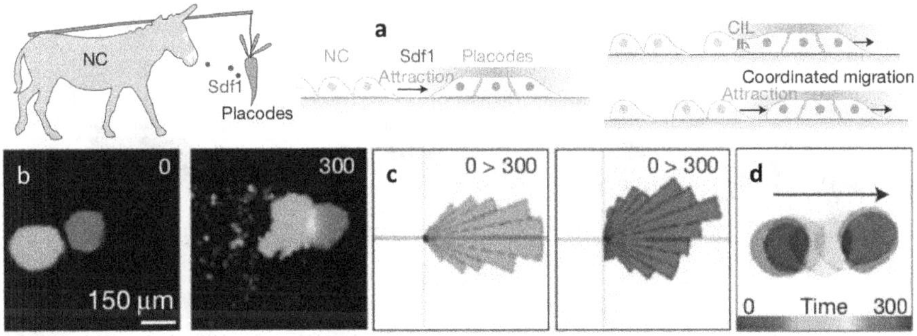

Fig. 7.21 Chase-and-run. (**a**) Sketch of the interaction between neural crest ("donkey") and placode ("carrot") cells. (**b**) Mobile NC cells (*green*) encounter immobile placodes and set them in motion. (**c**) Displacement maps for the two kinds of cells. (**d**) Change in the average position of the combined population with time (Theveneau et al, 2013)

7.5 Tumor Spreading

Motion and rearrangement of cells within a dense layer is essential for tissue remodeling in development; we will return to this topic in Sect. 8.6, but concentrate here on its detrimental role in tumor spreading. In the classical view of malignant transformation in the epithelium (Thiery, 2002), cells weaken their integrin-mediated interactions with the extracellular matrix. Adherens junctions that are crucial for cell–cell adhesion are subsequently lost, leading to the *epithelial–mesenchymal transition* (EMT), which renders cells more motile and invasive. Single cancerous cells that have lost contact with their surroundings can trigger metastatic spreading, and a reverse *mesenchymal–epithelial transition* (MET) leads to the formation of a macroscopic carcinoma. The entire sequence is sketched in Fig. 7.22.

Cancer cells disobey the rule of contact inhibition (Sect. 7.4) and do not stop dividing when they are surrounded by adjacent cells in a dense layer, but rather crawl over their neighbors, and continue to divide, producing cell clusters. Cancerous cells are often rounder in cross-section and capable to perform amoeboid movements. This behavior enables them to form metastases colonizing other body regions. Crossing the walls of blood vessels and invading solid tissues is facilitated by the ability of many malignant cells to produce enzymes that punch holes in the lamina of blood vessels and cut channels through connective tissues (Liotta et al, 1991).

Quite often, cancer cells possess *epithelial–mesenchymal plasticity*, which allows them to shift reversibly between adherent, static and detached, migratory states. This drives distinct stages of cancer progression, including invasiveness, dissemination,

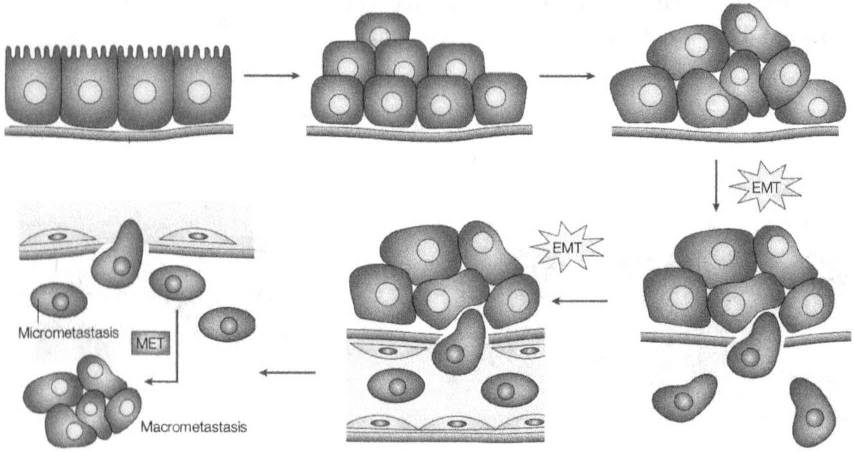

Fig. 7.22 Emergence and progression of carcinoma. Epithelial cells lose contact with a basement membrane, and genetic alterations lead to a carcinoma in situ, still outlined by an intact basement membrane. Further alterations induce EMT, and the basement membrane becomes fragmented. The cells spread to the extracellular matrix, and possibly penetrate into lymph or blood vessels, allowing their passive transport to distant organs. A macroscopic carcinoma forms through MET (Thiery, 2002)

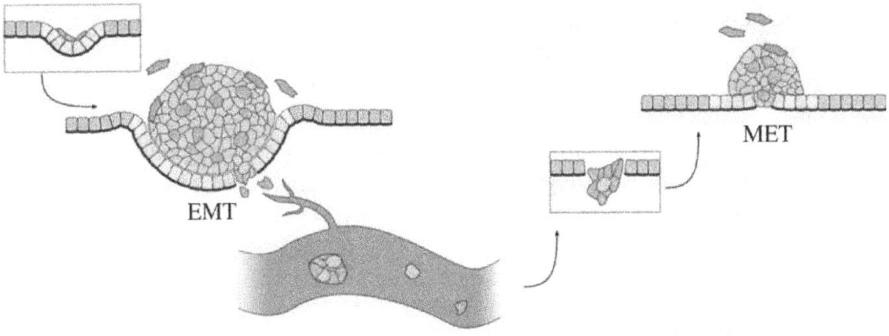

Fig. 7.23 Scheme of tumor spreading through repeated EMT and MET. Cancerous epithelial cells proliferate and can undergo partial or full EMT to form heterogeneous tumors (*colored*) within an otherwise non-cancerous tissue (*grey*). Tumors also recruit non-cancerous fibroblasts (*orange*) to facilitate growth and invasion. Migratory cancer cells escape the tissue via individual or collective migration and enter the circulatory system (*red*) as individual cells or as clusters, depending on their state, distinguished by color changing from *yellow* (epithelial) to *green* (mesenchymal). Metastatic colonization of secondary sites requires re-epithelialization of the tumor cells through MET, resulting in the formation of secondary tumors (Plygawko et al, 2020)

and metastatic colonization and outgrowth, as sketched in Fig. 7.23. The same feature in its benign role facilitates morphogenetic changes during normal development. However, although EMT exacerbates motility and invasiveness of many cell types, it is not a necessary prerequisite for tumor infiltration and metastasis, as some

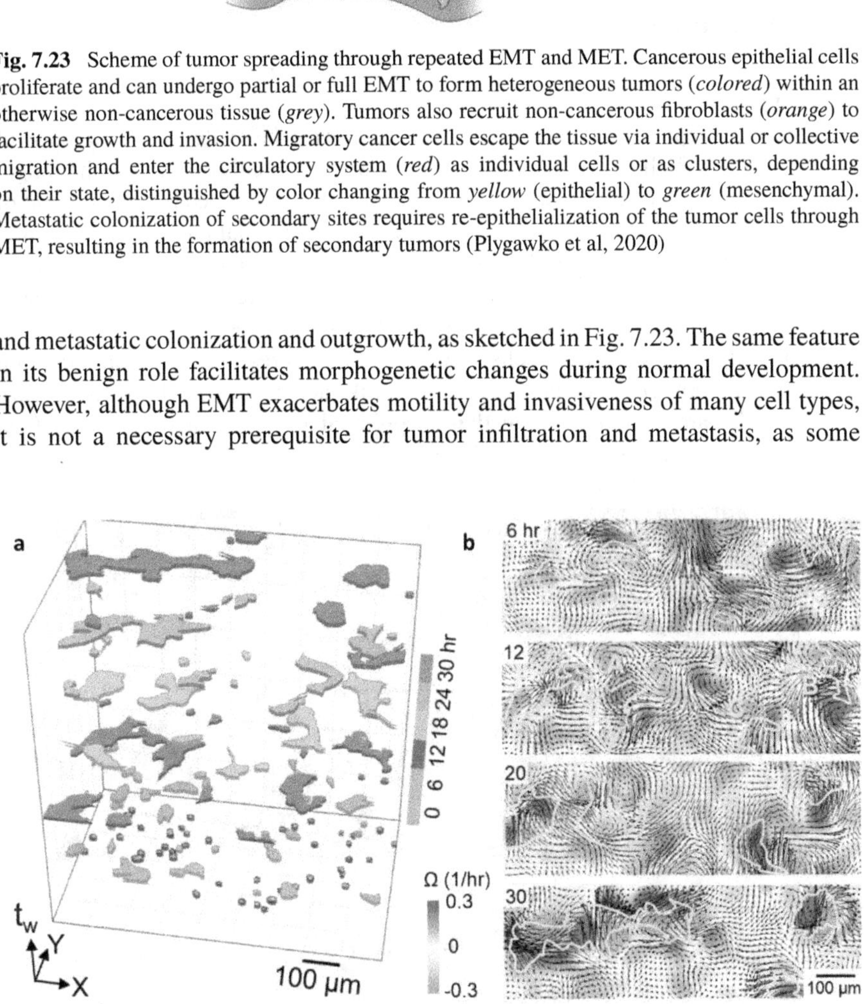

Fig. 7.24 (**a**) Spatiotemporal distribution of cancer cell clusters. (**b**) Sequential plots of velocity and color-coded vorticity fields, showing enhanced motion of both cancerous and nearby epithelial cells alongside the gradual aggregation of cancer cells into the clusters (A, B, C) with fluctuating boundaries, labeled by *yellow closed curves* (Chen et al, 2018)

carcinomas adopt mesenchymal features while still retaining common characteristics of epithelial cells (Christiansen and Rajasekaran, 2006).

Moreover, cancer cells are capable to energize epithelial layers as they proliferate and aggregate. Chen et al (2018) have found that the invasion of a small fraction of motile cancer cells induces turbulent cooperative motion, which intensifies with the increasing size of gradually aggregating cancer clusters and involves healthy cells through the disruption of cell–cell junctions (Fig. 7.24).

7.6 Polarization and Defects

In living tissues, polarization is expressed on a microscopic rather than a molecular scale and is evident in the shape of cells. Polarization can be related to the direction of stress bundles, which often, but by no means always, coincides with both the longer axis of the cell and the direction of motion. It may also be defined by the gradient of some chemical cues. A change of polarization in an epithelial layer from tangential to normal causes a conspicuous bending effect. Coupling of the polarization rotation to elasticity can give rise to symmetry-breaking instabilities that are apt to play a role in morphogenetic processes, as proposed by Belintsev et al (1987).

Elongated (spindle-shaped) cells form nematic textures that commonly contain half-charged topological defects (Sect. 2.4). Unlike active nematic fluids (Sect. 2.5), oriented cells are immobilized in dense epithelial layers, as activity is damped

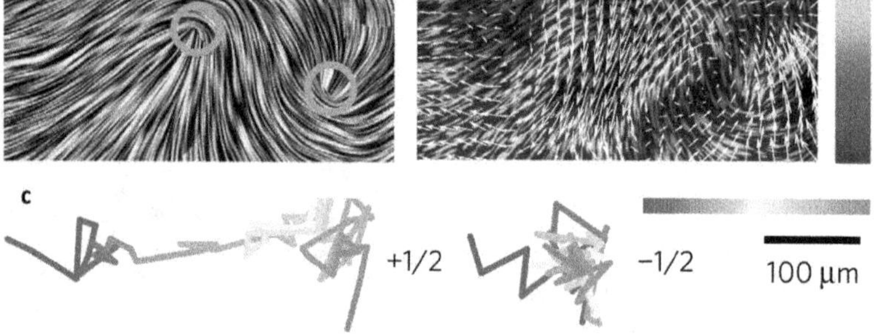

Fig. 7.25 (**a**) Contours of the orientation field. Colored circles emphasize $+1/2$ (*blue*) and $-1/2$ (*orange*) defects. (**b**) Velocity directions (*arrows*) and color-coded magnitudes, increasing from dark blue to dark red within the range from 0 to 2 μm/h. (**c**) Trajectories of $\pm 1/2$ defects color-coded with time increasing from blue to red during 60 hours (Duclos et al, 2017)

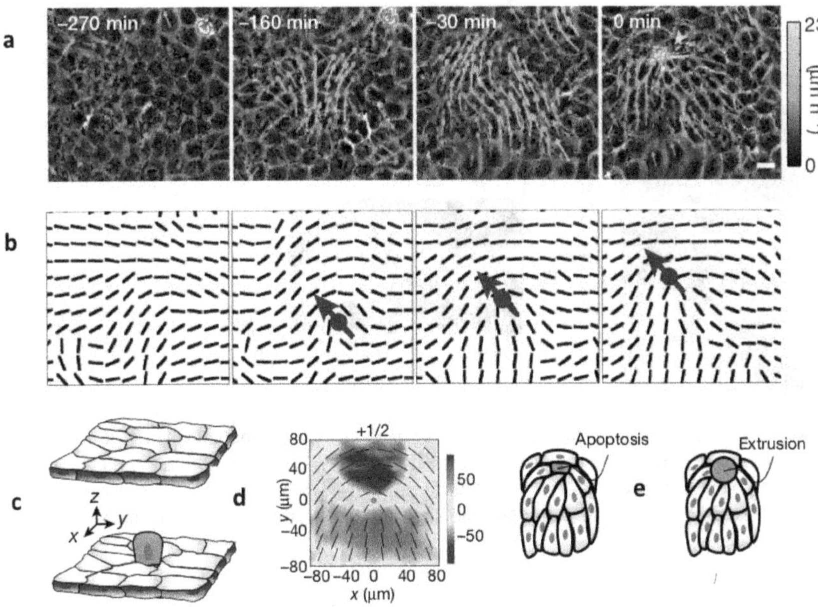

Fig. 7.26 (**a**) Monolayer dynamics before extrusion (indicated by the *yellow arrowhead*) at $t = 0$ min. Lengths of velocity vectors are proportional to their magnitude. (**b**) Schematized image showing the average local orientation of cells. (**c**) Confluent monolayer (*top*) and an extruded cell (*bottom*), colored *orange*. (**d**) Color-coded average isotropic stress (Pa µm) near a $+1/2$ defect. (**e**) Schematic views of apoptosis and extrusion at the location of a $+1/2$ defect (Saw et al, 2017)

by friction. Nevertheless, defects are mobile, as their interaction is controlled by the elastic nematic energy of the tissue responsible for the attraction and eventual pairwise annihilation of oppositely charged defects. Complex flow patterns develop in the vicinity of $+1/2$ defects propelled by their "comet tails", as described in Sect. 2.6. The nematic pattern and velocities are shown in Fig. 7.25a and b (Duclos et al, 2017). Note that motion is very slow, approaching a maximum of only 2 µm/h near $+1/2$ defects. The distinction between the motion of $+1/2$ defects, showing a net advance, and the disorderly displacements of $-1/2$ defects is clearly seen in Fig. 7.25c.

Motion along the "comet tails" toward a $+1/2$ defect leads to accumulated stress in its vicinity that causes cell extrusion or death (apoptosis), as demonstrated by Saw et al (2017) and illustrated in Fig. 7.26. The cells are not elongated in this experiment, and only average local orientations are shown in Fig. 7.26b. There is no available evidence as to whether or how these phenomena are related to typical penta-hepta defects in cellular patterns.

Polarization of cells determines the shape of epithelial shells, as illustrated by the distinction between the elongated form of a uniformly polarized eggshell and the round "kugel" mutant form with random cell orientation seen in Fig. 7.27a and b. The establishment of common polarity in cells of variegated shapes, often seen

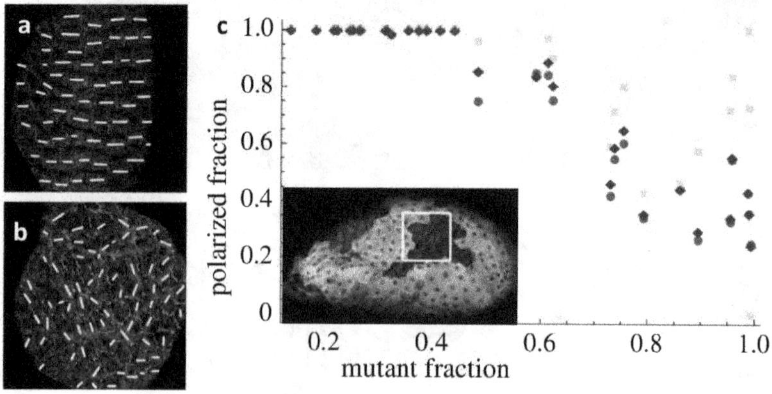

Fig. 7.27 (**a**) Elongated uniformly polarized eggshell. (**b**) Round mutant form with random cell orientation. (**c**) The fraction of cells with actin filaments polarized within ±10° deviation from the perpendicular to the long axis of the shell for all cells (*blue*), normal cells (*green*) and mutant cells (*red*) vs. mutant fraction. *Inset*: Mosaic shell mixing normal (*green*) and mutant (*red*) cells (Viktorinová et al, 2011)

in dense tissues, is not straightforward. It is believed that it depends on cadherin-mediated interactions through adherens junctions, but the mechanism still remains unclear. Viktorinová et al (2011) attempted to elucidate this question by studying *mosaic* shells mixing normal and mutant cells, like the one shown in the inset of Fig. 7.27c. Experiments and Monte Carlo simulations indicated that a signal establishing common polarity can propagate from its source only through a connected chain of normal cells. This is demonstrated by the dependence of the fraction of uniformly polarized cells on the mutant fraction in Fig. 7.27c. All cells are polarized with no more than ±10° deviation from the perpendicular to the long axis of the shell, as long as the mutant fraction does not exceed 1/2. This coincides exactly with the percolation limit on a triangular lattice (Feng et al, 2008), where each node has six neighbors, exactly like the number of neighbors of each cell in a hexagonal lattice. Beyond this critical point, the simulation results scatter.

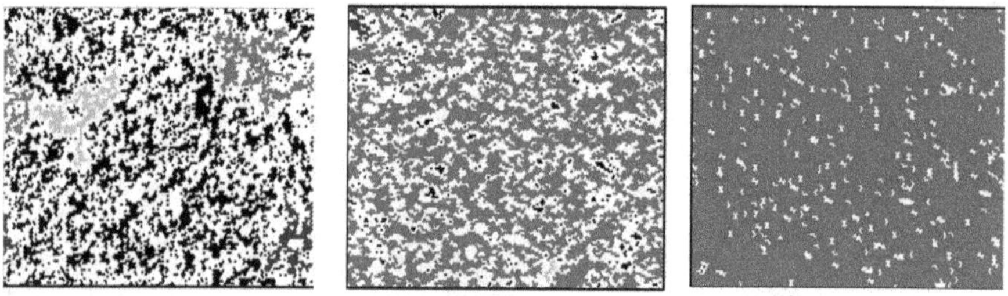

Fig. 7.28 Formation of polarized clusters (*colored*) and the establishment of a common polarization with increasing interaction strength (Chandrasekaran and Bose, 2019)

Chandrasekaran and Bose (2019) have also undertaken Monte Carlo simulations in order to elucidate the dependence of the establishment of uniform polarization on parameters that represent the relative strengths of cell interactions over the noise. As interactions strengthen, several polarized clusters form, distinguished by colors in Fig. 7.28, and a single cluster prevails following the percolation transition. These, as well as the above simulations, were carried out with periodic boundary conditions, which imply the topology of a torus, and do not therefore generate the defects that are necessarily formed on a spherical surface.

7.7 Bending and Folding

Bending and folding of epithelial sheets is an essential part of morphogenetic processes, the subject of the next chapter, and here we just concentrate on their general mechanisms, providing the most common path of the transition from 2D to 3D structures. The most straightforward way of bending a cellular sheet is to break the symmetry between the cells' *apical* and *basal* surfaces. This can be caused by *apical constriction*, initiated by protein-coding genes 'snail' and 'twist' and carried out by

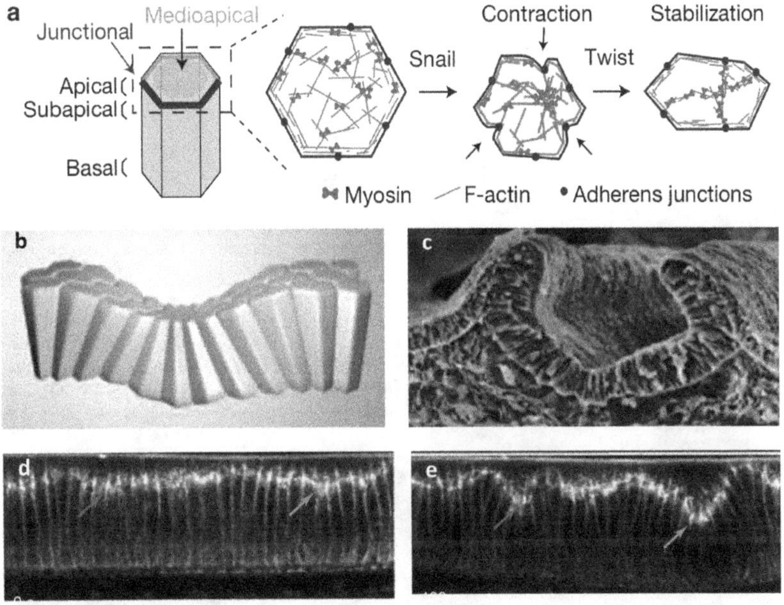

Fig. 7.29 (**a**) Scheme showing apical constriction (Mason et al, 2013). (**b**), (**c**) Bending of epithelial layer due to apical constriction of cells colored *pink* (Lecuit and Lenne, 2007). (**d**), (**e**) Bending by modulation of apical–basal polarity. *Arrows* show the anterior (*red*) and posterior (*blue*) junctions of cells initiating the folds (Wang et al, 2013)

enhanced actomyosin activity, as sketched in Fig. 7.29a. The resulting bending of the epithelial layer is schematized in Fig. 7.29b.

In a pioneering study of chemo-mechanical interactions, Odell et al (1981) postulated, without specifying a molecular mechanism, that a sufficiently large extension of the cellular cortex triggers restoring forces which are so strong that they cause it to contract beyond their equilibrium configuration. The coordination of epithelial cell shape changes is accompanied by the propagation of mechanical contraction waves. Beloussov and Mittenthal (1992) generalized this kind of response as *hyperrestoration*, which they put forward as a fundamental property of a developing and growing organism (although, of course, it is not expected to operate in all circumstances). This response is apt to generate a wave of cell shape change propagating along an epithelial layer. On this principle, Odell et al built up a 1D cell-based model of epithelial *invagination*, one of the most important morphogenetic processes (more on this in Sect. 8.6).

More complex 3D forms may be formed by apical constriction of a group of cells, as in Fig. 7.29c. Another possible cause of bending is modulation of apical–basal polarity (Fig. 7.29d and e) by shifting positions of adherens junctions (Wang et al, 2013).

A number of alternative bending mechanisms are sketched in Fig. 7.30. Some tissues use apoptosis to assist apical constriction. Mechanical forces that bend the epithelium in this case are thought to be produced by an apico-basally oriented actomyosin cable (colored blue in Fig. 7.30a) in the dying cell. Another possibility is "vertical telescoping", whereby the vertical shear between neighboring cells moves them relative to one another, assisted by basal protrusions (b) or apical depressions (c). A rather peculiar mechanism is "basal wedging" (d) in a layer of tightly packed

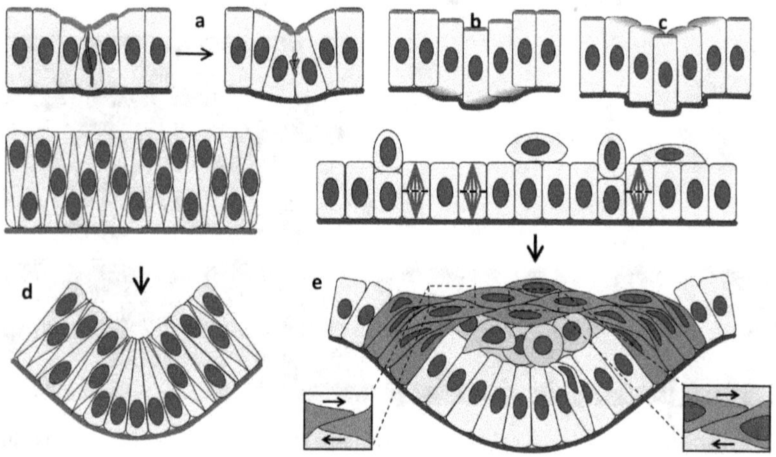

Fig. 7.30 Alternative bending mechanisms. (**a**) Apoptosis. (**b**), (**c**) Vertical telescoping. (**d**) Basal wedging. (**e**) Suprabasal intercalation. *Red*, actomyosin; *blue*, basal lamina; *orange*, cell protrusions; *purple*, nucleus. See the text for explanations (Pearl et al, 2017)

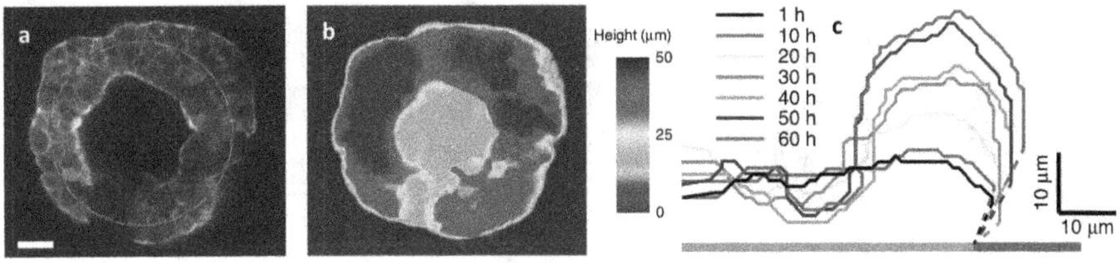

Fig. 7.31 (**a**) Confocal section 15 µm above the basal plane, showing a horizontal section of the rim. The *dashed line* is the limit of the adhesive patch in the substrate that limits the layer's spreading. (**b**) Color-coded 3D reconstruction of the same domain. Scale bar 20 µm. (**c**) Dynamics of the development of the rim. *Blue* and *red lines* show the adhesive and non-adhesive areas of the substrate (Deforet et al, 2013)

cells that bulge around their nuclei. Cells at the hinge point becoming wedge-shaped move the nuclei to the basal position, and this results in a fold. The process opposite to apoptosis is intercalation of extra cells (e), formed by division and creating the *placode* shown in its early stage in the upper picture. Cells at the placode's edges (orange) bend inwards and intercalate with more central cells, creating tension which leads to bending. Stratification creates suprabasal cells (pale and dark green), some of which intercalate (dark green cells), creating further tension to fully bend the epithelium. Boxes to the right show intercalating cells, where arrows indicate the direction of cell movement.

Constrained cell monolayers also develop 3D structures when a peripheral cell cord forms at the domain edge by differential extrusion of proliferating cells, as shown in Fig. 7.31. The rim location is a result of the additional degrees of freedom of the border cells (Deforet et al, 2013).

Extending the vertex model to 3D may be sufficient to generate complicated forms by defining appropriate rules for changing cell areas and folding at cell edges. Okuda et al (2018) based their approach on Turing's (1952) model of spatial patterning through combination of a slowly diffusing activator and rapidly diffusing inhibitor. Patterning led to deformation through the cell proliferation rate increasing with the activator concentration. Some dazzling structures (though unrelated to biological realities) generated in this way are shown in Fig. 7.32.

Misra et al (2017) generated 3D forms by folding a vertex model of a 2D sheet in a more realistic morphogenetic context. The starting point was the hypothesized genetic patterning of the *Drosophila* eggshell leading to the formation of two respiratory dorsal appendages (Simakov et al, 2012) shown in Fig. 7.33a. The origami-style folding of a patterned sheet was carried out by assigning different cell area and line tension energies to pre-patterned domains destined to form different parts of the structure and supplementing the 2D energy expression of Sect. 7.1 by an additional term that captures the distinction between apical and basal surfaces. Different domains, distinguished by color, bend in specific ways, as shown in Fig. 7.33b, with red and blue patches bulging, although the extended target form is not yet attained.

The bending algorithm can be made more precise if individual cells are represented by 3D elements. In the simplest case, it would be a 3D shape with trapezoidal

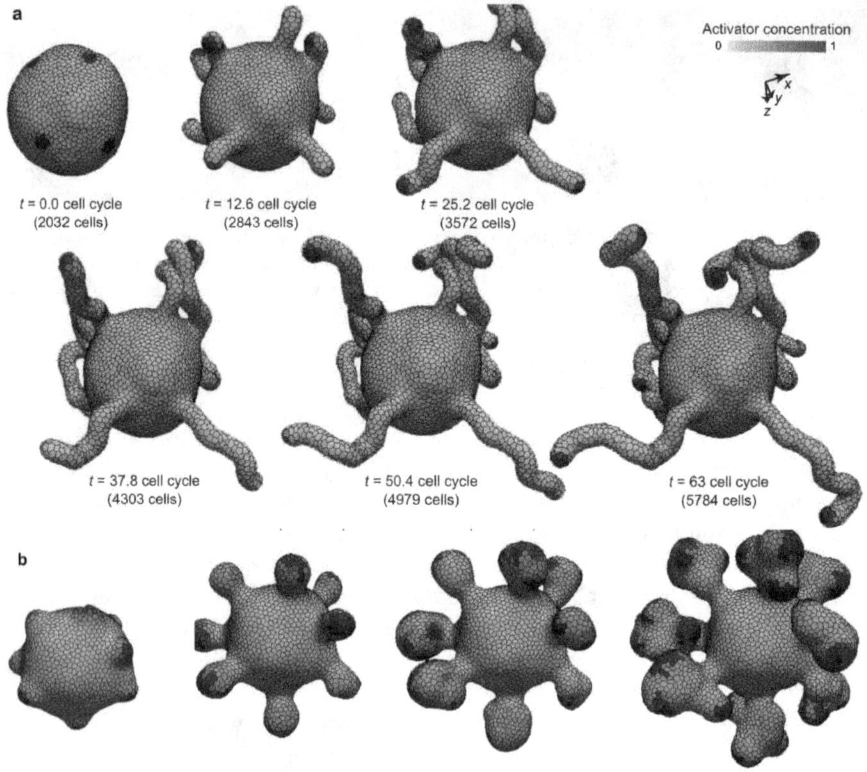

Fig. 7.32 Development of tubular (**a**) and branched (**b**) structures in the model by Okuda et al (2018). The number of cells growing with time measured in cell division cycles is omitted in the last sequence

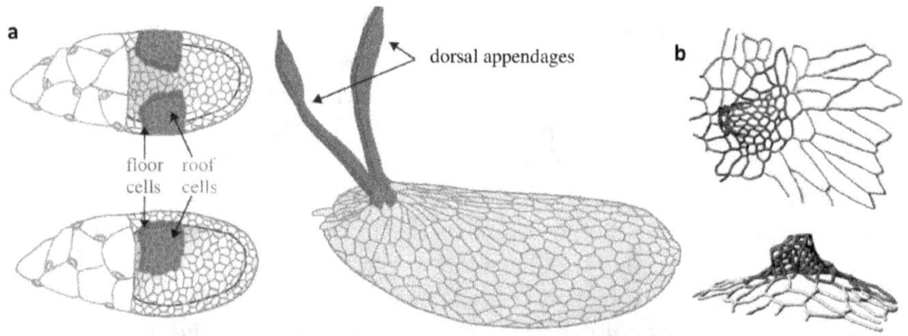

Fig. 7.33 (**a**) Schematic of cell types in the *Drosophila* egg chamber and the developing form. *Blue* highlights the roof cells, *red* the floor cells, *yellow* the midline cells, and the rest are main body cells. (**b**) Folding of different domains (Misra et al, 2017)

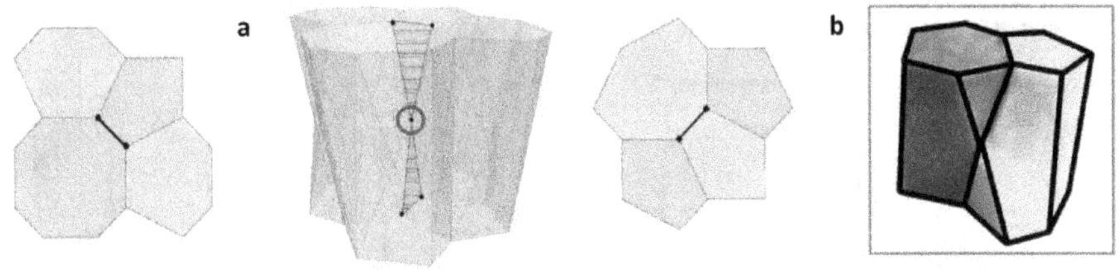

Fig. 7.34 (**a**) Trapezoidal elements with different connectivities at the apical and basal surfaces (Ioannou et al, 2020). (**b**) "Scutoidal" element (Gómez-Gálvez et al, 2018)

sides, which allows for different connectivities at the apical and basal surfaces (Fig. 7.34a). A still more sophisticated form, dubbed *scutoid* (Fig. 7.34b), includes an additional vertex along the apico-basal axis.

7.8 Plant Tissues

Cell walls in plants are stiff, and this impairs motor-driven deformations, but tissues can deform in other ways: by growth or changes in cell volume due to water intake or expulsion. This kind of motion, possible even in dead tissues, can be assigned to the category of *geometric* activity (Sect. 6.7). The increase in volume of stems and leaves does not occur through deformations of the existing material, but through the accretion of new material at its surface through growth, so that large strains of existing material are not necessary to achieve large volume changes, and this simplifies theoretical analysis.

Nonuniform growth causes plant tissues, not unlike hydrogels, to bend (Fig. 7.35a), and the resulting pattern depends on environmental conditions. A floating lotus

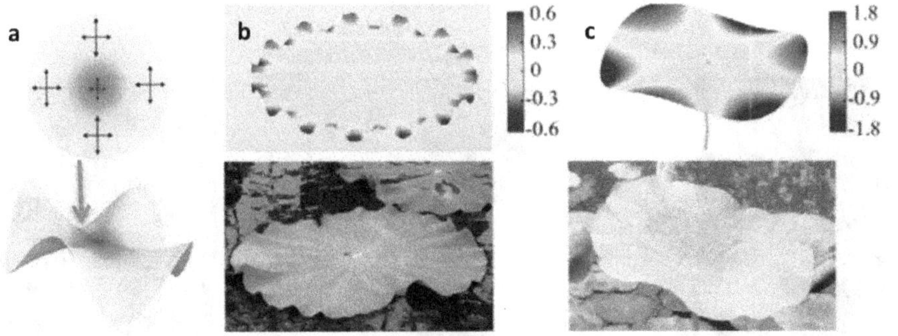

Fig. 7.35 (**a**) Ripples due to nonuniform growth (Sharon and Sahaf, 2018). (**b**), (**c**) Simulations (*top*) and natural shapes (*bottom*) of a lotus leaf growing while floating or suspended, respectively. In the simulation images, the relative deviation of the wavy edge from the central plane is color coded according to the adjacent scale (Xu et al, 2020)

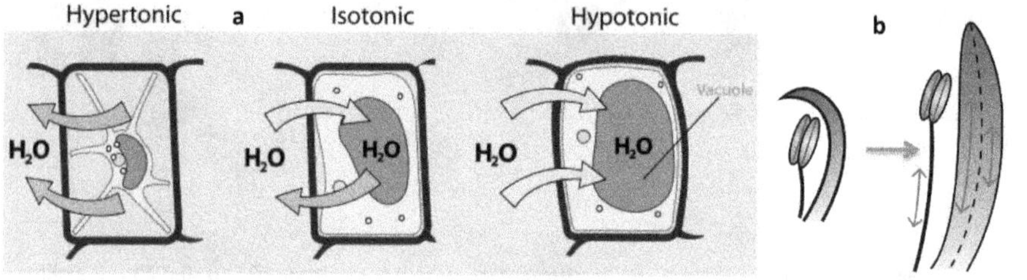

Fig. 7.36 (**a**) Water exchange in a plant cell. (**b**) Opening of a flower petal (Beauzamy et al, 2014)

leaf grows with short waves along the edge (Fig. 7.35b), while suspended leaves (Fig. 7.35c) develop long-wave ripples (Xu et al, 2020). Sharon et al (2007) found the buckling pattern of leaves near the edge to be similar to that of a torn plastic sheet in Fig. 6.24a. This causes polarized cell expansion in thin structures, such as stems and roots. The growth rate is largely controlled not by mechanics, but by a signaling molecule, auxin – more on this in Sect. 8.7.

Plants are in constant motion, which is required for breathing, access to light and water, avoiding hostile environments, and reproduction, but hydraulically driven motion is often imperceptibly slow. The water exchange mechanism in plants differs from that in Sect. 6.4, as *turgor pressure* is regulated in plant cells by *vacuoles*, membrane-bound organelles, which expand or shrink as water moves in and out (Fig. 7.36a). Although the turgor pressure is, by definition, isotropic, anisotropy may be imposed by the direction of cortical microtubules in cell walls. For example, flower petals can be caused to open either by differential growth or by different water intake at their sides, and anisotropy of deformation or growth leads to elongation (Fig. 7.36b).

Deformations in response to changes in relative humidity set the proper timing of seed dispersal. Pine cones open when it is dry, releasing seeds when the weather is more likely to disperse them, and close when it is damp; wheat awns behave in

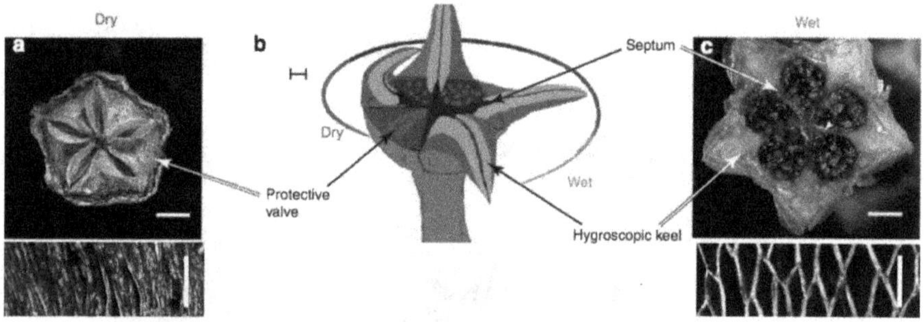

Fig. 7.37 Unfolding mechanism of the desert ice plant seed capsule (*top*) and the change in the keel tissue structure from the dry to the wet state (*bottom*). Scale bars are 2 mm in (**a**) and (**c**), 1 mm in (**b**), and 0.1 mm in the *lower panels* (Harrington et al, 2011)

the same way. Bending movement during drying and rewetting is achieved in both species by a bilayered structure in which one layer is soft and contracts lengthwise as it dries, while the other layer is stiff and resists contraction. The direction of bending is again controlled by the anisotropy of cell walls.

Desert plants, in contrast, need to scatter their seeds when more moisture is available and they have better chances of germinating, and for them it is a matter of life and death rather than just efficiency of dispersal. The desert *ice plant* seed has evolved a sophisticated mechanism, illustrated in Fig. 7.37 (Harrington et al, 2011). In the dry state, seed compartments, partitioned by *septa*, are covered by *keels* (bottom petals) that serve as protective valves preventing premature dispersion of seeds. When the hygroscopic *keel* tissue absorbs water and swells, as shown in the bottom panels, each valve, consisting of two halves separated in the dry state but coming into contact when swollen, unfolds in a way that depends on the anisotropic deformation of the keel cells.

Seed dispersal is most effective when motion is fast, which is not easy to achieve without muscles. Effective operation should rely on a mechanism that stores elastic energy gradually by integrating activities across different spatial scales, but releases it rapidly. Hofhuis et al (2016) thoroughly investigated such a mechanism in *C. hirsuta*, a plant commonly called popping cress, due to the explosive shatter of its fruit pods, which takes just a few milliseconds and fires seeds over a two-meter range (Fig. 7.38a–c). The firing structure of the two valves enclosing the seeds consists of three elastic layers, each with a different reference geometry: an active soft outer layer, the *exocarp*, a passive middle layer, and an inner layer, the *endocarp*, stiffened by lignin. They are attached together in such a way that, in the closed state, the outer layer is in tension and the other two are in compression.

During fruit maturation, the growing seeds deform the valve (Fig. 7.38d), so that its cross-section is not flat but bowed outward. Explosive opening is brought about

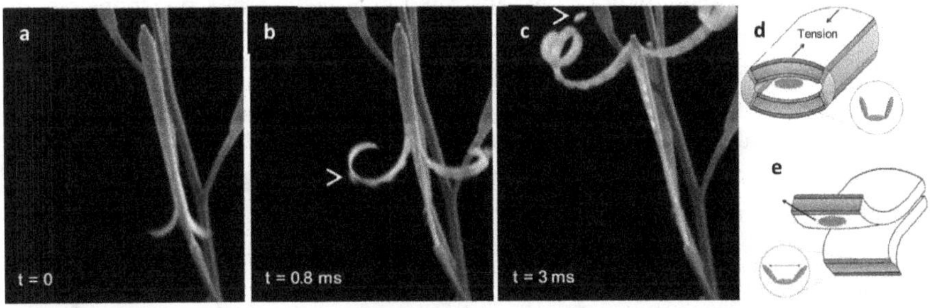

Fig. 7.38 (**a**)–(**c**) Explosive seed dispersal by *C. hirsuta*. The two valves detach from the fruit (**a**), curl back with seeds adhered to the inner valve surface (**b**), and launch seeds while coiling (**c**); *arrowheads* indicate seeds. (**d**), (**e**) Triggering the energy release. (**d**) The three-layered valves, with the exocarp, endocarp, and passive middle layer colored *red*, *blue*, and *green*, respectively, are curved in cross-section and build up tension while attached to the fruit (*brown*). Dehiscence zones (*orange*) form along the valve margins, weakening the attachment. (**e**) Valves flatten in cross-section via opening of the endocarp hinges, as shown in the *insets*, and release the tension by coiling (Hofhuis et al, 2016)

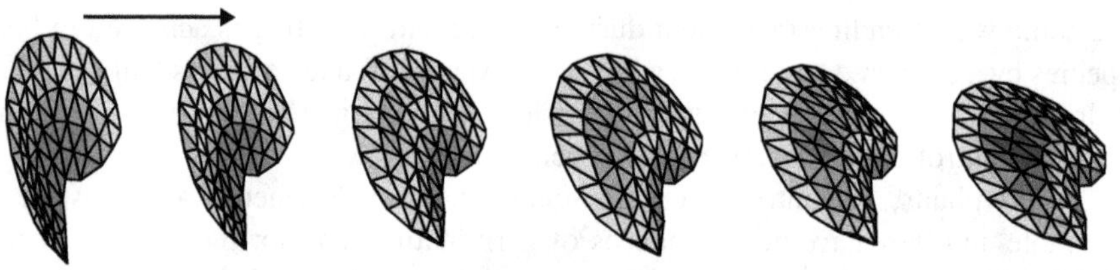

Fig. 7.39 Dynamic sequence of Venus flytrap leaf closure. The time between the images is 0.04 s. The local mean curvature is color coded, changing from *red* when convex to *blue* when concave (Forterre et al, 2005)

by the special stiff asymmetric fibrous texture of the cells forming the inside layer. In order to release valve tension by coiling lengthwise, the valve must first flatten, which requires either narrowing the exocarp or widening the endocarp layer. The latter is made possible by the structure of its cell walls, where lignin is deposited with subcellular precision to form three stiff rods connected by very thin hinges, as shown in the insets of Fig. 7.38d and e. Once sufficient tension is established along the length of the valve and the dehiscence zone at its margins weakens, these hinged cell walls open during explosion, allowing the valve to change rapidly from a curved to a flat cross-section and release the tension by coiling (Fig. 7.38e).

An exceptional example of rapid motion is the snapping shut of the Venus flytrap. Of course, it is not driven optically like its imitation in Fig. 6.33b, and the way it operates was long misunderstood. The trap closure is initiated by the mechanical stimulation of trigger hairs, which spread an electrochemical signal to the leaves. Forterre et al (2005) showed that the fast closure of the trap results from a snap-buckling instability, the onset of which is controlled actively by the plant. The leaf is curved outward (convex) in the open state and curved inward (concave) in the closed state. The snapping motion, with the local mean curvature changing as shown in Fig. 7.39, involves three phases, with the rapid intermediate phase responsible for 60% of the displacement in a tenth of a second, while the initial and final phases, lasting a third of a second each, are relatively slow. Strain measurements showed that closure is triggered primarily by differential strains in the direction perpendicular to the midrib. Later studies (Sachse et al, 2020) assert that the prerequisite for fast snapping is prestress due to the accumulation of internal turgor pressure, which is released after the trap is triggered.

Chapter 8
Morphogenesis

8.1 Approaches to Morphogenesis

Morphogenesis is the summit of the sophistication and accomplishments of active matter. It starts from an inseminated cell and ends with a miracle, a live creature. Perhaps we will never understand this process in full detail. Perhaps Artificial Intelligence will one day unravel the entire chain of genetic and metabolic interactions. Even then, receiving information from a black box will not qualify as *our* comprehension. The aim of our narrative is far more modest. We restrict to the observable and measurable outcomes of those hidden processes being uncovered step by step by biologists and biochemists.

Morphogenesis is directed genetically. In technical terms, the developing embryo is a computer-operated factory, but its architecture is unlike anything invented by ingenious primates. Unlike our most advanced computerized plants, the operation is carried out on the molecular level. Both "software" and "hardware" are polymer molecules, though of different kinds: nucleotides for information storage (DNA) and transfer (RNA), and proteins for information retrieval and production. Small organic molecules are used for some signaling functions.

The programming is stored in every processing unit. This might appear to be wasteful, but it is an evolutionary necessity that served to eliminate egotism and greed. Social insects eliminate conflicts in the same way, but it would not work with conscious animals, as even identical twins may think differently; egotism is quelled in our society in intricate ways, and not quite successfully. Cells do not think, and they did not choose the way they would evolve into multicellular organisms. Dumb replicators just multiply, producing more identical replicators of the same kind, and let less successful competitors die away. Diverse bacteria are capable of communal life in biofilms (Sect. 4.7) or in our guts, but their community is neither organized nor integrated. Enslavement is an alternative to the suppression of genetic diversity; this is what happened when an *archean* cell engulfed a bacterium, engaging it as its power unit, the mitochondrion, and thereby turning into a *eukaryotic* cell. Take

L. Pismen, *Active Matter Within and Around Us*, The Frontiers Collection,
https://doi.org/10.1007/978-3-030-68421-1_8

note of the prefix *eu*, the same as in *eugenic*, hinting at the aristocratic status of the enslaver.

All cells in a particular organism are clones, containing the same genome copied in consecutive divisions, and in a society of clones any individual can do any work. The same applies to social insects, but worker ants or bees remain undifferentiated. This is impossible in a complex society where diverse specialized functions need to be implemented. Therefore cells have to be directed as to which part of their genome to activate in order to carry out a particular assigned task. Once a cell turns into a narrow specialist, there is no way back. Only a few pluripotent *stem* cells remain as a reserve, idle for a time but capable of serving as a replacement whenever the need arises. There is no Central Committee assigning the tasks, but certain groups of cells, called *organizers* play this role by sending signals to somatic cells. Such an organizer establishing the major axis was first discovered by Spemann and Mangold (1924) in experiments on the *Hydra* embryo[1].

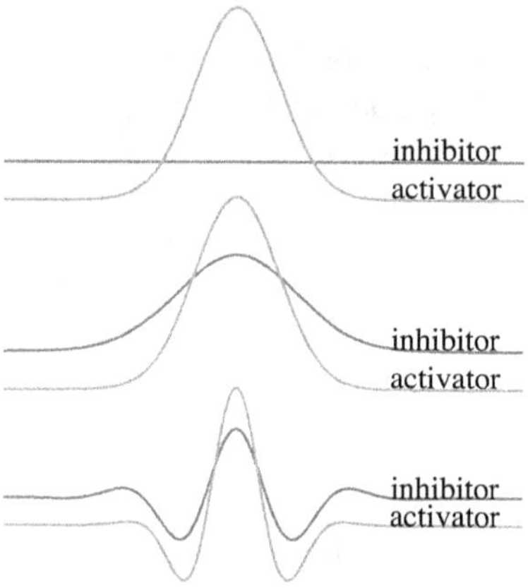

Fig. 8.1 Initiation of a pattern by Turing's symmetry-breaking mechanism

The sequence of nucleotides (*codons*) in DNA double strands is so far the best known element of the morphogenetic machinery. DNA sequencing is a perfect tool for getting incriminating evidence or proving parenthood, but a "gene" is still a vaguely defined string of codons responsible for this or that function, though capabilities for "editing" it have already been attained in some exceptional cases. The major part of the genome, noncoding sequences, used to be called "junk" DNA before realizing that they might be responsible for processing functions that we do not yet understand. Biologists are working hard to map the chemistry involved in the development of *Drosophila* and other "model animals", which may be irrelevant in some details, even for their close relatives. If we ever fully comprehend our own chemistry, *some* humans would be able to live forever (which, however, is a rather problematic prospect).

We are very far from this elusive aim, but the general principles of morphogenesis are gradually becoming transparent along the way, starting from far-reaching hopes and simplistic theories. The celebrated paper by Alan Turing (1952) bears a promising title but ends on a humble note acknowledging "the fact that biological phenomena are usually very complicated". The sound element of this long treatise is the symmetry-breaking mechanism in reaction–diffusion systems due to competing short-range activation and long-range inhibition. This is a perfect recipe for generating a great

[1] Spemann got a Nobel prize, but his reasoning and even priority are questioned (Sander, 1997).

Fig. 8.2 Simulated animal coats (Murray, 1980)

variety of patterns on a last-century computer, or even analytically (Pismen, 2006) if wide separation of the temporal and spatial scales of the reactants is assumed.

The mechanism of symmetry breaking sketched in Fig. 8.1 is utterly simple. A local upsurge of the activator also increases the concentration of the inhibitor, which spreads out suppressing the activator at neighboring locations. This, in turn, suppresses the inhibitor locally and, through inhibitor diffusion, enhances the activator further along the line, so that the inhomogeneous state spreads out. Murray (1980, 1989) employed models of this kind to imitate animal coats, achieving both visual semblance and variety by adjusting parameters and domain shapes (Fig. 8.2). Although far better endowed mathematically, this model is not much closer to the actual patterning mechanism than the Biblical story of Jacob inducing Laban's sheep to conceive speckled and spotted progeny, cited in Murray's comprehensive book.

The problem with symmetry-breaking theories is that, very early on, no symmetries remain to be broken in the developing embryo, as will be elaborated in the next section. Simple reaction–diffusion models of patterning are still applicable in combination with *asymmetry* due to an imposed gradient or growth. Starting from early 1970s, Meinhard (1982) devised several pattern-forming mechanisms of this kind, bringing Turing's scheme closer to biological reality. In a beautiful later book (Meinhard, 1995), he employed reaction–diffusion models to generate a plethora of sea shell patterns, one of which is shown in Fig. 8.3. These patterns are not really related to morphogenesis, where precision and functionality are a must. There is no selective pressure on a particular shell pattern, even less so than on the coloration of animal furs. As Meinhard observes, "the diversity indicates that it is possible to modify the pattern drastically without endangering a species. Nature is allowed to play". Patterning by growth is a mathematically rich system, including not just Turing's static symmetry breaking, but also oscillations and waves.

However, surveying these models diverts us from the actual mechanisms of development from a fertilized egg, where cells have to obey precise commands that would guide their differentiation. The general patterning model based on interpretation of signals was put forward by Wolpert (1969). It retains one basic feature of Turing's

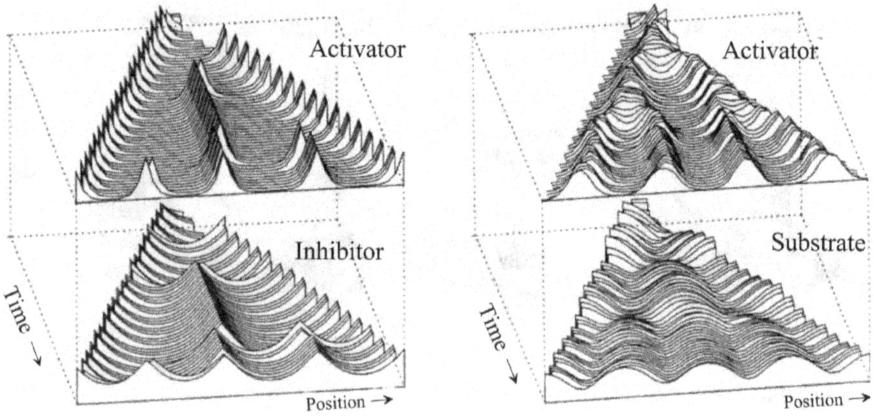

Fig. 8.3 Two models of the development of sea-shell pattern during growth (Meinhard, 1995). In the *left panel*, new regions are activated when the inhibitor level becomes too low to suppress autocatalytic action in the growing space between the maxima. On the *right*, existing activator maxima are shifted towards higher substrate concentrations, which leads to their splitting

model, combining activation with inhibition, but operates in an entirely different way. It presumes asymmetry sustained by chemical *gradients*, which have long been known to play a crucial role in development (Sander, 1997). In Wolpert's scheme the activating and repressing actions are tied into the *feed-forward* motif, $S \rightarrow P$, $S \rightarrow T$, $P \dashv T$ (Fig. 8.4), which includes two activating ($\rightarrow$) links with different thresholds initiated by the same signal S induced by a morphogen diffusing from a signaling source, and an inhibiting ($\dashv$) link from the intermediate protein P to the target T. The inhibiting link has a higher activation threshold, so that, as the morphogen level decreases, it is switched off. This generates the classical "French flag" pattern shown in the lower part of Fig. 8.4, with the target T expressed in the middle ("white") interval, where the signal level is below the higher threshold of the link to the protein P and above the lower threshold of the direct link to the target.

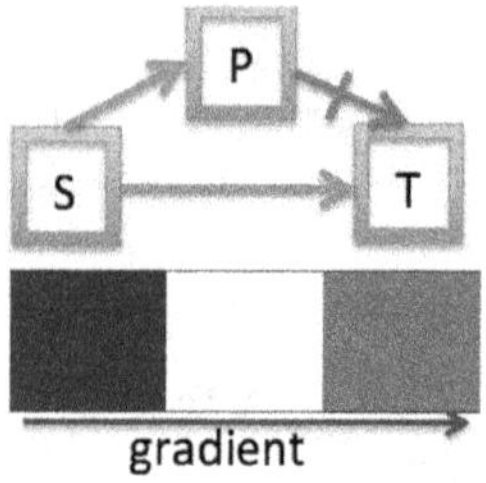

Fig. 8.4 The feed-forward motif and Wolpert's French flag

This basic scheme has been further extended, and its abstract links filled by specific interactions of specific chemicals, as will be elaborated in Sect. 8.3. Very readable expositions of the development processes, including molecular details of genes and signaling proteins (avoided here), were published by some of the most prominent figures in the field: Wolpert (2002) and Nüsslein-Volhard (2006). Murray (1989) criticized the notion of positional information for its allegedly *static* character, but this limitation fades in realistic complex networks involving more than just a single morphogen. The early dynamic hypothesis (Goodwin and Cohen, 1969) related the patterning to the phase difference of two morphogenetic waves, propagating with different velocities. This opens rich possibilities for activating cellular processes within phase intervals with the "right" combination of chemicals, which can be easily shifted by manipulating the signaling waves. We will return to modern implementations of these ideas in Sect. 8.3.

8.2 Oogenesis

The starting point of the development is *meiosis*, the first division of a *diploid zygote*, an inseminated cell containing a double set of chromosomes. Meiosis differs from common division of somatic cells (mitosis) by proceeding in two stages required to separate two sets of chromosomes, maternal and paternal. Paradoxically, this essential proliferation step is dubbed by the term derived from the Greek word for decreasing, lessening, which refers to the halving of the number of chromosomes in one of the two cell divisions. Like in mitosis, replicated chromosomes are separated into pairs of identical (save mutations) sister *chromatids* (Fig. 8.5a). *Crossing over* may happen at this stage, with genetic material exchanging between parental chromatids, as indicated by the color changes in the picture.

Separation is facilitated in animal cells by the *centrosome* consisting of two *centrioles*. When the cell starts to divide, microtubules are nucleated at the centrioles, and, as they elongate, push the centrioles apart, forming a mitotic spindle (Fig. 8.5b), which extends to move the centrioles to opposite poles of the cell. Chromatids, set free by a dissolving nuclear membrane, attach to the strings of the spindle, and are pulled apart, concentrating in the two hemispheres. A cleavage develops between them, and an actomyosin ring formed in the equatorial plane constricts to pull the

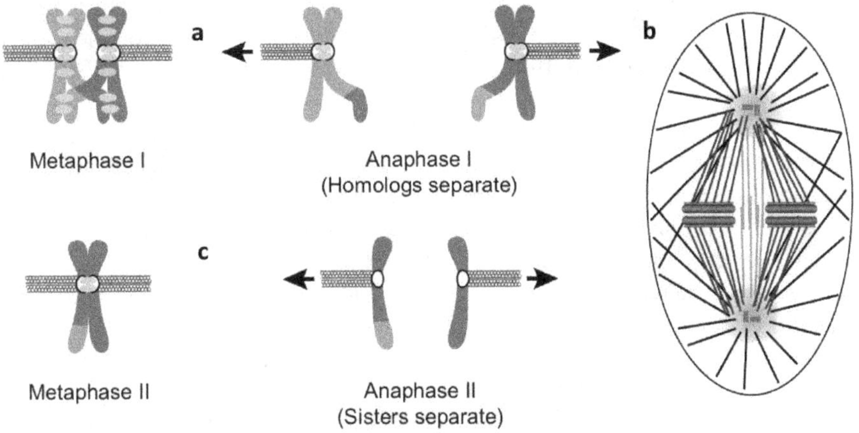

Fig. 8.5 (**a**) First stage of cell division. (**b**) Mitotic spindle. The centriole pair is shown in *red*; variously colored lines are microtubules, chromatids are *dark green*. (**c**) Second stage of meiotic division (Severson et al, 2016)

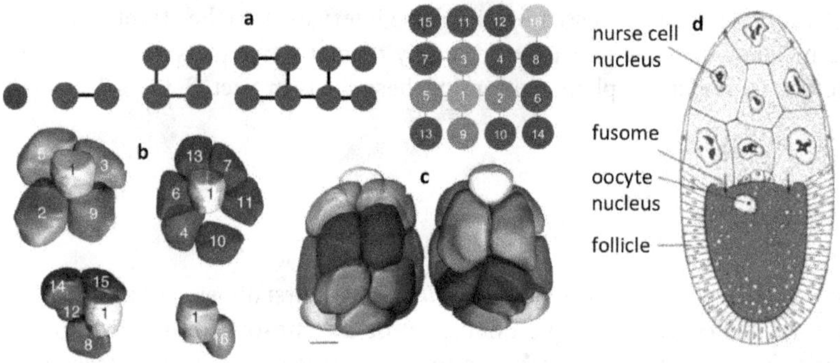

Fig. 8.6 (**a**) Sequence of cell division leading to 16 cells, numbered according to their order of appearance. (**b**) Cells colored according to the number of ring canals (1, *blue*; 2, *red*; 3, *green*; 4, *yellow*) separating them from the oocyte (*gray*). (**c**) Layered spatial organization of nurse cells. Scale bar 10 μm (Alsous et al, 2017). (**d**) Sketch of a mature oocyte (Müller, 1997)

daughter cells apart. As this still leaves a double set of sister chromosomes in each of them, the second stage, specific to meiosis, follows (Fig. 8.5c).

The remaining four cells with the ordinary set of chromosomes further multiply by the usual mitosis. In the most intensively studied "model animal", *Drosophila*, 16 cells connected by thin tubes, called ring canals, form after two more divisions (Fig. 8.6a and b). Four divisions prior to the onset of specialization is a conserved feature among animals; also in mice, far ahead on the evolution ladder and a proxy for humans in genetic studies, roles of cells begin to diversify from the 12 to16 cell stage (Müller, 1997).

Cells compete for the role of the *oocyte*, the locus of further development, while the rest are relegated to the role of *nurse* cells supplying the oocyte with building material through ring canals. The choice is determined by dynamic localization of oocyte-specification factors (Lin et al, 1994), and commonly the winner is the most connected cell, one of the first in the division cycle. Further cell proliferation leads to the formation of the *blastula*, where somatic *follicle* cells form an undifferentiated layer enclosing the inner cavity rich in cytoplasm and yolk. A highly branched organelle, *fusome*, forms a network extending through the ring canals.The name implies that, after each cell division, a new material *fuses* with the existing structure. The oocyte remains transcriptionally inactive, while nurse cells become polyploid (containing more than two paired sets of chromosomes) and synthesize proteins, messenger RNA molecules, organelles, and nutrients transported to the oocyte throughout its development.

The oocyte develops through several numbered stages (Fig. 8.7). The picture of stage 9 shows a mesh of filaments facilitating cargo transport from nurse cells to growing and multiplying follicle cells. The activity of motors running on these filaments also excites cytoplasmic flow (Sect. 5.8), shown on the cartoon of the next stage, which supplements the traffic of messenger RNA and proteins.

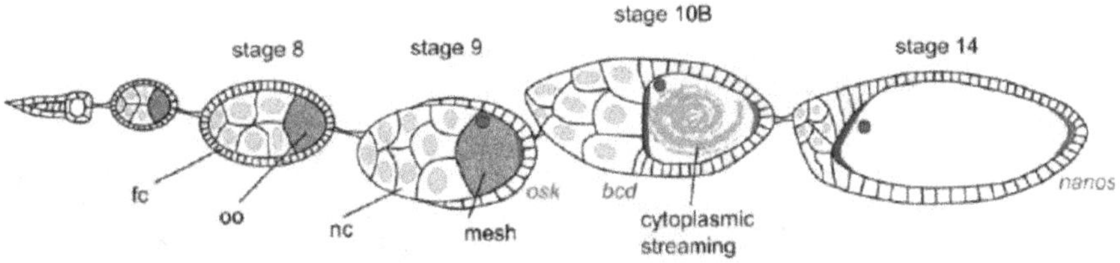

Fig. 8.7 Stages of *Drosophila* oocyte growth. *Colored letters* denote the localization of various messenger RNA molecules, which encode addressing for the delivery of particular proteins along the network of microtubules; the *black dot* is the nucleus (Bor et al, 2014)

The oocyte becomes polarized early on, as determinants of the anterior and posterior ends (indicated in Fig. 8.7) appear at the respective locations specifying the major body axis of the future fly. This used to be the only axis of ancient radially symmetric creatures. The appellations of particular messenger RNA molecules bringing instructions to set this axis are of no importance to us: they are specific to *Drosophila*, the darling of all geneticists, and are not present even in the genomes of other flies. The anterior end is called the *animal* pole and its opposite, the *vegetal* pole. Although establishing this axis is fundamental to embryo patterning, it is not conserved among different species, which may use different strategies to set it up. In amphibians, represented by their own model animal, the *Xenopus* frog, the egg possesses a distinct polarity even before it is fertilized.

Next comes the *dorso-ventral* axis. It first appeared in bilateral animals, which emerged in the Cambrian Explosion. Lewis Wolpert (2002) writes: "We are much more like flies in our development than you might think". But not quite: there is an evolutionary surprise here, first suggested by Geoffroy Saint-Hilaire (1822). What is dorsal in *Drosophila* and its *arthropod* relatives, is ventral in our relatives, *chordates*. The way this axis is set up in a radially symmetric egg appears to be of no consequence. Thus, in *Xenopus* it is determined by the sperm's entry point (Wolpert, 2002). There are, of course, less trivial distinctions between the development of arthropods and chordates.

8.3 Cell Fates

Cell fates have to be fixed early in the development, before the embryo has grown too large to be reached by morphogenetic signals. The outcomes are decisive: transplanted cells maintain the fate bestowed on them by their original position. In this way, experimentalists have caused limbs and appendages to grow in the wrong places. Francis Crick (1970), supporting Wolpert's idea of positional information, estimated that a morphogen profile could be established by diffusional transport in a reasonable time, within a few hours, in a millimeter-sized embryo, but would take a full day in a centimeter-sized animal. Although diffusion is not the only transport mechanism

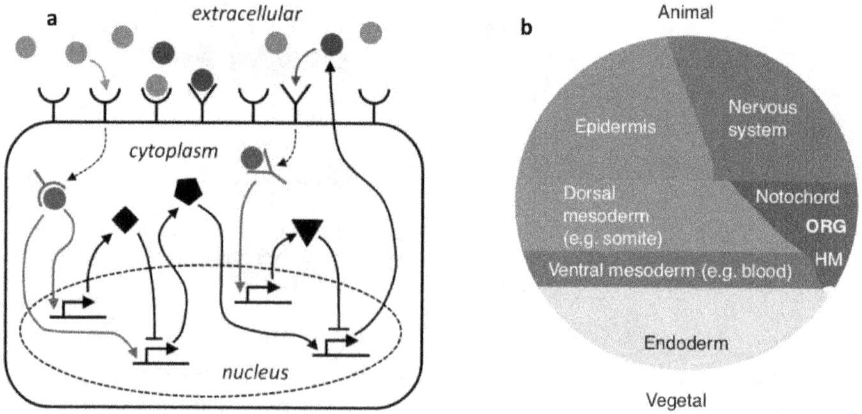

Fig. 8.8 (**a**) Cartoon of intracellular interactions and signaling. *Circles* denote diffusive ligands, and polygons, various intracellular proteins produced by and affecting the expression of various genes. Intracellular proteins, signals, morphogens, and ligands are colored, respectively, *black*, *red*, *green*, and *blue* (Pismen and Simakov, 2011). (**b**) Cell fate map of *Xenopus*. Abbreviations: ORG, organizer; HM, head mesoderm; LTM, lateral tail muscle (Kourakis and Smith, 2005)

in the oocyte, neither advection nor delivery by molecular motors are suitable for creating a smooth gradient.

The idea remained speculative before Frohnhöfer and Nüsslein-Volhard (1986) identified the first protein morphogen, *Bicoid*, forming a gradient along the antero-posterior axis in the *Drosophila* embryo. Morphogens may spread along a cellular layer in different ways: via active transport through dedicated channels in cell junctions, along the surface of a cellular tissue, or by extracellular diffusion. Gradients are established because signaling molecules are unstable and decay as they spread from the source. Differences in morphogen diffusivities and decay rates set the locations of activating and repressing thresholds, but do not affect the pattern in a qualitative way. Morphogens bind to specific cellular receptors and induce intracellular signaling cascades, which may be further enhanced by intercellular communication (Fig. 8.8a). A morphogen spreading from the anterior pole may interact with another one initiated at the opposite end. In a complex expression scheme, different concentrations of a single morphogen may determine a variety of different developmental states.

However, patterning in a single direction is not sufficient. Another gradient might run along the dorsal–ventral axis. Approximate cell fate maps shaped by the crossed gradients, such as the one in Fig. 8.8b, are known for thoroughly studied model animals. Wolpert's flag can be extended to 2D as the "Franco-German flag" drawn in Fig. 8.9a. Its number of distinct domains even exceeds that in Fig. 8.8b. The square tessellation is of no significance: it can be modified by taking point rather than line morphogen sources and shifting their locations, as well as thresholds of activating and repressing links.

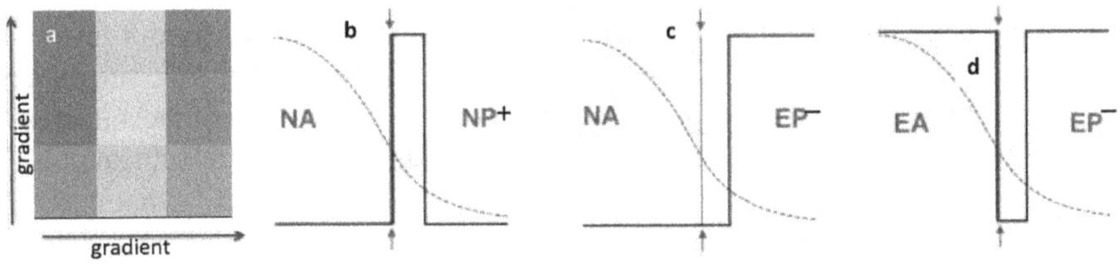

Fig. 8.9 (**a**) Opaque superposition of French and German flags. (**b**)–(**d**) Illustration of the action of an autocrine ligand on neighboring domains. Stationary target and ligand concentration distributions are shown by *black solid* and *blue dashed lines*, respectively. The domain boundary is indicated by a *thin line* emphasized by *arrows*. See the text for further explanation (Pismen and Simakov, 2011)

Patterning becomes more variegated if external (*paracrine*) signals are complemented by *autocrine* morphogenetic signaling initiated within the embryonic tissue by proteins whose expression is in turn determined by local morphogen levels. Meinhard (1992) explored mechanisms for finer patterning due to short-range communication of adjacent cells. Pismen and Simakov (2011) described the general scheme allowing for variegated patterning of a "target" T. The scheme includes a locally active autocrine ligand L expressed under the action of the same pair of morphogens M_1, M_2 within a common scheme, which may include several activating and repressing links and several intermediate proteins P_i. The state of the system in each of the nine squares of the composite "flag" is now characterized by expression

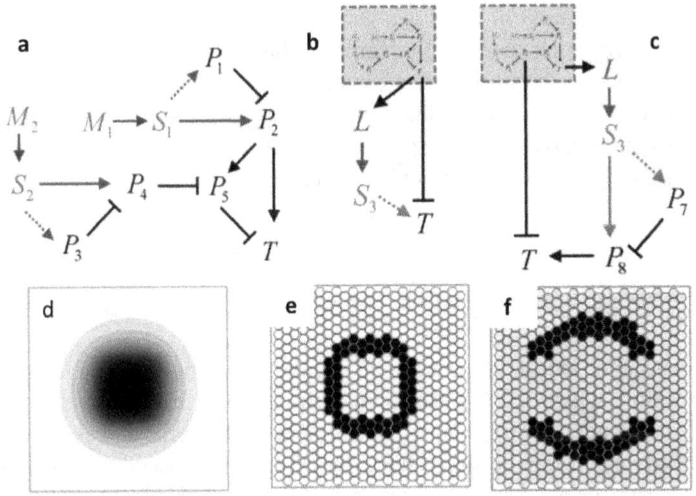

Fig. 8.10 (**a**) Example of a target expression scheme. (**b**), (**c**) The same scheme (drawn as the common block in the dashed frame) with the target replaced by a ligand-induced scheme. (**d**) Concentration distribution of the ligand expressed in the central (*pink*) square of the "Franco-German flag". (**e**), (**f**) Computed target expression for schemes (**b**) and (**c**) (Pismen and Simakov, 2011). The distinction between the morphogens M_i and the ligand L, on the one hand, and the respective signals S_i, on the other hand, is retained, as in the original publication, but can be ignored

of both the target and the ligand, dependent on the structure of the overall interaction scheme. If there is a single target gene and a single ligand, there are altogether sixteen combinations of their expression in the presence or absence of the autocrine signal. The number of combinations grows exponentially as $2^{2(n+1)}$ with the number n of autocrine ligands, leading to a great variety of expression domains that may be generated by the same intrinsic genetic scheme.

The autocrine ligand's signal is assumed to be short-range, so that it affects expression of the target only in the vicinity of the domain where the ligand is expressed. In the three sketches in Fig. 8.9, the letters E and N show whether the target is expressed or not in the absence of the ligand. The letter A denotes expression of the ligand, which can affect expression in a neighboring domain, either promoting it (P^+, as in Fig. 8.9b) or repressing it (P^-, as in Fig. 8.9c and d). An example of expression patterns generated in this way by two versions of a simple interaction scheme is shown in Fig. 8.10. Some expression schemes generate oscillatory patterns that may lead to propagating waves (Pismen and Simakov, 2011). This approach has been applied to locate a narrow strip destined to develop into *Drosophila*'s dorsal appendages (Sect. 7.7), but it is too abstract to attract biologists' attention, which is concentrated on particular genes and proteins, even though they may be specific to certain model animals.

A distinct example of the action of autocrine signaling is the development of stripes in *Drosophila* (Nüsslein-Volhard, 2006). The area where these stripes emerge is defined by *gap gene* expression induced by a gradient along the antero-posterior axis. Each gap gene encodes a diffusive signaling protein which represses other gap genes wherever its concentration is above some threshold. As its concentration decreases, activation of other gap genes becomes possible, creating a periodic pattern in a way similar to Turing's scenario illustrated in Fig. 8.1. As commonly happens in studies of development, this mechanism was validated by observing the effects of mutations (Nüsslein-Volhard and Wieschaus, 1980). Molecular details of patterning (including oscillatory expression) via local cell–cell interactions in other development processes have been reviewed recently by Boareto (2019).

8.4 Dynamic Patterning

In vertebrates, the axis is segmented in another way, by a propagating wave of gene expression, which is closer (but not identical) to the dynamic mechanism by Goodwin and Cohen (1969). Somites, precursors of vertebrae and skeletal muscles, form successively, at regular time intervals, as regularly spaced subdivisions along the antero-posterior axis, as it progressively elongates posteriorly. In view of the inhomogeneities in living tissue, it is impossible to maintain phase coherence on the long scale required to generate a fixed number of vertebrae in this sequence. Such precision requires a mechanism whereby the number of cells per somite adjusts to the overall size of the embryo. Several alternative oscillation-based models have been proposed, but the winning mechanism is the *clock and wavefront* model by

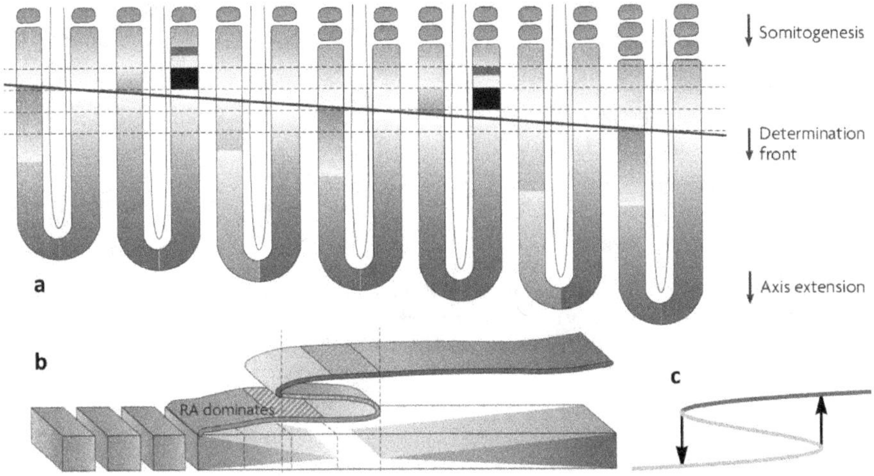

Fig. 8.11 (**a**) Time sequence of somitogenesis. The *black line* shows the shifting front position dividing the regions dominated by retinoic acid (*green*) and the growth factor (*purple*). The wave of cyclic gene expression controlled by the segmentation clock oscillator is shown in *orange* on the left side, and a differentiation marker activated during the oscillation cycle, in *black*. (**b**) Model for segment determination showing the S-shaped oscillatory region bounded by the *dashed lines*; the anterior is on the left. A cohort of cells, the future segment, exposed to retinoic acid (RA) signaling is formed in the area *hatched orange*, and the next cohort of cells to be simultaneously determined to form the future segment is *hatched blue* (Dequéant and Pourquié, 2008). (**c**) Cartoon of the relaxation oscillation cycle

Cooke and Zeeman (1975), which was later propped up by elucidating its molecular framework, as reviewed by Dequéant and Pourquié (2008).

Somitogenesis proceeds, as sketched in Fig. 8.11a, by gradual advance of the determination front combined with the extension of the axis. The front position (black line) is determined by the opposing gradients of a growth factor (purple) and retinoic acid (green). A bistability region, shown as the S-shaped region in Fig. 8.11b, is formed under conditions set by the intersection of these gradients, and relaxation oscillations arise between the levels dominated by either morphogen (Fig. 8.11c). Somites form successively in the oscillatory region as it progresses along the extending axis.

Coordination between positional information and growth should also be essential for Wolpert's patterning mechanism. Cell fates have to be scaled by the size of the embryo, as sketched in Fig. 8.12a, rather than by morphogen diffusivities and decay rates. There have been various attempts, some of them naive, to rectify this contradiction. Doubling a signal by either a counter-propagating signal or a sink at the opposite edge would become forbiddingly clumsy in the case of two-dimensional patterning and in the presence of several morphogens. Gregor et al (2005) judged that variation in the morphogen lifetime was the only mechanism consistent with their data, but have not suggested any way it might work. Making decay rates of a

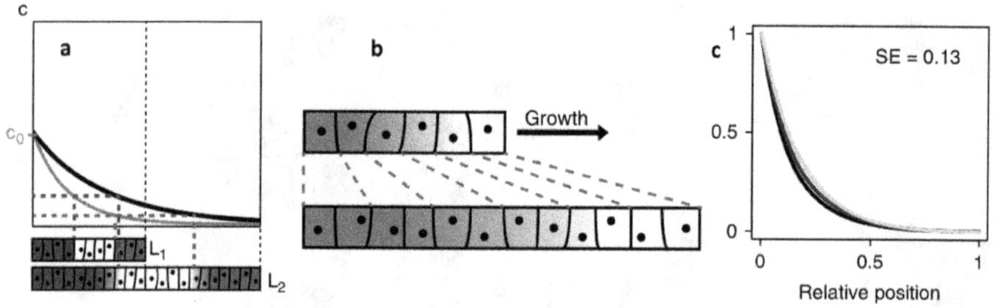

Fig. 8.12 (**a**) Gradients have to lengthen to maintain proportions on longer domains. (**b**) Cell-bound morphogens are advected passively as cells are pushed out during tissue growth. (**c**) The normalized simulated gradient profiles overlay on the rescaled domain with a slight scaling error SE (Fried and Iber, 2014)

morphogen dependent on its concentration levels would not help: it just deforms the morphogen profile.

Some kind of global control could solve the problem. For example, having morphogen degradation depend on some chemical species present in a fixed amount and uniformly distributed in a developing embryo would automatically make the morphogen gradients scale-invariant. In the model by Ben-Zvi and Barkai (2010), the role of a global regulator was played by an "expander" molecule that broadens the morphogen distribution but is repressed by the morphogen, so that, when the morphogen spreads over the entire domain, the production of the expander stops and a size-dependent morphogen distribution is established. However, global agents require a fast mechanism for sustaining their uniform concentration, and this is unlikely to exist in real tissues.

Since differentiation of tissues takes place during growth, advection and dilution due to division and migration of cells should be important factors. In experiments with the developing imaginal disk[2] of the *Drosophila* wing, Wartlick et al (2011) found that cells divide when they feel an increase in the morphogen level, which automatically flattens the gradient. These observations led to a debate, but a significant piece of supporting evidence is that the measured gradients overlay on a domain rescaled by the current tissue length. This overlay was reproduced (Fig. 8.12c) by Fried and Iber (2014) in computations based on a dynamical model in which the morphogenetic gradient never reaches a steady state, but is permanently shifted and adjusted by advection of a cell-bound morphogen in a growing tissue (Fig. 8.12b). In continuation of these studies, Aguilar-Hidalgo et al (2018) identified a critical point of self-organized feedback dynamics that leads to spatially homogeneous growth and proportional scaling of patterns with tissue length.

Although this mechanism is persuasive, it has been experimentally justified in only a single development system. Čapek and Müller (2019), reviewing it alongside mechanisms discussed critically above, concluded that there is no reason to believe that only one of these models is applicable to every case. Indeed, development pro-

[2] Imaginal disks are precursors of adult organs that remain inactive at the larval stage.

cesses are versatile, but no viable alternative to self-organized dynamically adjusted scaling has been suggested so far. However, interactions between morphogenetic gradients and growth may be implemented in a variety of ways. Vollmer et al (2017) reviewed a range of mechanisms of this kind ensuring patterning robustness, which depend heavily on molecular details of both signaling and growth regulation. These mechanisms answer Murray's (1989, 2002/3) complaint that "morphogen prepattern models are not dynamic, in the sense that once the pattern is laid down it cannot be changed by the dynamic behavior, for example, domain growth".

8.5 Mechanotransduction

Molecular details of cellular and developmental processes, avoided in our narration, came onto the central stage in biological research. More than a half of Nobel Prizes in *Chemistry* awarded in this century have been awarded for *biochemical* studies. Meanwhile, the alternative direction emphasizing *mechanical* interactions was evolving, starting from the early 1980s. It never became prominent in studies of development, but interest is reviving lately. The key is *mechanotransduction*, defined as a force-induced process initiating biochemical responses, first of all, changes in gene expression and protein synthesis.

Murray (1989, 2002/3) discusses at length earlier work by George Oster, himself, and their coauthors, one of which, an innovative biomechanical model of apical constriction (Odell et al, 1981), was mentioned in Sect. 7.7. The motivation was that "the very notion of morphogenesis (*morphos* = shape, genesis = change) implies motions – the motions that shape the embryo". Murray goes on to write:

> All motions require forces to generate them. It is surprising that this fundamental law of nature has largely been ignored by embryologists and cell biologists. Very few books on embryology even mention forces. There may be good reasons for this, for only recently has it become possible to actually measure mechanical forces at the cell and tissue level. And what good to ponder immeasurable quantities while the sirens of chemistry and genetics beckon with tangible rewards? [...] It is certainly possible to construct an organism by first laying down a chemical prepattern, and then have the cells execute their internally programmed instructions for mechanical behavior (for example, shape change) according to the chemically specified recipe. In this view, mechanics is simply a slave process to chemistry. Indeed, there is a large number of biologists who think that embryogenesis works in just this way.

Murray further argues that any such process would be unstable without mechanical feedback. This an exaggeration, and so is the ease in unfolding the interaction network of genes and proteins. Measurements of mechanical forces on the cell and tissue level described in Sect. 7.2 are sophisticated and precise, but they could be carried out only *in vitro*. Gauging forces in a live developing embryo has become possible by means of single-molecule fluorescence force spectroscopy, which combines confocal scanning fluorescence microscopy with optical tweezers (Hohng et al, 2007). Recent applications of this method are reviewed by Liu (2020). *Optogenetic* technologies make possible manipulation of morphogens and molecular motors (Herrera-Perez and Kasza, 2019). However, operation of single-molecule devices already comes closer to chemistry than to continuum mechanics, and would not contribute to continuous chemo-mechanical models of the kind described by Murray, as well as in Chap. 6 of this book.

Concurrently with Murray and Oster, Donald Ingber was advancing the ideas, already mentioned in Sect. 6.6, of mechanotransduction affecting the biochemistry of cells and tissues, recently reviewed by himself (Ingber, 2018), and Larry Taber started a long series of morphomechanical studies (Taber, 2009). On the other side of the Iron Curtain and for a while outside the international discourse, Lev Beloussov and his coworkers paid particular attention to mechanical effects in embryonic development, also reviewed by himself (Beloussov, 2018). Beloussov's attitude mirrors that of Murray:

> The embryological textbooks are either completely descriptive, or enumerate one by one the instructions, "ordering" a given piece of embryonic tissue to develop in this or that direction. [...] Most modern embryological texts give an impression that an embryo is a mere toy of its genes, so that the ultimate and sole task of investigating development is in enumerating, one after another, the expressed genes.

However, not all biologists share this attitude, and there are clear signs of a revival of interest in physical factors of a general nature in this field dominated by the particularities of proteomics and genetics. The changing attitude, confirmed by numerous citations, is expressed in a review coauthored by a large group of biologists (Paluch et al, 2015):

> Despite our often detailed understanding of the biochemical reactions that control cellular fates, or maybe because of it, we may overlook the fact that mechanical forces are a powerful means to modulate or override many of these biochemical reactions.

Established cases of stress-induced gene expression have been known at least since studies by Farge (2003), who proved this by submitting the early *Drosophila* embryo

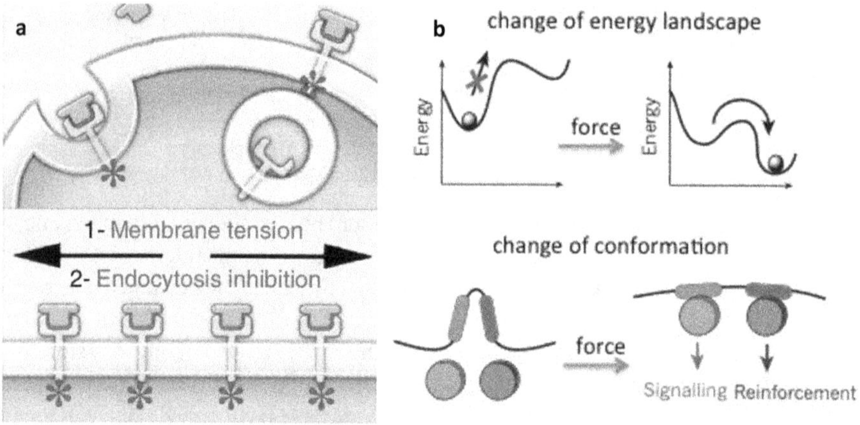

Fig. 8.13 (**a**) Modulation of signaling protein transfer by membrane stretching (Farge, 2011). (**b**) Change of energy landscape and the resulting conformation under applied force (Hoffman et al, 2011)

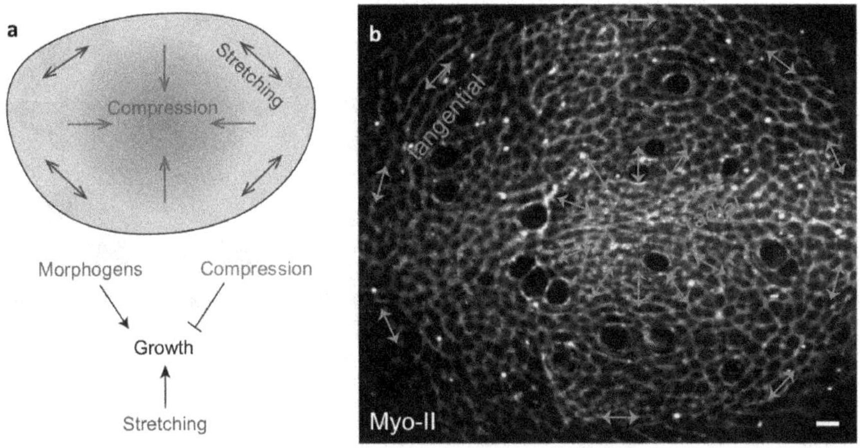

Fig. 8.14 (**a**) Model of wing disc growth. The morphogen gradient (*green*) from the center to the periphery promotes cell proliferation in the center, which causes stretching forces (*blue*) in the periphery promoting proliferation there and compressive forces (*red*) inhibiting proliferation in the center. (**b**) Wing disc marked for myosin. *Blue arrows* show the orientation of cell divisions. Scale bars 10 μm (Chanet and Martin, 2014)

to a transient lateral deformation; more examples are found in the reviews by Farge (2011), Hoffman et al (2011), and Chanet and Martin (2014). A possible mechanism is the interference with trafficking of a signaling protein, which takes place when flattening of the cell membrane by mechanical tension inhibits endocytosis, as shown in Fig. 8.13a. The mechanical action is also tied up with endogenous bioelectricity, as mechanosensitive ion channels, gated by membrane tension, modulate ion fluxes and shift the membrane potential (Leronni et al, 2020). On a still deeper level, forces transmitted through adherens junctions and focal adhesions may change the energy landscape on the molecular scale, breaking weak (noncovalent) molecular bonds and thereby modifying protein conformations (Fig. 8.13b).

Mechanical stresses are unavoidable in growing and deforming embryonic tissues, and shape them in joint action with morphogenetic signaling. For example, growth of the wing imaginal disk is regulated by a combination of biochemical and mechanical signals. While secreted growth factors are stronger in the center, mechanical feedback enhancing proliferation in the periphery and suppressing it in the center ensures uniform growth (Fig. 8.14a). Compression in the center leads to small cell areas, whereas cells are stretched in the periphery. The stress pattern controls the orientation of cell divisions: radial in the center and tangential in the periphery, as shown in Fig. 8.14b.

Chemo-mechanical instabilities can induce Turing-like patterning due to feedback interactions between elastic stresses in epithelium and intercellular processes. In the model by Brinkmann et al (2018), a morphogen induces an apical or a basal constriction locally, causing the epithelial layer to bend, while the tissue stretching induces production of the morphogen. The resulting shapes shown in Fig. 8.15 are

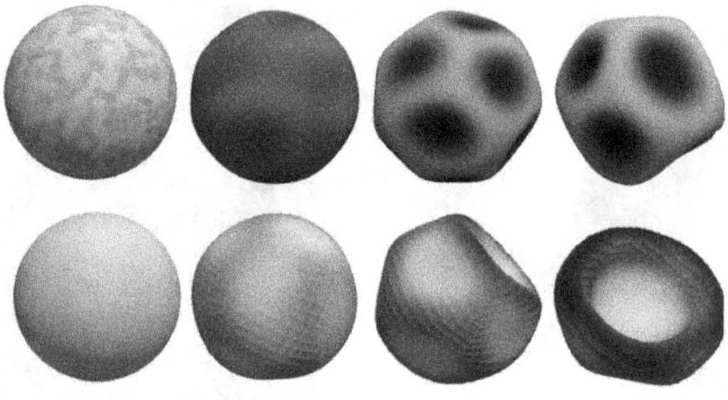

Fig. 8.15 Simulation snapshots of the development of a chemo-mechanical instability due to a mechanochemical feedback loop including basal (*top*) and apical (*bottom*) constriction. *Darker shades* correspond to higher morphogen concentrations (Brinkmann et al, 2018)

more realistic that those in Fig. 7.32, but are no more closely related to actual embryonic development than Turing's original model.

A more persuasive piece of evidence is the emergence of a follicle pattern in the avian skin (Shyer et al, 2017). Based on their observations, the authors questioned the earlier notion of the guiding role of a molecular prepattern, which might be established by the Turing instability, similar to Murray's (1980) simulations of animal fur patterns (Sect. 8.1). Instead, the mechanism turned out to be mechanical, which should have pleased Murray, whose works on mechanically-induced morphogenetic patterning coauthored with Oster featured in the citation list. In accordance with the mechanical models, the follicle structure was proven to be initiated by compression due to spontaneous mesenchymal cell aggregation within a skin layer. Moreover, it turned out that cellular self-organization conveyed via mechanosensation directly affects genetic expression.

The influence of nematic orientation on developing shapes is another variety of mechanotransduction; one example was given in Sect. 7.6. A spectacular effect

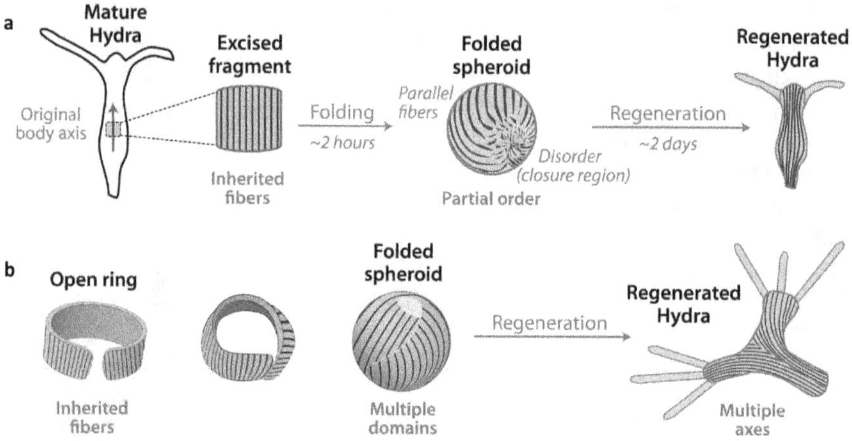

Fig. 8.16 Regeneration of *Hydra* from a closed (**a**) and open (**b**) ring of excised tissue (Braun and Keren, 2018)

of cell polarization was observed by Braun and Keren (2018) in experiments with *Hydra* regeneration. This simple radially symmetric aquatic animal has been the subject of numerous studies because of its ability to regenerate from almost any piece of tissue; it owes its name to its resemblance to the mythical Lernaean Hydra, which would regrow two heads for every head chopped off. Braun and Keren started with a fragment of excised tissue that inherited a parallel array of actin fibers from the parent *Hydra*. The tissue folded into a hollow spheroid, as shown in Fig. 8.16. The parallel fiber array was partially retained in the spheroid but a perfect alignment without defects is impossible on a spherical surface, so the area near the closure region became disordered. When the initial fragment was a closed ring, the residual actin fiber alignment in the folded spheroid induced order across the entire system and led to the formation of a regenerated *Hydra* with a nematic order of fibers throughout the animal, in the direction of the inherited fibers (Fig. 8.16a). But when the regeneration started from an open ring, multiple domains appeared with different fiber alignments, leading to the regeneration of a monstrous animal with multiple axes (Fig. 8.16b).

8.6 Remodeling

The attention to early patterning of the embryo in Sect. 8.3 may appear excessive in view of the diverse ultimate destinations of cells with different fates. Embryonic tissue undergoes a topological transformation, so that cells in adjacent regions of the fate map in Fig. 8.8b may end up in different *germ layers* of the final *triploblastic* body plan. This extensive remodeling through collective cell migration has to be no less precise than the primary plan. Whereas key genes and local cellular processes in the development of model animals are quite well understood, very little is known about global coordination of remodeling that would ensure the precision necessary to deliver cells with designated fates to proper locations.

The routes of migration should follow genetically defined chemical cues. Cells may navigate along pre-patterned chemoattractant gradients, but more likely they are able to autonomously generate local gradients that travel with them, a strategy allowing for self-determined directionality (Rørth, 2011). Remodeling, like any motion, involves mechanical forces, which are far less evident than the mere kinematics

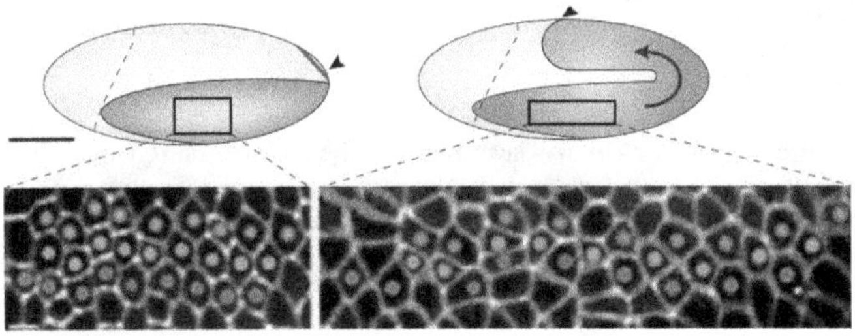

Fig. 8.17 Elongation of the *Drosophila* germ band through intercalation, as emphasized by *colored dots* in the blowup of deforming segments (Lecuit and Lenne, 2007)

of remodeling. There are some speculations about the way precision is attained in addressing and the way errors are dealt with. Misdirected cells may commit suicide, through apoptosis, or may retrain for another job.

The mechanisms of cell rearrangement and sorting, their collective migration, and folding of epithelial layers in the embryo do not differ in principle from those discussed in Chap. 7, but they have to follow specific genetic cues, likely self-sustained by the cells involved. Thus, genetic mechanisms regulating intercalation in *Drosophila* were uncovered by Irvine and Wieschaus (1994). This is an efficient tool for tissue elongation (accompanied by lateral contraction), as illustrated in Fig. 8.17. Stiffening was observed in collectively migrating cells (Barriga et al, 2018), which may be a sign of a supracellular cytoskeletal organization between multiple cells allowing them to function as a single unit. Yet, mechanics of cellular flow remains poorly understood due to difficulties in measuring driving forces.

As noted in Sect. 7.1, intercalation, as well as mitosis, increases the disorder of a cellular layer. In some organs, where a regular arrangement is essential, the hexagonal tiling is restored at a later stage. The patterning of *Drosophila*'s wings is already regularized on the adult stage, at the start of bristle growth (Classen et al, 2005), and requires the rearrangement of intercellular junctions, also guided genetically. Order is essential in retinal patterning. Insect's eyes contain an ordered array of hexagonal units called ommatidia. Their assembly is a complex process (Voas and Rebay, 2004), where ordering is established by an autoregulatory feedback loop between neighboring cells, disrupted by mutations.

Cell sorting serves to maintain boundaries between cell patches with different fates. The most common mechanism is based on differential adhesion properties (Sect. 7.1). Another model states that specific mechanical properties of cells at the boundaries prevent cell mixing through the formation of a stiff barrier that cells cannot cross. Both mechanisms are regulated genetically and are responsible for surface tension that prevents the addition and stabilization of new cells at the boundary and reduces the area of contact between compartments. The role of cell

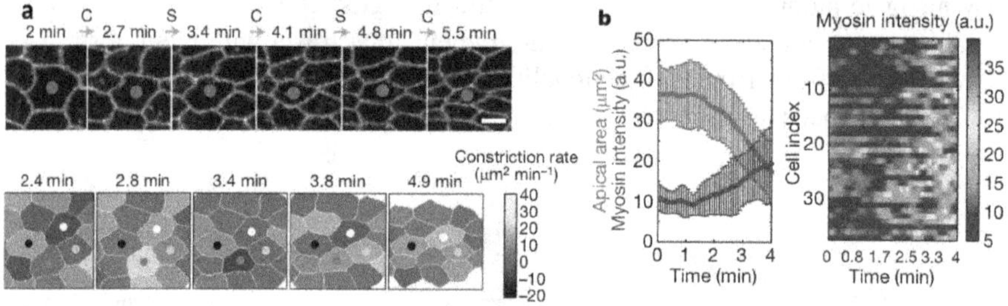

Fig. 8.18 (**a**) *Top*: Time-lapse images of a cellular layer marking the constriction (C) and stabilization (S) phases of the cell marked by the *red dot*. *Bottom*: Color-coded constriction rates at consecutive time moments. Scale bar 4 μm. (**b**) *Left*: Change in the mean apical area and myosin intensity with time. *Right*: Color-coded myosin intensity of individual cells (Martin et al, 2009)

surface mechanics and cell sorting in development is reviewed by Lecuit and Lenne (2007) and Krens and Heisenberg (2011).

The crucial event requiring radical remodeling of the embryonic tissue is *gastrulation*, turning a plain hollow sheath of cells, *blastula*, into a layered structure, *gastrula*. Lewis Wolpert called it "truly the most important time in your life". The site of invagination and internalization of the cells driven to inner layers is designated by a furrow formed by apical constriction of ventral cells. This is a standard mechanism of epithelial folding that involves, as noted in Sect. 7.7, enhanced actomyosin activity. However, the development of the ventral furrow in *Drosophila* does not proceed continuously, but involves repeated pulsed constrictions, which are asynchronous between neighboring cells (Martin et al, 2009). As shown in Fig. 8.18, individual cells go through the stages of constriction and stabilization, while the average apical area of a band of cells oscillates as well. These pulsations are accompanied by oscillations of myosin intensity, both on the average and in individual cells.

Gastrulation is necessarily accompanied by global tissue deformations. No *in vivo* force measurements are available as yet, but Streichan et al (2018) have mapped cellular flow together with the accompanying myosin intensity, as an indicator of the forces at work. Low-resolution patterns of cellular flow in the *Drosophila* oocyte at three moments before and following the first occurrence of the cephalic furrow, taken as $t = 0$, are shown in Fig. 8.19a. A flow pattern with a dorsal sink and ventral source emerges well before the ventral furrow forms. No cells are internalized during this flow, but rather cells reduce their cross-section on the dorsal side. As the furrow forms, the source and sink swap sides, and a large group of cells internalize on the ventral side. At a later stage, during the germ band extension, the flow pattern exhibits two saddles on the dorsal and ventral sides, as well as four vortices, two at the posterior and two at the anterior end.

The myosin distribution shown in Fig. 8.19b and c is in a rough correspondence with the flow pattern. Myosin is enriched at the basal side near the sink prior to the gastrulation onset, and there is a pronounced dorso-ventral asymmetry at this stage. The myosin intensity spreads far from the flow singularity due to the mechanical

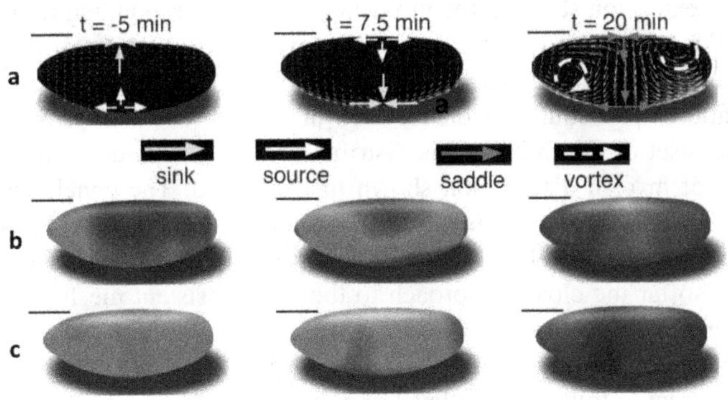

Fig. 8.19 Cellular low field (**a**) and myosin intensity at the basal (**b**) and apical (**c**) cell surface in the *Drosophila* embryo during gastrulation. Anterior is to the left, dorsal up, and scale bars represent 100 μm (Streichan et al, 2018)

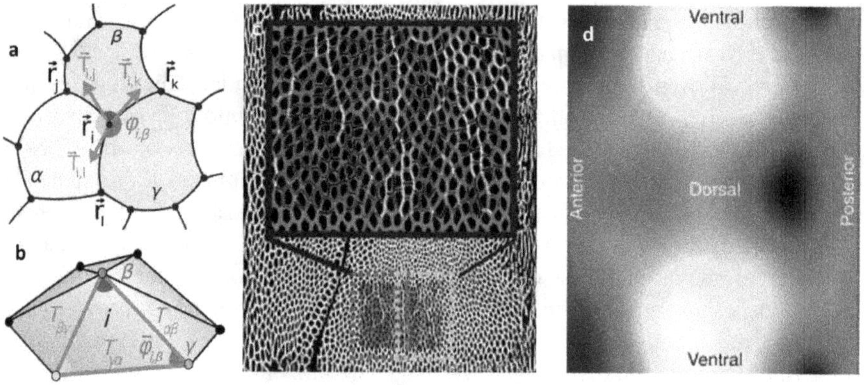

Fig. 8.20 (**a**) Polygonal tiling with circular arcs. (**b**) Tension triangles at vertices. (**c**) Two overlapping image frames shown on a cylindrical projection with the ventral line cut and mapped onto the top and bottom edges of the image with the dorsal side along the midline. *Inset*: Color-coded inferred edge tension (darker when low). (**d**) Mesoscopic anisotropy of myosin intensity (Noll et al, 2020)

interaction of cells. There is no clear reversal when the source and sink locations interchange, but myosin accumulates on the apical side during the late stage.

These data are still insufficient for the dynamic description of cellular flow, as long as there are no direct measurements of stress *in vivo*. In the subsequent work by the same group (Noll et al, 2020), stresses were inferred by measuring the curvatures of cell boundaries. The procedure involves the extension of the standard cellular model (Sect. 7.1), with straight lines connecting vertices replaced by circular arcs. In mechanical equilibrium, the curvature of each edge is controlled by pressure differences between adjacent cells. Force balance at each vertex requires the three tension vectors tangent to each edge, marked by green arrows in Fig. 8.20a, to sum up to zero. As adjacent vertices share edges, the tension triangles of each vertex form a triangulated surface (Fig. 8.20b), the dual representation of force balance, with triangular faces corresponding to each vertex.

Noll et al analyzed images of fluorescent-labeled membranes and myosin intensity during the germ band extension stage, at the moment roughly corresponding to the rightmost images in Fig. 8.19. They used the data to approximate the apical geometry of an epithelial tissue by a circular arc tiling and inferred from its structure through a sophisticated variational procedure the distribution of equilibrium edge tension, such as shown in the inset of Fig. 8.20c. This distribution was compared with the measured anisotropy of myosin distribution shown in Fig. 8.20d. The conclusion was that most of the myosin activity is involved in a static internal force balance within the epithelial layer. The correlation between inferred and measured data was imperfect, but this is so far the closest approach to the self-consistent mechanical description of cellular flow *in vivo*.

Complex species-specific processes take place during specialization of the inner layers. They are still harder to follow on the level of a dynamical description and lie

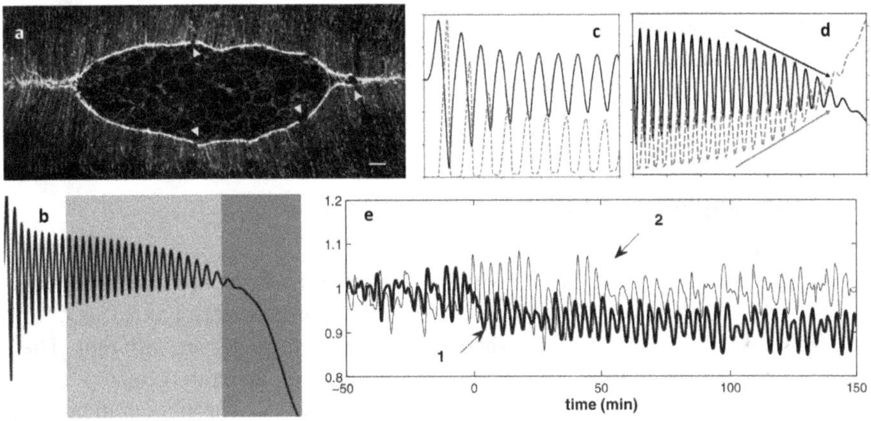

Fig. 8.21 (**a**) The dorsal gap, showing the actin cable, with interruptions indicated by *yellow triangles*, surrounding the amnioserosa. The protein specifically required for actin cable formation is marked *green*, and actin is marked *magenta*. Scale bar 10 μm (Ducuing and Vincent, 2016). (**b**) Change in the area of a representative cell through the three phases of dorsal closure, distinguished by *shading*. (**c**), (**d**) Dynamics of the normalized area of a representative cell (*black solid line*) and its myosin content (*red dashed line*) during the early (**c**) and slow/fast (**d**) phases (Durney et al, 2018). (**e**) Area oscillations of two neighboring cells. Zero time corresponds to the formation of the actin cable (Wang et al, 2012)

beyond the scope of this book. A thoroughly investigated post-gastrulation process, still observable from outside, is *dorsal closure* in the developing *Drosophila* embryo, during which a transient dorsal gap, left upon invagination and covered by an extra-embryonic tissue, the *amnioserosa*, is closed. The closure begins, similarly to some wound-healing processes (Sect. 7.2), with the appearance of a supracellular actin cable that surrounds the opening and provides a contractile force. It is complemented by a pulsed force pulling on the surrounding tissue (Solon et al, 2009). Later studies (Ducuing and Vincent, 2016) showed that, as in wound healing, the force provided by the cable is dispensable, and the cable itself is not a continuous structure (Fig. 8.21a); however, the cable helps to keep the leading edge straight.

Dorsal closure proceeds through several phases (Fig. 8.21b). The early phase, prior to the formation of the actin cable, is characterized by persistent oscillation of amnioserosa cells with no net contraction. The cyclic apical constriction correlates with the assembly and disassembly of myosin condensates. During the following slow phase, cell oscillations decrease in amplitude and period, and a net shrinkage in cell area is observed. The area contraction accelerates markedly during the fast phase, and most cells contract consistently with little fluctuation.

Cell-based simulations (Wang et al, 2012; Durney et al, 2018) taking into account myosin-generated forces and the dynamics of myosin attachment/detachment, and in the later publication also the dynamics of other proteins, reproduced pulsations in amnioserosa, which arise, similarly to other tissue remodeling processes, as a result of mechanical coupling among neighboring cells. Fluctuations of the cell area are in antiphase with the myosin intensity, as shown in Fig. 8.21c and d. Solon et al (2009) reported antiphase correlations between neighbors, which are also reproduced in the simulations (Fig. 8.21e). Sustained oscillations are possible even under conditions when a single cell does not oscillate.

8.7 Morphogenesis in Plants

Plants, like animals, start their life with the merger of an egg and a sperm cell, and meiosis. As an animal oocyte polarizes to form the antero-posterior axis (Sect. 8.2), a plant seed polarizes to form its shoot and root – but plants are different. Their morphogenesis never stops: *phyllotaxis*, the process of generating new *phylla* – leaves, roots, stalks, florets – continues while they are alive. Throughout their lives, they retain embryonic tissues, *meristems* (Fig. 8.22a), nucleating *primordia* of these repetitive structures, starting out as small undifferentiated bumps on the surface of the plant. Growth is promoted in plants by a small but ubiquitous molecule – *auxin*.

Phyllotactic patterning in early land plants, such as mosses and ferns, is lineage-dependent, caused by the patterned cell divisions of a single apical cell. The invasion of land was enabled by acquiring the ability for cell cleavage in three dimensions, facilitating the formation of bushy upright body plans.The daughters of the initial cell in a moss shoot (Fig. 8.22b) adopt the same initial fate as parts of a leaf, with the exception of the hair cell (marked by the asterisk) resulting from the first cleavage.

In flowering plants, the spacing of primordia evolved from hereditary to position-dependent. It is determined by the inhibitory action of existing phylla, which is reminiscent of Turing's symmetry-breaking mechanism (Sect. 8.1), but was understood by botanists (Schoute, 1913) long before Turing. However, as in the morphogenetic processes discussed in Sect. 8.6, mechanical interactions compete with chemical inhibition.

In many plants, phylla form a pattern of two mutually intersecting spirals, called *parastichy*. Through the ages, both botanists and mathematicians were intrigued by the relation of this pattern with Fibonacci numbers, the sequence originating from a book published in 1202 by Leonardo of Pisa, *filius Bonacci*. It starts with 1 and 1, and

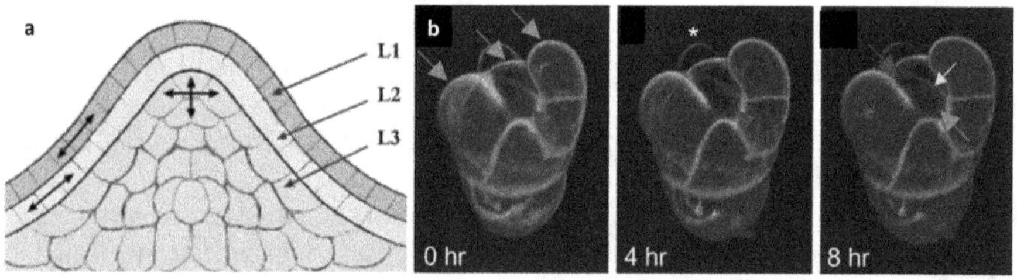

Fig. 8.22 (**a**) Apical meristem of a growing tip with the outer epidermal (L1) and subepidermal (L2) layers and the inner volume L3 (CC). (**b**) Early growth of a moss shoot. Three presumptive leaves (*gray arrows*) flank a single apical initial cell (*red arrow*). Its self-renewing cleavage at 8 hr is indicated by the *yellow arrow* (Harrison et al, 2009)

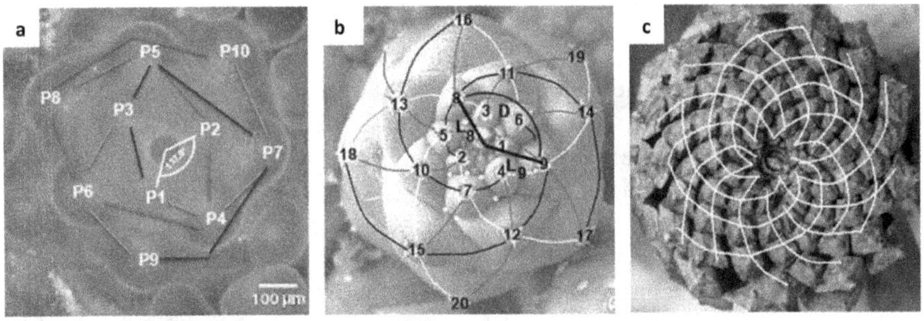

Fig. 8.23 (**a**) A 2/3 parastichy superimposed on the image of newly forming buds of *S. muticum* alga (Linardić and Braybrook, 2017). The spirals connecting buds with birth moments differing by three are colored *blue* and *green*, and the spirals connecting buds with birth moments differing by two, *red* and *purple*. (**b**) 3/5 parastichy on the top of a cactus. The family of 3 spirals is drawn in *black* and the family of 5 spirals, in *white*. (**c**) 8/13 parastichy on a pine cone (Shipman and Newell, 2005)

each subsequent member of the sequence is the sum of the two preceding numbers. The history of numerous wrong theories has been narrated by Adler et al (1996). A proper explanation depends on two long overlooked physical factors: interaction energy and growth rate. Douady and Couder (1992) simulated the formation of spiral patterns assuming that primordia form at a fixed distance R from the apex and, due to the shoot's growth, move away from the center with a radial velocity V, which may depend on their radial location; new primordia are formed at regular time intervals T at an angular location minimizing their interaction energy with the existing ones. The simulations were supported by a non-botanical experiment with deposition of magnetically interacting droplets.

The most interesting result was not just reproducing the parastichy governed by Fibonacci numbers, but locating *bifurcations* between different pairs of spirals with changing growth rate. In fact, this qualitatively important result can be obtained without computing energies (which need not be the same for plants and magnetic droplets anyway), but just by following the old recipe due to Hofmeister (1868): an incipient primordium forms in the largest available space left by the previous ones. In this case, bifurcations are located by a simple algebraic calculation.

The growth rate V can be made dimensionless by taking R and T as length and time units. When growth is very fast, $V \geqslant 2$, it is sufficient to take into account the repulsion by the single immediate forerunner, so that successive new phylla move away in opposite directions. In the interval $2 > V \geqslant 1$, the angular position of a new primordium depends on the two preceding ones. Here the first spiral pattern appears, characterized by the Fibonacci numbers 2 and 3: two spirals connecting phylla with the birth moments differing by 2 and three spirals connecting phylla with the birth moments differing by 3. They are indicated by different colors on the image of newly forming buds of *S. muticum* alga in Fig. 8.23a. They don't look as neat as in simulations, and the mechanism of the bud formation differs from that in land plants (Linardić and Braybrook, 2017), but they signify position-dependent patterning.

Higher-order parastichies in Fig. 8.23b and c are more distinct. For $1 > V \geqslant 2/3$, three precursors determine the angular position of a new primordium generating intersecting spirals characterized by the Fibonacci numbers 3 and 5, as shown in Fig. 8.23b. For $2/3 > V \geqslant 1/2$, the location depends on four precursors and the parastichy includes 5 and 8 spirals connecting phylla with the differing birth moments. Further bifurcations leading to sequential Fibonacci numbers follow at decreasing intervals.

This model catches the essence of phyllotaxis, but in reality everything is not that simple, since signaling and mechanical forces certainly play a role in this, as in all other morphogenetic processes. The growth hormone auxin is instrumental in initiating the formation of primordia, affecting the spatial and temporal intervals and the growth rate, which may be variable. The underlying geometry may be spherical or cylindrical rather than flat, and compressive stresses may lead to buckling of the plant surface. Some of these factors were taken into account by Shipman and Newell (2005) and in their later work. They asserted that Fibonacci patterns are persistent among different growth scenarios, but may come out with topological defects when bifurcations become less sharp, so that different parastichies coexist within a certain parametric interval.

Development of differently shaped leaves follows the same rule due to Hofmeister, now in the 1D geometry of a growing edge, with new primordia emerging at locations that are sufficiently distant from the nearest primordia formed previously. Growth directions are aligned with main veins, which is an evolutionary inheritance from ancestral branching structures. Veins are formed and differentiate through positive feedback, which creates narrow canals of auxin transport in a manner analogous to the carving of rivers by flowing water. Following these rules, the leaf develops as shown in Fig. 8.24a and b (Runions et al, 2017). Inhibited intervals are marked in red, and the shapes become more angular when the inhibitor deepens the troughs. The shapes in Fig. 8.24c differ by the insertion angle of new veins, and the shapes in Fig. 8.24d, by the growth rate at the vein tip, producing broader leaves as it increases.

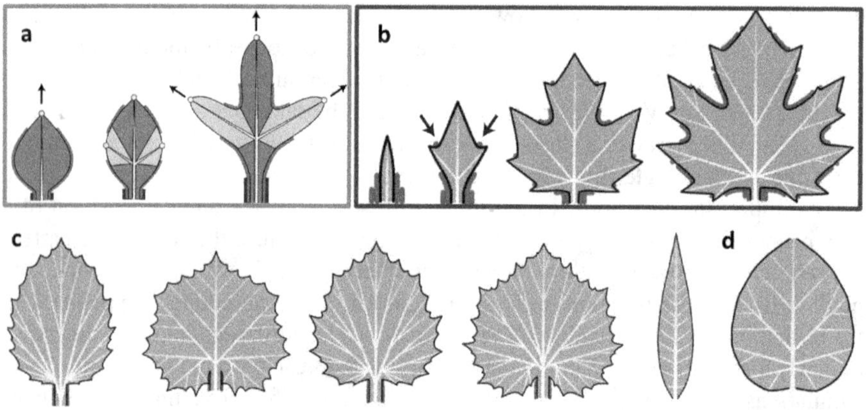

Fig. 8.24 Simulations of growing leaves (Runions et al, 2017). See the text for explanation

Fig. 8.25 Simulations of composite leaves (Runions et al, 2017). See the text for explanation

The compound leaf in Fig. 8.25a may form by exaggerating the tendency in Fig. 8.24b. Another mechanism, illustrated in Fig. 8.25b, involves two morphogen species: the one marked purple inhibits branching and the one marked cyan delimits leaflets at their base.

As in 2D growth, mechanical interactions complement chemical signaling; they are also responsible for finer effects (Boudaoud, 2010). Slowly and rapidly growing tissues exert mutual stresses. Veins are stiffer than the surrounding leaf tissue and grow more slowly, which subjects them to a tensional stress. This affects vein junctions, so that three veins of similar thickness tend to form angles of 120°. Enhanced growth at the periphery causes a compressive stress that leads to buckling, as already noted in Sect. 7.8.

Auxin plays contradictory roles in the overall plant development. Thus, it promotes growth of roots but suppresses shoot branching, competing with other hormones which take over when the apex of a growing plant, where auxin is produced, is cut or when downward transport of auxin is suppressed. An experiment demonstrating apical dominance and the inhibiting role of auxin is illustrated in Fig. 8.26 (Leyser

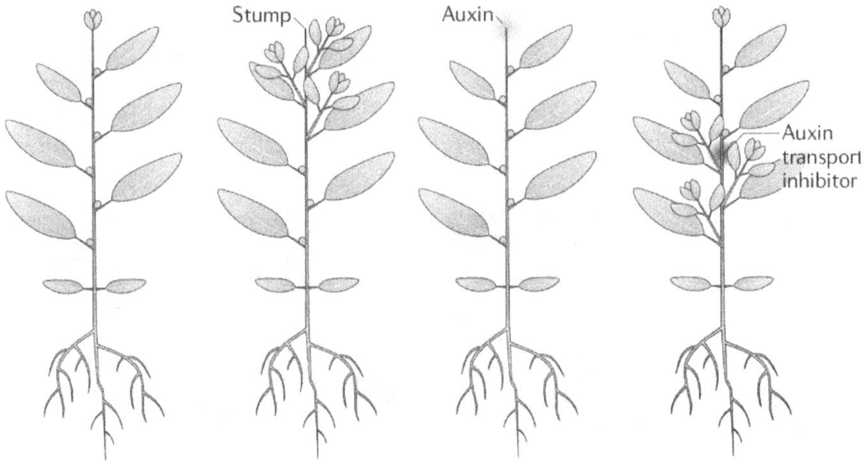

Fig. 8.26 Demonstration of branching inhibition by auxin (Leyser and Domagalska, 2011). See the text for explanation

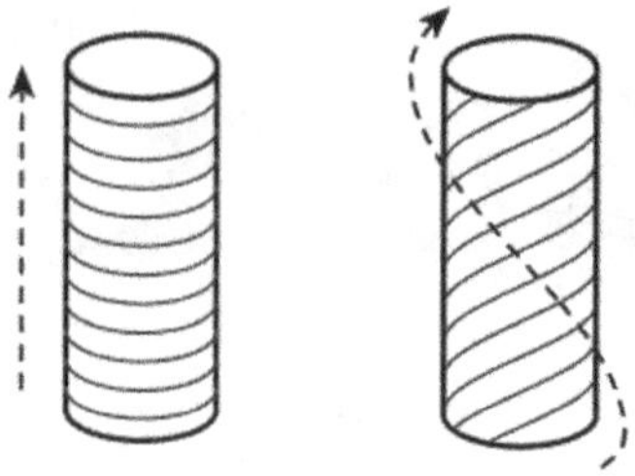

Fig. 8.27 Linear growth directed by transverse alignment of the microtubular cortex and growth with chirality opposite to that of the cortex, as indicated by the *dashed line* (Wada and Matsumoto, 2018)

and Domagalska, 2011). Removing the shoot apex results in activation of buds, indicating that the shoot apex inhibits the formation of the buds below. When the missing apex is replaced with an auxin source, branching inhibition is restored, but applying an auxin transport inhibitor leads to bud outgrowth at nodes below the application site.

Plant tissues may grow not only through cell division, but by cell enlargement, and in linear structure elements, such as stalks and roots, the expansion of cells is highly anisotropic. An individual cell in a root may grow at a rate of a few μm/min, with no appreciable radial expansion. The latter is restricted by a cortex formed by bundles of microtubules oriented transversely to the growth direction. The chirality of such a bundle causes chiral growth with the opposite twist (Fig. 8.27).

8.8 Biomorphs and Biohybrids

In Sects. 6.7 and 7.8, we noted a similarity between plant tissues and synthetic *geometrically active* materials. However, there is a considerable difference between our traditional fabrication methods, by assembling parts or removing excess material, and the way Nature actually operates, by growth and diversification. This contrast has been mitigated with the spread of additive manufacturing through 3D printing, guided by programming in lieu of genetic induction. Sophisticated microfluidic

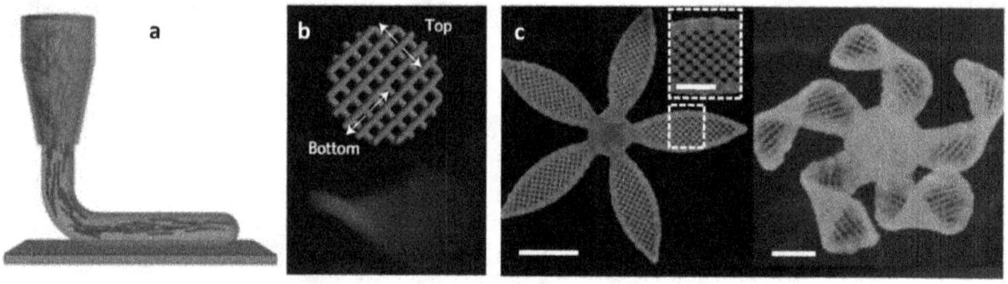

Fig. 8.28 (**a**) Alignment of fibrils in the printing nozzle. (**b**) Grid pattern with different placement of swelling layers deforming into a saddle shape. (**c**) Printed (*left*) and actuated (*right*) flower-like shapes. Scale bar 10 mm. The *inset* shows the grid orientation in the printed petals (Gladman et al, 2016)

techniques make it possible to integrate different materials into emerging forms, which can be as fantastic as any designer could imagine.

Incorporating geometrically active soft materials into a 3D-printed form led to what is called 4D printing, adding time as the fourth dimension – of course, not on par with spatial dimensions in relativity theory, since only the spatial form is printed, but by allowing controllable reshaping under actuation. This method is most suitable for fabricating layered and anisotropic structures. Gladman et al (2016) printed patterns using hydrogel ink with imbedded fibrils, which aligned when passing the deposition nozzle, as shown in Fig. 8.28a. This makes the material anisotropic, so that it swells along the filament length rather than uniformly in all directions, imitating the effect of the directional orientation of microtubules within plant cell walls (Fig. 8.27). In a bilayer system, differential swelling between the top and bottom layers induces curvature with the direction of bending determined by the way the swelling layer is placed. In Fig. 8.28b, it is placed on the top or bottom sides of the two sets of filaments, oriented perpendicularly to one another. Upon actuation, this leads to a saddle-like form. In this way, a variety of shapes can be created, like the flower-like forms in Fig. 8.28c.

The contraptions featured in Sect. 6.8, though imitating worms and caterpillars, are biomorphic in a very restricted sense, as are the many versatile bilayer or patterned hydrogel structures reviewed by Erol et al (2019), but advanced engineering design is bringing forward more sophisticated soft robots complete with autonomous power sources, sensors, and onboard processors, and partly 3D-printed. The *octobot* (Wehner et al, 2016) presented in all its glory in Fig. 8.29a, is an "animal" of this kind moving by actuating its legs pneumatically as shown in Fig. 8.29b by gases produced in the built-in catalytic reactor. Lu et al (2020) constructed a soft millirobot (Fig. 8.29c) integrating the power generation and actuation functions in a multilayer thin film less than 0.5 mm thick fitting on a finger. It can sense the terrain, walk on multiple tapered feet, and communicate remotely by coupling magnetic and piezo-

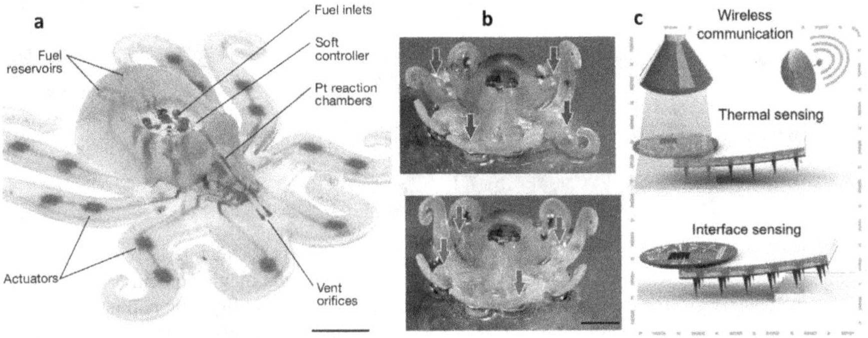

Fig. 8.29 (**a**) Octobot; fluorescent dyes have been added to assist in visualizing internal features. (**b**) Octobot autonomously alternating between "blue" and "red" actuation states. Scale bar 10 mm (Wehner et al, 2016). (**c**) Multilegged, battery-less, wireless sensing soft millirobot by Lu et al (2020)

electric effects. So far, these are just toys, but soft robotics is a hot topic. The number of reviews alone exceeded 200 in the last five years, and whatever could be written here will soon become obsolete.

A notable trend is combining natural and synthetic components in a *biohybrid* device. Molecular motors are most efficient microscopic tools converting chemical to mechanical energy, and they are simpler to operate when they remain within intact cells and make use of their locomotive machinery. This suggests employing bacteria as movers (Behkam and Sitti, 2007). They are not picky: a simple nutrient such as glucose would suffice, there is no need for toxic fuels or catalysts, and benign bacteria can be used for such fine tasks as intravenous drug delivery. The problem is how to control directionality. Stanton et al (2016) suggested attaching *E. coli* (found naturally in human guts) to Janus particles (featured in Sect. 3.2), where the bacterium preferentially adheres to the metal cap (Fig. 8.30a). Iron capping gives the advantage of magnetic guidance: orienting the particle by external magnetic field forces the bacteria to swim along a guided route (Fig. 8.30b).

Vizsnyiczai et al (2017) constructed a micromotor powered by bacteria (Fig. 8.30c). The outer rim of its rotor features a number of microchambers, each capable of accommodating the body of a single cell, leaving the entire flagellar bundle outside for maximal propulsion. The torque exerted by each cell increases with the tilt, but the number of chambers that can be accommodated along the circumference decreases, so the 45° tilt is the optimal choice. Self-assembly of the hybrid system was completed when bacteria were captured by suspended micromotors. Earlier, Sokolov et al (2010) induced freely moving bacteria to rotate asymmetric microscopic gears

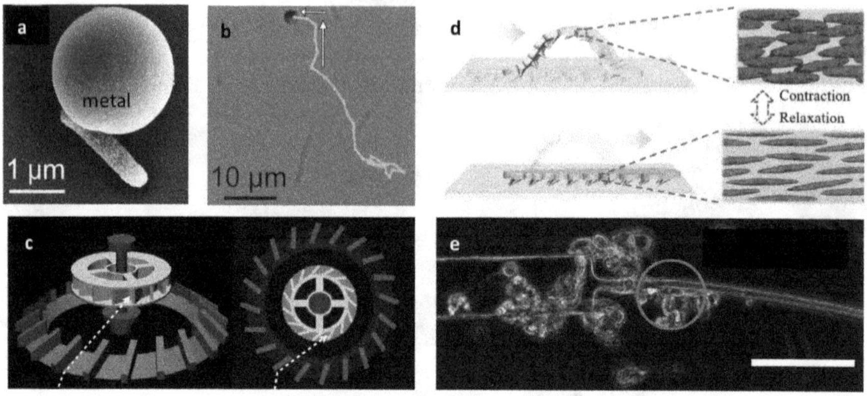

Fig. 8.30 (**a**) *E. coli* attached to the metal-covered part of a Janus particle. (**b**) Trajectory of the swimmer with the iron-covered Janus particle oriented by an external magnetic field (Stanton et al, 2016). (**c**) Scheme of a micromotor powered by bacteria. Colours highlight the component parts: ramp (*red*), axis (*blue*), and rotor (*green*). The *dashed white line* depicts the trajectory of a cell guided by the ramp structure into a rotor microchamber (Vizsnyiczai et al, 2017). (**d**) Crawling "caterpillar" powered by cardiac muscle (Sun et al, 2020). (**e**) Flagellar swimmer powered by cardiac muscle cells (the region near the head of the filament is shown). The contractile cells are circled in *red*. Scale bar 0.4 mm (Williams et al, 2014)

through a "Brownian ratchet" effect. This did not require confining microbes as slaves on an oared galley, but the energy they generate in this way would be hard to harvest.

A cardiac muscle is another tool for locomotion driven by molecular motors. Sun et al (2020) assembled a "caterpillar" (Fig. 8.30d) crawling, like its better equipped counterpart in Fig. 8.29d, on tapered feet and advancing by alternately bending and flattening its soft body (like its light-driven cousin in Fig. 6.31b) in response to muscle contraction and relaxation. Williams et al (2014) used a cardiac muscle to power a flagellar swimmer. The head of a flexible filament is attached to the load, while its tail is left free. The muscle, which is only allowed to attach near the head region of the filament (Fig. 8.30e), wiggles the filament while contracting and relaxing, thereby setting the swimmer into motion, though slowly compared with its sperm lookalike, and not well controlled.

3D printing spreads out to living tissues. Noor et al (2019) 3D-printed a small-scaled cellularized human heart (Fig. 8.31) using fat cells reprogrammed to become pluripotent stem cells and differentiated to cardiac tissues. The publication is marked as a "hot paper", having received more than a hundred citations in just over a year. *Bioprinting* opens the way to replacements of tissues and organs and studies of their *in vitro* models (Mota et al, 2020).

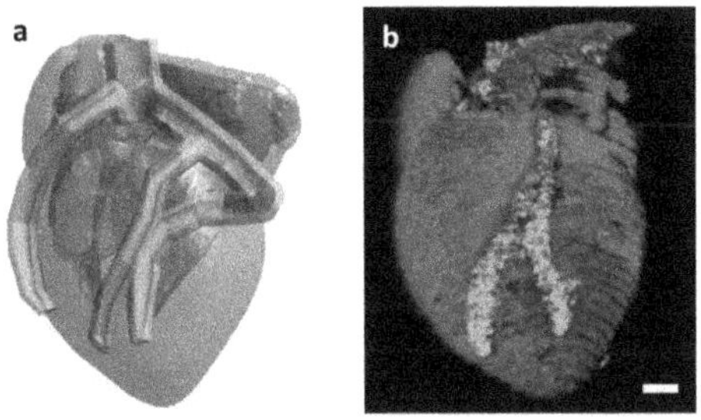

Fig. 8.31 Computer model (**a**) and confocal image (**b**) of a 3D-printed heart. Cardiac muscles are in *pink*, and endothelial cells that form the blood vessels, in *orange*. Scale bar 1 mm (Noor et al, 2019)

But this is not the end of the road. An earlier "hot paper" (Deglincerti et al, 2016) reports an *in vitro* system for studying the post-implantation development of the human embryo. Laurent et al (2017), reviewing technical progress in bioprinting, set the aim of directed multicellular self-organization into higher-order large-scale structures, mimicking advanced stages of embryo development.

Our narrative has to stop at this point, leaving us to contemplate the wider prospects of saving the critically ill or creating monsters.

References

Abercrombie M and Heaysman JEM, 1953. Observations on the social behaviour of cells in tissue culture. I. Speed of movement of chick heart fibroblasts in relation to their mutual contacts. Exp. Cell Res. 5: 111–131

Abkenar M, Marx K, Auth T, and Gompper G, 2013. Collective behavior of penetrable self-propelled rods in two dimensions, Phys. Rev. E 88: 062314

Adler I, Barabe D, and Jean RV, 1997. A history of the study of phyllotaxis, Ann. Botany 80: 231–244

Agostinelli D, Cerbino R,Del Alamo JC, DeSimone A, Höhn S, Micheletti C, Noselli G, Sharon E, and Yeomans J, 2019. MicroMotility: State of the art, recent accomplishments and perspectives on the mathematical modeling of bio-motility at microscopic scales Math. in Eng. 2: 230–252

Aguilar-Hidalgo D, Werner S, Wartlick O, González-Gaitán M, Friedrich BM, and Frank Jülicher F, 2018. Critical Point in Self-Organized Tissue Growth, Phys. Rev. Lett. 120: 198102

Alberts B, Bray D, Lewis J, Raff M, Roberts K, and Watson JD, 2002. *Molecular Biology of the Cell*, Garland Science Publishing, New York

Alsous JI, Villoutreix P, Berezhkovskii AM, and Shvartsman SY, 2017. Collective Growth in a Small Cell Network, Curr. Biol. 27: 2670–2676

Alvarado J, Sheinman M, Sharma A, MacKintosh FC, and Koenderink GH, 2013. Molecular motors robustly drive active gels to a critically connected state, Nat. Phys. 9: 591–597

Ambrosio A, Marrucci L, Borbone F, Roviello A, and Maddalena P, 2012. Light-induced spiral mass transport in azo-polymer films under vortex-beam illumination. Nat. Commun. 3: 989

Anderson JL, 1989. Colloidal transport by interfacial forces, Annu. Rev. Fluid Mech. 21: 61–99

Aranson IS and Tsimring LS, 2003. Model of coarsening and vortex formation in vibrated granular rods, Phys. Rev. E 67: 021305

Aranson IS, Volfson D, and Tsimring LS, 2007. Swirling motion in a system of vibrated elongated particles, Phys. Rev. E 75: 051301

Arciero J, Mi Q, Branca MF, Hackam DJ, and Swigon D, 2011. Continuum model of collective cell migration in wound healing and colony expansion, Biophys. J. 100: 535

Atis S, Weinstein BT, Murray AW, and Nelson DR, 2019. Microbial Range Expansions on Liquid Substrates, Phys. Rev. X 9: 021058

Attanasi A, Cavagna A, et al, 2014 Information transfer and behavioural inertia in starling flocks, Nat. Phys. 10: 691–696

L. Pismen, *Active Matter Within and Around Us*, The Frontiers Collection, https://doi.org/10.1007/978-3-030-68421-1

Ballerini M, ..., Zdravkovic V (12 coauthors), 2008. Interaction ruling animal collective behavior depends on topological rather than metric distance: Evidence from a field study. Proc. Natl. Acad. Sci USA 105: 1232–1237

Banerjee S and Marchetti MC, 2019. Continuum Models of Collective Cell Migration, in *Cell Migrations: Causes and Functions*, eds. La Porta CAM and S. Zapperi S, Springer, Cham

Bär M, Großmann R, Heidenreich S, and Peruani F, 2020. Self-Propelled Rods: Insights and Perspectives for Active Matter, Annu. Rev. Condens. Matter Phys. 11: 441–466

Barnhart EL, Lee K-C, Keren K, Mogilner A, and Theriot JA, 2011. An Adhesion-Dependent Switch between Mechanisms That Determine Motile Cell Shape, PLoS Biology 9: e1001059

Barriga EH, Franze K, Charras G, and Mayor R, 2018. Tissue stiffening coordinates morphogenesis by triggering collective cell migration in vivo, Nature 554: 523–527

Barton DL, Henkes S, Weijer CJ, and Sknepnek R, 2017. Active vertex model for cell-resolution description of epithelial tissue mechanics. PLoS Comput. Biol. 13: e1005569

Batchelor GK, 1976. Brownian diffusion of particles with hydrodynamic interaction, J . Fluid Mech. 74: 1–29

Bäuerle T, Löffler RC, and Bechinger C, 2020. Formation of stable and responsive collective states in suspensions of active colloids, Nat. Commun. 11: 2547

Bausch AR and Kroy K, 2006. A bottom-up approach to cell mechanics, Nat. Phys. 2: 231–238

Beauzamy L, Nakayama N, and Boudaoud A, 2014. Flowers under pressure: ins and outs of turgor regulation in development, Ann. Botany 114: 1517–1533

Be'er A, Ilkanaiv B, Gross R, Kearns BD, Heidenreich S, Bär M, and Ariel G, 2020. A phase diagram for bacterial swarming, Commun. Phys. 3: 66

Behkam B and Sitti M, 2007. Bacterial flagella-based propulsion and on/off motion control of microscale objects, Appl. Phys. Lett. 90: 023902

Belintsev BN, Beloussov LV, and Zaraisky AG, 1987. Model of pattern formation in epithelial morphogenesis. J. Theor. Biol. 129: 369–394

Beloussov LV and Mittenthal JE, 1992. Hyperrestoration of mechanical stresses as a possible driving mechanism of morphogenesis, Zh. Obshch. Biol. 53: 797–807 (in Russian)

Beloussov LV, 2018. Morphomechanical rules of embryonic development, BioSystems 173: 10–17

Ben-Jacob E, Cohen I, Shochet O, Tenenbaum A, Czirók A, and Vicsek T, 1995. Cooperative Formation of Chiral Patterns during Growth of Bacterial Colonies, Phys. Rev. Lett. 75: 2899–2902

Ben-Jacob E, Cohen I, and Czirók A, 1997. Smart Bacterial Colonies, in *Physics of biological systems: from molecules to species*, Flyvbjerg H et al (Eds.) pp. 307–324, Springer-Verlag, Berlin

Ben-Zvi D and Barkai N, 2010. Scaling of morphogen gradients by an expansion-repression integral feedback control, Proc. Natl. Acad. Sci. USA, 107: 6924–6929

Berg HC, 2004 *E. coli in Motion*, Springer, New York

Bergert M, Erzberger A, Desai RA, Aspalter IM, Oates AC, Charras G, Salbreux G, and Paluch EK, 2015. Force transmission during adhesion-independent migration, Nat. Cell Biol. 17: 524–529

Beroz F, Yan J, Meir Y, Sabass B, Stone HA, Bassler BL, and Wingreen NS, Verticalization of bacterial biofilms, Nat. Phys. 14: 954–960

Bertin E, Droz M, and Grégoire G, 2006. Boltzmann and hydrodynamic description for self-propelled particles, Phys. Rev. E 74: 022101

Bertin E, Chaté H, Ginelli F, Mishra S, Peshkov A, and Ramaswamy S, 2013. Mesoscopic theory for fluctuating active nematics, New J. Phys. 15: 085032

Bi D, Lopez JH. Schwarz JM, and Manning ML, 2015. A density-independent rigidity transition in biological tissues, Phys. Rev. X 6: 021011, Nat. Phys. 11: 1074–1079

Bi D, Yang X, Marchetti MC, and Manning ML, 2016. Motility-Driven Glass and Jamming Transitions in Biological Tissues, Phys. Rev. X 6: 021011

Bialek W, Cavagna A, Giardina I, Mora T, Silvestri E, Viale M, and Walczak AM, 2012. Statistical mechanics for natural flocks of birds, Proc. Natl. Acad. Sci USA 109: 4786–4791

Biot MA, 1941. General Theory of Three-Dimensional Consolidation, J. Appl. Phys. 12: 155–164

Blair DL, Neicu T, and Kudrolli A, 2003. Vortices in vibrated granular rods, Phys. Rev. E 67: 031303

Blaschke J, Maurer M, Menon K, Zöttl A, and Stark H, 2016. Phase separation and coexistence of hydrodynamically interacting microswimmers, Soft Matter, 12: 9821–9831

Boareto M, 2019. Patterning via local cell–cell interactions in developing systems, Develop. Biol. 460: 77–85

Bonelli F, Gonnella G, Tiribocchi A, and Marenduzzo D, 2016. Spontaneous flow in polar active fluids: the effect of a phenomenological self propulsion-like term, Eur. Phys. J. E 39: 1

Bonelli F, Carenza NL, Gonnella G, Marenduzzo D, Orlandini E, and Tiribocchi A, 2019. Lamellar ordering, droplet formation and phase inversion in exotic active emulsions, Sci. Rep. 9: 2801

Bor B, Bois JS, and Quinlan ME, 2015. Regulation of the Formin Cappuccino is Critical for Polarity of *Drosophila* Oocytes, Cytoskeleton 72: 1–15

Boudaoud A, 2010. An introduction to the mechanics of morphogenesis for plant biologists, Trends Plant Sci. 15: 353–360

Braun E and Keren K, 2018. Hydra Regeneration: Closing the Loop with Mechanical Processes in Morphogenesis, BioEssays 40: 1700204

Bricard A, Caussin J-B, Desreumaux N, Dauchot O, and Bartolo D, 2013. Emergence of macroscopic directed motion in populations of motile colloids, Nature 503: 95–98

Bricard A, Caussin J-B, Das D, Savoie C, Chikkadi V, Shitara K, Chepizhko O, Peruani F, Saintillan D, and Bartolo D, 2015. Emergent vortices in populations of colloidal rollers, Nat. Commun. 6: 7470

Brinkmann F, Mercker M, Richter T, and Marciniak-Czochra A, 2018. Post-Turing tissue pattern formation: Advent of mechanochemistry, PLoS Comput. Biol. 14: e1006259

Brugués A, Anon E, Conte V, Veldhuis JH, Gupta M, Colombelli J, MuŨoz JJ, Brodland GW, Ladoux B, and Trepat X, 2014. Forces driving epithelial wound healing, Nat. Phys. 10: 683–689

Brumley DR, Wan KY, Polin M, Goldstein RE, and Nelson WJ, 2014. Flagellar synchronization through direct hydrodynamic interactions, eLife 3: e02750

Brumley DR, Polin M, Pedley TJ, and Goldstein RE, 2015. Metachronal waves in the flagellar beating of Volvox and their hydrodynamic origin, J. Roy. Soc. Interface 12: 20141358

Burnette DT, Manley S, Sengupta P, Sougrat R, Davidson MW, Kachar B, and Lippincott-Schwartz J, 2011. A role for actin arcs in the leading-edge advance of migrating cells, Nat. Cell Biol. 13: 371–382

Buttinoni I, Bialké J, Küňmmel F Loẅen H, Bechinger C, and Speck T, 2013. Dynamical Clustering and Phase Separation in Suspensions of Self-Propelled Colloidal Particles, Phys. Rev. Lett. 110: 238301

Camacho-Lopez M, Finkelmann H, Palffy-Muhoray P, and Shelly M, 2004. Fast liquid-crystal elastomer swims into the dark, Nat. Mater. 3: 307–310

Camley BA, Zimmermann J, Levine H, and Rappel W-J, Collective Signal Processing in Cluster Chemotaxis: Roles of Adaptation, Amplification, and Co-attraction in Collective Guidance, PLoS Comput. Biol. 12: e1005008

Campbell AI, Ebbens SJ, Illien P, and Golestanian R, 2019. Experimental observation of flow fields around active Janus spheres, Nat. Commun. 10: 3952

Čapek D and Müller P, 2019. Positional information and tissue scaling during development and regeneration, Development 146: 177709

Case LB and Waterman CM, 2015. Integration of actin dynamics and cell adhesion by a three-dimensional, mechanosensitive molecular clutch, Nat. Cell Biol. 17: 955–963

Cates ME, Wittmer JP, Bouchaud J-P, and Claudin P, 1998. Jamming, Force Chains, and Fragile Matter, Phys. Rev. Lett. 81: 1841–1844

Cates ME and Tailleur J, 2015. Motility-induced phase separation, Annu. Rev. Condens. Matter Phys. 6: 219–244

Chandrasekaran K and Bose B, 2019. Percolation in a reduced equilibrium model of planar cell polarity, Phys. Rev. E 100: 032408

Chanet S and Martin AC, 2014. Mechanical Force Sensing in Tissues, Prog. Mol. Biol. Transl. Sci. 126: 317–352

Chapman S and Cowling TG, 1970. *The Mathematical Theory of Non-Uniform Gases*, Cambridge University Press

Charras GT, Yarrow JC, Horton MA, Mahadevan L, and Mitchison TJ, 2005. Non-equilibration of hydrostatic pressure in blebbing cells, Nature 435: 365–369

Charras G and Paluch E, 2008. Blebs lead the way: How to migrate without lamellipodia. Nat. Rev. Molec. Cell Biol. 9: 730–736

Chaté H, Ginelli F, Grégoire F, Peruani F, and Raynaud F, 2008. Modeling collective motion: variations on the Vicsek model, Eur. Phys. J. B 64: 451–456

Chaté H, 2020. Dry Aligning Dilute Active Matter, Annu. Rev. Condens. Matter Phys. 11: 189–212

Chaudhuri O, Parekh SH, and Fletcher DA, 2007. Reversible stress softening of actin networks, 2007. Nature 445: 295–298

Chen H-Y, Hsiao Y-T, Liu S-C, Hsu T, Woon W-Y, and I L, 2018. Enhancing Cancer Cell Collective Motion and Speeding up Confluent Endothelial Dynamics through Cancer Cell Invasion and Aggregation, Phys. Rev. Lett. 121: 018101

Christiansen JJ and Rajasekaran AK, 2006. Reassessing Epithelial to Mesenchymal Transition as a Prerequisite for Carcinoma Invasion and Metastasis, Cancer Res. 66: 8319–8326

Cicuta P, 2020. The use of biophysical approaches to understand ciliary beating, Biochem. Soc. Trans. 48: 221–229

Cisneros LH, Cortez R, Dombrowski C, Goldstein RE, and Kessler JO, 2007. Fluid dynamics of self-propelled microorganisms, from individuals to concentrated populations, Exp. Fluids 43: 737–753

Classen AK, Anderson KI, Marois E, and Eaton S, 2005. Hexagonal packing of *Drosophila* wing epithelial cells by the planar cell polarity pathway. Dev. Cell 9: 805–817

Cogan NG and Guy RD, 2010. Multiphase flow models of biogels from crawling cells to bacterial biofilms. HFSP J. 4: 11–25

Cohen I, Brenner MP, Eggers J, and Nagel SR, 1999. Two Fluid Drop Snap-Off Problem: Experiments and Theory, Phys. Rev. Lett. 83: 1147–1150

Colin R, Drescher K, and Sourjik V, 2019. Chemotactic behaviour of *Escherichia coli* at high cell density, Nat. Commun. 10: 5329

Coussy O, 2004. *Poromechanics*, 2nd ed. Chichester, England; Wiley, Hoboken NJ

Cox RG, 1970. The motion of long slender bodies in a viscous fluid, J. Fluid Mech. 44: 791–810

Crick F, 1970. Diffusion in Embryogenesis, Nature 225: 420–422

Czajkowski M, Sussman DM, Marchetti MC, and Manning ML, 2019. Glassy dynamics in models of confluent tissue with mitosis and apoptosis, Soft Matter 15: 9133–9149

Dahmann C, Oates AC, and Brand M, 2011. Boundary formation and maintenance in tissue development, Nat. Rev. Genetics 12: 43–55

Danuser G, Allard J, and Mogilner A, 2013. Mathematical Modeling of Eukaryotic Cell Migration: Insights Beyond Experiments, Annu. Rev. Cell and Develop. Biol. 29: 501–528

Darnton NC and Berg HC, 2007. Force-Extension Measurements on Bacterial Flagella: Triggering Polymorphic Transformations, Biophys. J. 92: 2230–2236

Das R, Kumar M, and Mishra S, 2017. Order-disorder transition in active nematic: A lattice model study, Sci. Rep. 7: 7080

Das S, Garg A, Campbell AI, Howse J, Sen A, Velegol D, Golestanian R, and Ebbens SJ, 2015. Boundaries can steer active Janus spheres, Nat. Commun. 6: 8999

Dearden PK, 2019. Hourglass or Twisted Ribbon? in Evo-Devo: Non-model Species in Cell and Developmental Biology, eds Tworzydlo W and Bilinski SM, Springer, Cham

Deforet M, Hakim V, Yevick HG, Duclos G, and Silberzan P, 2013. Emergence of collective modes and tri-dimensional structures from epithelial confinement, Nat. Commun. 5: 3747

de Gennes PG, 1975. One type of nematic polymers, C. R. Acad. Sci. Ser. B **281**, 101–103

de Gennes PG and Prost J, 1995. *The Physics of Liquid Crystals*, Oxford University Press

Dell'Arciprete D, Blow ML, Brown AT, Farrell FDC, Lintuvuori JS, McVey AF, Marenduzzo D, and Poon WCK, 2018. A growing bacterial colony in two dimensions as an active nematic, Nat. Commun. 9: 4190

Dequéant M-L and Pourquié O, 2008. Segmental patterning of the vertebrate embryonic axis, Nat. Rev. Genetics 9: 370–381

Dembo M and Harlow F, 1986. Cell motion, contractile networks, and the physics of interpenetrating reactive flow. Biophys. J. 50: 109–121

Dickinson RB and Purich DL, 2002. Clamped-Filament Elongation Model for Actin-Based Motors, Biophys. J. 82: 605–617

Dietrich K, Renggli D, Zanini M, Volpe G, Buttinoni I, and Isa L, 2017. Two-dimensional nature of the active Brownian motion of catalytic microswimmers at solid and liquid interfaces, New J. Phys. 19: 065008

Dogterom M and Koenderink GH, 2019, Actin-microtubule crosstalk in cell biology, Nat. Rev. Molec. Cell Biol. 20: 39–54

Deglincerti A, Croft GF, Pietila LN, Zernicka-Goetz M, Siggia ED, and Ali H. Brivanlou AH, 2016. Self-organization of the in vitro attached human embryo, Nature 533: 251–254

Doostmohammadi A, Ignés-Mullol J, Yeomans JM, and Sagués F, 2018. Active Nematics, Nat. Commun. 9: 3246

Douady S and Couder Y, 1992. Phyllotaxis as a physical self-organized process. Phys. Rev. Lett. 68: 2098–2101

Dreher A, Aranson IS, and Kruse K, 2014. Spiral actin-polymerization waves can generate amoeboidal cell crawling, New J. Phys. 16: 055007

Duclos G, Erlenkämper C, Joanny J-F, and Silberzan P, 2017. Topological defects in confined populations of spindle-shaped cells, Nat. Phys. 13: 58–62

Ducuing A and Vincent S, 2016. The actin cable is dispensable in directing dorsal closure dynamics but neutralizes mechanical stress to prevent scarring in the *Drosophila* embryo, Nat. Cell Biol. 18: 1149–1160

Durney CH, Harris TJ, and James J. Feng, 2018. Dynamics of PAR Proteins Explain the Oscillation and Ratcheting Mechanisms in Dorsal Closure, Biophys. J. 115: 2230–2241

Dyer JRG, Johansson A, Helbing D, Couzin ID, and Krause J, 2009. Leadership, consensus decision making and collective behaviour in humans, Phil. Trans. R. Soc. B 364: 781–789

Echarri A, ..., Del Pozo MA (16 coauthors), 2019. An Abl-FBP17 mechanosensing system couples local plasma membrane curvature and stress fiber remodeling during mechanoadaptation, Nat. Commun. 10: 5828

Einstein A, 1905. On the Movement of Small Particles Suspended in Stationary Liquids Required by the Molecular-Kinetic Theory of Heat, Ann. Phys. 17: 549–560

Elgeti J, Winkler RG, and Gompper G, 2015. Physics of microswimmers – single particle motion and collective behavior: a review, Rep. Prog. Phys. 78: 056601

Engelmann TW, 1883. Ein Beitrag zur vergleichenden Physiologie des Licht- und Farbensinnes, Pflügers Arch. 30: 95–124

Ennomani H, Letort G, Guerin C, Martiel JL, Cao W, Nédélec F, De la Cruz EM, Thery M, and Blanchoin L, 2016. Architecture and Connectivity Govern Actin Network Contractility, Curr. Biol. 26: 616–626

Erol O, Aishwarya Pantula A, Liu W, and Gracias DH, 2019. Transformer Hydrogels: A Review, Adv. Mater. Technol. 4: 1900043

Etournay R, Merkel M, Popović M, Brandl H, Dye NA, Aigouy B, Salbreux G, Eaton S, and Jülicher F, 2016. TissueMiner: A multiscale analysis toolkit to quantify how cellular processes create tissue dynamics, eLife 5: e14334

Faraday M, 1831. On a peculiar class of acoustical figures; and on certain forms assumed by groups of particles upon vibrating elastic surfaces, Phil. Trans. Roy. Soc. London 121: 299–318

Farge E, 2003. Mechanical Induction of Twist in the *Drosophila* Foregut/Stomodeal Primordium, Curr. Biol. 13: 1365–1377

Farge E, 2011. Mechanotransduction in Development, Curr. Top. Develop. Biol. 95: 243–265

Farhadifar R, Röper JC, Aigouy B, Eaton S, and Jülicher F, 2007. The influence of cell mechanics, cell–cell interactions, and proliferation on epithelial packing. Curr. Biol. 17: 2095

Feinerman O, Pinkoviezky I, Gelblum A, Fonio E, and Gov NS, 2018. The physics of cooperative transport in groups of ants, Nat. Phys. 14: 683–692

Feng X, Deng Y, and Blote HW, 2008. Percolation transitions in two dimensions. Phys. Rev. E. 78, 031136

Finkelmann H, Ringsdorf H, and Wendorff JH, 1978. Model considerations and examples of enantiotropic liquid crystalline polymers, Makromol. Chem. 179: 273–276

Fletcher DA and Mullins D, 2010. Cell mechanics and the cytoskeleton, Nature 463: 485–492

Forterre Y, Skotheim JM, Dumais J, and Mahadevan L, 2005. How the Venus flytrap snaps, Nature 433: 421–425

Fried P and Iber D, 2014. Dynamic scaling of morphogen gradients on growing domains, Nat. Commun. 5: 5077

Friedrich B, 2016. Hydrodynamic synchronization of flagellar oscillators, Eur. Phys. J. Spec. Top. 225: 2353–2368

Frohnhöfer and Nüsslein-Volhard, 1986. Organization of anterior pattern in the *Drosophila* embryo by the maternal gene bicoid, Nature 324: 120–125

Fu X, Kato S, Long J, Mattingly HH, He C, Vural DC, Zucker SW, and Emonet T, Spatial self-organization resolves conflicts between individuality and collective migration, Nat. Commun. 9: 2177

Fukuda J-i, 1998. Effect of hydrodynamic flow on kinetics of nematic-isotropic transition in liquid crystals, Eur. Phys. J. B 1: 173–177

Fürthauer, S, Strempel M, Grill SW, and Jülicher F, 2012. Active chiral fluids, Eur. Phys. J. E 35: 89

Ge F and Zhao Y, 2019. Microstructured Actuation of Liquid Crystal Polymer Networks, Adv. Func. Mater. 30: 1901890

Geiger B, Spatz JP, and Bershadsky AD, 2009. Environmental sensing through focal adhesions, Nat. Rev. Molec. Cell Biol. 10: 21–33

Genkin MM, Sokolov A, Lavrentovich OD, and Aranson IS, 2017. Topological defects in a living nematic ensnare swimming bacteria, Phys. Rev. X 7: 011029

Geyer D, Martin D, Tailleur J, and Bartolo D, 2019. Freezing a Flock: Motility-Induced Phase Separation in Polar Active Liquids, Phys. Rev. X 9: 031043

Geyer VF, Jülicher F, Howard J, and Friedrich BM, 2013. Cell-body rocking is a dominant mechanism for flagellar synchronization in a swimming alga, Proc. Natl. Acad. Sci. U.S.A. 110: 18058–18063

Ginelli F and Chaté H, 2010. Relevance of Metric-Free Interactions in Flocking Phenomena, Phys. Rev. Lett. 105: 168103

Ginelli F, 2016. The Physics of the Vicsek model, Eur. Phys. J. Spec. Top. 225: 2099–2117

Ginot F, Theurkauff I, Detcheverry F, Ybert C, and Cottin-Bizonne C, 2018. Aggregation-fragmentation and individual dynamics of active clusters, Nat. Commun. 9: 696

Giomi L, Bowick MJ, Ma X, Marchetti MC, 2013. Defect annihilation and proliferation in active nematics, Phys. Rev. Lett. 110, 228101

Giomi L and DeSimone A, 2014. Spontaneous Division and Motility in Active Nematic Droplets, Phys. Rev. Lett. 112: 147802

Gladman AS, Matsumoto EA, Nuzzo RG, Mahadevan L, Lewis JA, 2016. Biomimetic 4D printing, Nat. Mater. 15: 413–418

Goldenfeld N, 1992. *Lectures On Phase Transitions And The Renormalization Group*, Taylor & Francis, Boca Raton, FL

Goldstein RE and van de Meent J-W, 2015. A physical perspective on cytoplasmic streaming, Interface Focus 5: 20150030

Goldstein RE, 2016. Fluid dynamics at the scale of the cell, J. Fluid Mech. 807: 1–39

Goldstein RE, Lauga E, Pesci AI, and Proctor MRE, 2016. Elastohydrodynamic synchronization of adjacent beating flagella, Phys. Rev. Fluids 1: 073201

Goley ED and Welch MD, 2006. The ARP2/3 complex: an actin nucleator comes of age, Nat. Rev. Molec. Cell Biol. 7: 713–726

Golovin AA and Ryazantsev YS, 1990. Drift of a reacting droplet due to the chemoconcentration capillary effect, Fluid Dynamics 25: 370–378

Golovin AA, Nir A, and Pismen LM, 1995. Spontaneous Motion of Two Droplets Caused by Mass Transfer, Ind. Eng. Chem. Res. 34: 3278–3288

Gómez-Gálvez P, . . . , Escudero LM (16 coauthors), 2018. Scutoids are a geometrical solution to three-dimensional packing of epithelia, Nat. Commun. 9: 2960

Gomez-Solano JR, Samin S , Lozano C, Ruedas-Batuecas P, van Roij R and Bechinger C, 2017. Tuning the motility and directionality of self-propelled colloids, Sci. Rep. 7: 14891

Gompper G et al (37 coauthors), 2020. The 2020 motile active matter roadmap, J. Phys. Condens. Matter 32: 193001

Gong A, Rode S., Kaupp UB, Gompper G, Elgeti J, Friedrich BM, and Alvarez L, 2019. The steering gaits of sperm, Trans. Roy. Soc. B 375: 20190149

Goodwin BC and Cohen HM, 1969. A phase-shift model for the spatial and temporal organization of developing systems. J. Theor. Biol. 25: 49–107

Goudarzi M, Boquet-Pujadas A, Olivo-Marin J-C, and Raz E, 2019. Fluid dynamics during bleb formation in migrating cells *in vivo*, PLoS One 14: e0212699

Graner F and Glazier JA, 1992. Simulation of biological cell sorting using a 2-dimensional extended Potts model, Phys. Rev. Lett. 69: 2013Đ2017

Grégoire G, Chaté H, and Tu Y, 2003. Moving and staying together without a leader, Phys. D 181: 157–170

Gregor T, Bialek W, de Ruyter van Steveninck RR, Tank DW, and Wieschaus EF, 2005. Diffusion and scaling during early embryonic pattern formation, Proc. Natl Acad. Sci. USA 102: 18403–18407

Griniasty I, Aharoni H, and Efrati E, 2019. Curved Geometries from Planar Director Fields: Solving the Two-Dimensional Inverse Problem, Phys. Rev Lett. 123: 127801

Großmann R, Romanczuk P, Bär M, and Schimansky-Geier L, 2015. Pattern formation in active particle systems due to competing alignment interactions, Eur. Phys. J. Spec. Top. 224: 1325–1347

Gruler H, Schienbein M, Franke K, and de Boisfleury-Chevance A, 1995. Migrating Cells: Living Liquid Crystals, Mol. Cryst. Liq. Cryst. 260: 565–574

Guo Y, Jiang M, Peng C, Sun K, Yaroshchuk O, Lavrentovich OD, and Wei Q-H, 2016. High-Resolution and High-Throughput Plasmonic Photopatterning of Complex Molecular Orientations in Liquid Crystals, Adv. Mater. 28: 2353

Hakim V and Silberzan P, 2017. Collective cell migration: a physics perspective, Rep. Prog. Phys. 80: 076601

Harrington MJ, Razghandi K, Ditsch F, Guiducci L, Rueggeberg M, Dunlop JWC, Fratzl P, Neinhuis C, Burgert I, 2011. Origami-like unfolding of hydro-actuated ice plant seed capsules, Nat. Commun. 2: 337

Harrison CJ, Roeder AHK, Meyerowitz EM, and Langdale JA, 2009. Local Cues and Asymmetric CellDivisions Underpin Body Plan Transitions in the Moss *Physcomitrella patens*, Curr. Biol. 19: 461–471

Hartmann R, Singh PK, Pearce P, Mok R, Song B, Díaz-Pascual F, Dunkel J, and Drescher K, 2018. Emergence of three-dimensional order and structure in growing biofilms, Nat. Phys. 15: 251–256

Harvey CW, Alber M, Tsimring LS, and Aranson IS, 2013. Continuum modeling of myxobacteria clustering, New J. Phys. 15: 035029

Haupt M and Hauser MJB, 2020. Effective mixing due to oscillatory laminar flow in tubular networks of plasmodial slime moulds, New J. Phys. 22: 053007

Helbing D and Molnár P, 1995. Social force model for pedestrian dynamics, Phys. Rev. E 51: 4282–4286

Helbing D, Farkas I, and Vicsek T, 2000. Simulating dynamical features of escape panic, Nature 407: 487–490

Herant M and Dembo M, 2010. Form and function in cell motility: from fibroblasts to keratocytes, Biophys. J. 98: 1408–1417

Herrera-Perez R and Kasza KE, 2019. Manipulating the Patterns of Mechanical Forces That Shape Multicellular Tissues, Physiology 34: 381–391

Hoffman BD, Grashoff C, and Schwartz MA, 2011. Dynamic molecular processes mediate cellular mechanotransduction, Nature 475: 316–323

Hofhuis H, . . . , Hay A (15 coauthors), 2016. Morphomechanical Innovation Drives Explosive Seed Dispersal, Cell 166: 222–233

Hofmeister WFB, 1868. *Allgemeine Morphologie der Gewächse*, Handbuch der Physiologischen Botanik 1: 394–553, Engelmann, Leipzig

Hohenberg PC and B. I. Halperin PI, 1977. Theory of Dynamic Critical Phenomena, Rev. Mod. Phys. 49: 435 –479

Hohng S, Zhou R, Nahas MK, Yu J, Schulten K, Lilley DMJ, and Ha T, 2007. Fluorescence-force spectroscopy maps two-dimensional reaction landscape of the Holliday junction, Science 318: 279–283

Holmes WR and Edelstein-Keshet L, 2012. A comparison of computational models for eukaryotic cell shape and motility, PLoS Comput. Biol. 8: e1002793

Holtfreter JF, 1988. A New Look at Spemann's Organizer, Developmental Biology, ed. Browder LW, 5: 127–150, Plenum Press, New York

Howard J, 2001. *Mechanics of Motor Proteins and the Cytoskeleton*, Sinauer Associates, Sunderland MA

Hu J, Yang M, Gompper G, and Winkler RG, 2015. Modelling the mechanics and hydrodynamics of swimming *E. coli*, Soft Matter, 11: 7867–7876

Huang C, Lv J-a, Tian X, Wang Y, Yu Y, and Jie Liu J, 2015. Miniaturized Swimming Soft Robot with Complex Movement Actuated and Controlled by Remote Light Signals, Sci. Rep. 5: 17414

Ibrahim Y and Liverpool TB, 2016. How walls affect the dynamics of self-phoretic microswimmers, Eur. Phys. J. Spec. Top. 225: 1843–1874

Ideses Y, Brill-Karniely Y, Haviv L, Ben-Shaul A, and Bernheim-Groswasser A, 2008. Arp2/3 Branched Actin Network Mediates Filopodia-Like Bundles Formation In Vitro, PLos One 3: e3297

Ideses Y, Erukhimovitch V, Brand R, Jourdain D, Salmeron Hernandez J, Gabinet UR, Safran SA, Kruse K, and Bernheim-Groswasser A, 2018. Spontaneous buckling of contractile poroelastic actomyosin sheets, Nat. Commun. 9: 2461

Ihle T, 2011. Kinetic Theory of Flocking: Derivation of Hydrodynamic Equations, Phys. Rev. E 83: 030901

Ihle T, 2014. Towards a quantitative kinetic theory of polar active matter, Eur. Phys. J. Spec. Top., 223: 1293–1314

Ihle T and Chou Y-L, 2014. Discussion on Ohta et al, "Traveling bands in self-propelled soft particles", Eur. Phys. J. Spec. Top. 223: 1409–1415

Illukkumbura R, Bland T, Goehring NW, 2020. Patterning and polarization of cells by intracellular flows' Curr. Opinion Cell Biol. 62: 123–134

Ingber DE, 1997. Tensegrity: The architectural basis of cellular mechanotransduction, Annu. Rev. Physiol. 59: 575–599

Ingber DE, 2018. From mechanobiology to developmentally inspired engineering, Phil. Trans. R. Soc. B 373: 20170323

Ioannou F, Dawi MA, Tetley RJ, Mao Y, and Muñoz JJ, 2020. Development of a New 3D Hybrid Model for Epithelia Morphogenesis, Front. Bioeng. Biotechnol. 8: 405

Ionov L, 2013. Biomimetic Hydrogel-Based Actuating Systems, Adv. Funct. Mater. 23, 4555–4570

Irvine K and Wieschaus E, 1994. Cell intercalation during *Drosophila* germ-band extension and its regulation by pair-rule segmentation genes, Development 120: 827–841

Ishikawa T and Pedley TJ, 2008. Coherent Structures in Monolayers of Swimming Particles, Phys. Rev. Lett. 100: 088103

Jemielita M, Wingreen NS, and Bassler BL, 2018. Quorum sensing controls *Vibrio cholerae* multicellular aggregate formation, eLife 7: e42057

Jikeli JF, . . . , Kaupp UB (10 coauthors), 2015. Sperm navigation along helical paths in 3D chemoattractant landscapes, Nat. Commun. 6: 7985

Joanny, J-F, Jülicher F, Kruse K, and Prost J, 2007. Hydrodynamic theory for multi-component active polar gels, New J. Phys. 9: 422

Jülicher F and Prost J, 2009. Generic theory of colloidal transport, Eur. Phys. J. E 29: 27–36

Kaiser A, Sokolov A, Aranson IS, and Löwen H, 2015. Motion of two micro-wedges in a turbulent bacterial bath, Eur. Phys. J. Spec. Top. 224: 1275–1286

Kaiser A, Snezhko A, and Aranson IS, 2017. Flocking ferromagnetic colloids, Sci. Adv. 3: e1601469

Karsenti E, Nédélec F, and Surrey F, 2006. Modelling microtubule patterns, Nat. Cell Biol. 8: 1204–1211

Katyal N, Dey S, and Das D, 2020. Coarsening dynamics in the Vicsek model of active matter, Eur. Phys. J. E 43: 10

Kaupp UB and Alvarez L, 2016. Sperm as microswimmers – navigation and sensing at the physical limit, Eur. Phys. J. Spec. Top. 225: 2119–2139

Keller EF and Segel LA, 1971. Traveling Bands of Chemotactic Bacteria: A Theoretical Analysis, J. Theor. Biol. 30: 235–248

Khoromskaia D and Alexander GP, 2017. Vortex formation and dynamics of defects in active nematic shells, New J. Phys. 19: 103043

Kim JH, . . . , Fredberg JJ (13 coauthors), 2013. Propulsion and navigation within the advancing monolayer sheet, Nat. Mater. 12: 856–863

Kléman M and Lavrentovich OD, 2003. *Soft Matter Physics: An Introduction*, Springer-Verlag, New York

Klughammer N, Bischof J, Schnellbächer ND, Callegari A, Péter Lénáńrt P, and Schwarz US, 2018. Cytoplasmic flows in starfish oocytes are fully determined by cortical contractions, PLoS Comput. Biol. 14: e1006588

Koenderink GH and Paluch EK, 2018. Architecture shapes contractility in actomyosin networks, Curr. Opinion Cell Biol. 50: 79–85

Komareji M and Bouffanais R, Resilience and Controllability of Dynamic Collective Behaviors, PLoS One 8: e82578

Köpf MH and Pismen LM, 2013a. Non-equilibrium patterns in polarizable active layers, Physica D 259: 48–54

Köpf MH and Pismen LM, 2013b. A continuum model of epithelial spreading, Soft Matter 9: 3727–3734

Köpf MH and Pismen LM, 2013c. Phase separation and disorder in doped nematic elastomers, Eur. Phys. J. E 36: 121

Kosterlitz JM and Thouless DJ, 1973. Ordering, metastability and phase transitions in two-dimensional systems, J. Phys. C: Solid State Phys. 6: 1181–1203

Kourakis MJ and Smith WC, 2005. Did the first chordates organize without the organizer? Trends in Genetics 21: 506–510

Krens SFG and Heisenberg C-P, 2011. Cell Sorting in Development, Current Topics in Developmental Biology, 95: 189–213

Kroy K, Chakraborty D, and Cichos F, 2016. Hot microswimmers, Eur. Phys. J. Spec. Top. 225: 2207–2225

Kruse K, Joanny, J-F, Jülicher F, Prost J and Sekimoto K, 2004. Asters, vortices, and rotating spirals in active gels of polar filaments, Phys. Rev. Lett. 92: 078101

Kruse K, Joanny, J-F, Jülicher F, Prost J and Sekimoto K, 2005. Generic theory of active polar gels: a paradigm for cytoskeletal dynamics, Eur. Phys. J. E 16: 5–16

Kruse K, 2016. Cell Crawling Driven by Spontaneous Actin Polymerization Waves, in *Physical Models of Cell Motility*, ed. Aranson IS, pp. 69–93, Springer, Cham

Kuksenok O, Yashin VV, and Balazs AC, 2013. Three-dimensional model for chemoresponsive polymer gels undergoing the Belousov–Zhabotinsky reaction, Phys. Rev. E 78: 041406

Kuksenok O and Balazs AC, 2013. Modeling the Photoinduced Reconfiguration and Directed Motion of Polymer Gels, Adv. Funct. Mater. 23: 4601–4610

Kuron M, Kreissl P, and Holm C, 2018. Toward Understanding of Self-Electrophoretic Propulsion under Realistic Conditions: From Bulk Reactions to Confinement Effects, Acc. Chem. Res. 51: 2998–3005

Ladoux B and Mège R-M, 2017. Nat. Rev. Molec. Cell Biol. 18: 743–757

Lämmermann T, Bader BL, Monkley SJ, Worbs T, Wedlich-Söldner R, Hirsch K, Keller M, Förster R, Critchley DR, Fässler R, and Sixt M, 2008. Rapid leukocyte migration by integrin-independent flowing and squeezing, Nature 453: 51–55

Landau LD, Lifshits EM, and Pitaevskii LP, 1980. *Statistical Physics*, Pergamon, Oxford

Landau LD and Lifshits EM, 1986. *Theory of Elasticity*, Pergamon, Oxford

Lång E, ..., Bøe SO (13 coauthors), 2018. Coordinated collective migration and asymmetric cell division in confluent human keratinocytes without wounding, Nat. Commun. 9: 3665

Lauga E and Powers TR, 2009. The hydrodynamics of swimming microorganisms Rep. Prog. Phys. 72: 096601

Lauga E, 2016. Bacterial Hydrodynamics, Annu. Rev. Fluid Mech. 48: 105–130

Laurent J, Blin G, Chatelain F, Vanneaux V, Fuchs A, Larghero J and Théry M, 2017. Convergence of microengineering and cellular self-organization towards functional tissue manufacturing, Nat. Biomed. Eng. 1: 939–956

Leaver M, Domínguez-Cuevas P, Coxhead JM, Daniel RA, and Errington J, 2009. Life without a wall or division machine in *Bacillus subtilis*, Nature 457: 849–853

Lecuit T and Lenne P-F, 2007. Cell surface mechanics and the control of cell shape, tissue patterns and morphogenesis, Nat. Rev. Molec. Cell Biol. 8: 633–644

Lee JM, 2013. *The Actin Cytoskeleton and the Regulation of Cell Migration*, Morgan & Claypool

Lee P and Wolgemuth CW, 2011. Crawling cells can close wounds without purse strings or signaling. PLoS Comput. Biol. 7: e1002007

Leronni A, Bardella L, Dorfmann L, Pietak A, and Levin M, 2020. On the coupling of mechanics with bioelectricity and its role in morphogenesis, J. R. Soc. Interface 17: 20200177

Levich VG, 1962. *Physicochemical hydrodynamics*, Prentice Hall, Englewood Cliffs NJ

Lewis OL, Zhang S, Guy RD, and del Álamo J, 2014. Coordination of contractility, adhesion and flow in migrating *Physarum* amoebae, J. R. Soc. Interface 12: 20141359

Leyser O and Domagalska MA, 2011. Signal integration in the control of shoot branching, Nat. Rev. Molec. Cell Biol. 12: 211–221

Li X, Das A, and Bi D, 2019. Mechanical Heterogeneity in Tissues Promotes Rigidity and Controls Cellular Invasion, Phys. Rev. Lett. 123: 058101

Lighthill J, 1975. *Mathematical Biofluiddynamics*, SIAM, Philadelphia

Lin H, Yue L, and Spradling AC, 1994. The *Drosophila* fusome, a germline-specific organelle, contains membrane skeletal proteins and functions in cyst formation, Development 120: 947–956

Linardić M and Braybrook SA, 2017. Towards an understanding of spiral patterning in the *Sargassum muticum* shoot apex, Sci. Rep. 7: 13887

Liotta LA, Steeg PS, and Stetler-Stevenson WG, 1991. Cancer metastasis and angiogenesis: An imbalance of positive and negative regulation. Cell 84: 327–336

Liu AJ and Nagel SR, 1998. Jamming is not just cool any more, Nature 396: 21–22

Liu J, Prindle A, Humphries J, Gabalda-Sagarra M, Asally M, Lee DD, Ly S, Garcia-Ojalvo J, and Süel GM, 2015. Metabolic co-dependence gives rise to collective oscillations within biofilms, Nature 523: 550–554

Liu J, 2020. Intracellular Force Measurements in Live Cells With Förster Resonance Energy Transfer-Based Molecular Tension Sensors, in Mechanobiology From Molecular Sensing to Disease, ed. Niebur GL, pp. 161–171, Elsevier

Löber J, Ziebert F, and Aranson IS, 2014. Collisions of deformable cells lead to collective migration, Sci. Rep. 5: 9172

Loewe B, Chiang M, Marenduzzo D, and Marchetti MC, 2020. Solid-Liquid Transition of Deformable and Overlapping Active Particles, Phys. Rev. Lett. 125: 038003

López HM, Gachelin J, Douarche C, Auradou H, and Clément E, 2015. Turning Bacteria Suspensions into Superfluids, Phys. Rev. Lett. 115: 028301

Lovelock J, 1979. *Gaia: A New Look at Life on Earth*, Oxford University Press

Lu H, Hong Y, Yang Y, Yang Z, and Shen Y, 2020. Battery-Less Soft Millirobot That Can Move, Sense, and Communicate Remotely by Coupling the Magnetic and Piezoelectric Effects, Adv. Sci. 2000069

MacKintosh FC, Käs J, and Janmey PA, 1995. Elasticity of Semiflexible Biopolymer Networks, Phys. Rev. Lett. 75: 4425–4428

Malinverno C, . . . , Scita G (23 coauthors), 2017. Endocytic reawakening of motility in jammed epithelia, Nat. Mater. 16: 587–596

Marchetti MC, Joanny J-F, Ramaswamy S, Liverpool TB, Prost J, Rao M, and Simha RA, 2013. Hydrodynamics of soft active matter, Rev. Mod. Phys. 85: 1143–1189

Marenduzzo D, 2016. An introduction to the statistical physics of active matter: motility-induced phase separation and the "generic instability" of active gels, Eur. Phys. J. Spec. Top. 225: 2065–2077

Marth W, Praetorius S, and Voigt A, 2015. A mechanism for cell motility by active polar gels. J. R. Soc. Interface 12: 20150161

Martin AC, Kaschube M, and Wieschaus EF, 2009. Pulsed contractions of an actinÐmyosin network drive apical constriction, Nature 457: 495–499

Martin P and Lewis J, 1992. Actin cables and epidermal movement in embryonic wound healing, Nature 360: 179–183

Mason FM, Tworoger M, and Martin AC, 2013. Apical domain polarization localizes actin-myosin activity to drive ratchet-like apical constriction, Nat.Cell Biol. 15: 926–936

Matsumoto K, Takagi S, and Nakagaki T, 2008. Locomotive mechanism of *Physarum* plasmodia based on spatiotemporal analysis of protoplasmic streaming, Biophys. J. 94: 2492–2504

Meinhardt H, 1982. *Models of Biological Pattern Formation*, Academic Press, London

Meinhardt H, 1995. *The algorithmic beauty of sea shells*, Springer-Verlag, Berlin (4th ed., 2009)

Meredith CH, Moerman PG, Groenewold J, Chiu Y-J, Kegel WK, van Blaaderen A, and Zarzar LD, 2020. PredatorÐprey interactions between droplets driven by non-reciprocal oil exchange, Nat. Chem. 12: 1136–1142

Michelin S and Lauga E, 2015. Autophoretic locomotion from geometric asymmetry, Eur. Phys. J. E 38: 7

Miklius MP and Hilgenfeldt S, 2011. Epithelial tissue statistics: Eliminating bias reveals morphological and morphogenetic features, Eur. Phys. J. E 34: 50

Misra M, Audoly B, and Shvartsman SY, 2017. Complex structures from patterned cell sheets. Phil. Trans. R. Soc. B 372: 20150515

Moeendarbary E, Valon E, Fritzsche M, Harris AR, Moulding DA, Thrasher AJ, Stride E, Mahadevan L, and Charras GT, 2013. The cytoplasm of living cells behaves as a poroelastic material, Nat. Mater. 12: 253–261

Moerman PG, Moyses HW, van der Wee EB, Grier DG, van Blaaderen A, Kegel WK, Groenewold J, and Brujic J, 2017. Solute-mediated interactions between active droplets, Phys. Rev. E 96: 032607

Mogilner A and Oster G, 2006. Cell motility driven by actin polymerization, Biophys. J. 71: 3030–3045

Mogilner A and Manhart A, 2018. Intracellular Fluid Mechanics: Coupling Cytoplasmic Flow with Active Cytoskeletal Gel, Annu. Rev. Fluid Mech. 50: 347–370

Moran JL and Posner JD, 2017. Phoretic Self-Propulsion, Annu. Rev. Fluid Mech. 49: 511–540

Morozov KI and Pismen LM, 2011. Cytoskeleton fluidization versus resolidification: Prestress effect, Phys. Rev. E 83: 051920

Mota C, Camarero-Espinosa S, Baker MB, Wieringa P, and Moroni L, 2020. Bioprinting: From Tissue and Organ Development to in Vitro Models, Chem. Rev. 120: 10547–10607

Mueller J, . . . , Small, JV (13 coauthors), Electron Tomography and Simulation of Baculovirus Actin Comet Tails Support a Tethered Filament Model of Pathogen Propulsion, PLoS Biology, 12: e1001765

Müller WA, 1997. *Developmental Biology*, Springer-Verlag, New York

Mullins WW and Sekerka RF, 1963. Morphological stability of a particle growing by diffusion or heat flow, J. Appl. Phys. 34: 323–329

Murray JD, 1980. A pattern formation mechanism and its application to mammalian coat markings, Lect. Notes in Biomathematics, 39: 360–399, Springer-Verlag, Berlin

Murray JD, 1989. *Mathematical Biology*, Springer, Berlin (3rd ed., 2002/2003)

Murrell M, Oakes PW, Lenz M, and Gardel ML, 2015. Forcing cells into shape: the mechanics of actomyosin contractility, Nat. Rev. Molec. Cell Biol. 16: 486–498

Nadell CD, Foster KR, and Xavier JB, 2010. Emergence of spatial structure in cell groups and the evolution of cooperation, PLoS Comput. Biol. 6: e1000716

Nagai K, Sumino Y, and Yoshikawa K, 2007. Regular self-motion of a liquid droplet powered by the chemical Marangoni effect, Coll. Surf. B: Biointerfaces 56: 197–200

Nagai T, Kawasaki K, and Nakamura K, 1988. Vertex Dynamics of Two-Dimensional Cellular Patterns, J. Phys. Soc. Jpn. 57: 2221–2224

Nagai T and Honda H, 2001. A dynamic cell model for the formation of epithelial tissues, Phil. Mag. 81: 699–719

Nagy M, Akos Z, Biro D, and Vicsek T, 2010. Hierarchical group dynamics in pigeon flocks. Nature 464: 890–894

Narayan V, Menon N, and Ramaswamy S, 2006. Nonequilibrium steady states in a vibrated-rod monolayer: tetratic, nematic, and smectic correlations, J. Stat. Mech. Theor. Exp. P01005

Narayan V, Ramaswamy S, and Menon N, 2007. Long-Lived Giant Number Fluctuations in a Swarming Granular Nematic, Science 317: 105

Nasouri B and Golestanian R, 2020. Exact Phoretic Interaction of Two Chemically Active Particles, Phys. Rev. Lett. 124: 168003

Nédélec F, Surrey T, Maggs AC, and Leibler S, 1997. Self-organization of microtubules and motors, Nature 389: 305–308

Needleman D and Dogic Z, 2017. Active matter at the interface between materials science and cell biology, Nature Rev. Mater. 2: 17048

Nikolić DL, Boettiger AN, Bar-Sagi D, Carbeck JD, and Shvartsman SY, 2006. Role of boundary conditions in an experimental model of epithelial wound healing, Am. J. Physiol. 291: C68–C75

Nishiguchi D, Aranson IS, Snezhko A, and Sokolov A, 2018. Engineering bacterial vortex lattice via direct laser lithography, Nat. Commun. 9: 4486

Noll N, Streichan SJ, and Boris I. Shraiman BI, 2020. Method for Image-Based Inference of Internal Stress in Epithelial Tissues, Phys. Rev. X 10: 011072

Noor N, Shapira A, Edri R, Gal I, Wertheim L, and Dvir T, 2019. 3D Printing of Personalized Thick and Perfusable Cardiac Patches and Hearts, Adv. Sci. 6: 1900344

Notbohm J, Banerjee S, Utuje KJ, Gweon B, Jang H, Park Y, Shin J, Butler JP, Fredberg JJ, and Marchetti MC, 2016. Cellular contraction and polarization drive collective cellular motion, Biophys. J. 110: 2729–2738

Nüsslein-Volhard C and Wieschaus E, 1980. Mutations affecting segment number and polarity in *Drosophila*, Nature 287: 795–801

Noll N, Streichan SJ, and Shraiman BI, 2020. Variational Method for Image-Based Inference of Internal Stress in Epithelial Tissues, Phys. Rev. X 10: 011072

Nüsslein-Volhard C, 2006. *Coming to Life: How Genes Drive Development*, Kales Press, San Diego CA

Odell GM, Oster G, Alberch P, and Burnside B, 1981. The mechanical basis of morphogenesis. I. Epithelial folding and invagination. Dev. Biol. 85: 446–462

Ohm C, Brehmer M, and Zentel R, 2012. Applications of Liquid Crystalline Elastomers, Adv. Polym. Sci. 250: 49–94

Okuda S, Miura T, Inoue Y, Adachi T, and Eiraku M, 2018. Combining Turing and 3D vertex models reproduces autonomous multicellular morphogenesis with undulation, tubulation, and branching, Sci. Rep. 8: 2386

Oparin AI, 1924. *Proiskhozhdenie zhizni* (in Russian). Izd. Moskovskii Rabochii, Moscow. Transl. 1938, *The Origin of Life*, Macmillan, New York

Oscurato SL, Salvatore M, Maddalena P, and Ambrosio A, 2018. From nanoscopic to macroscopic photo-driven motion in azobenzene-containing materials, Nanophotonics 7: 1387–1422

Oster GF and Odell GM, 1984. The mechanochemistry of cytogels, Physica D 12: 333–350; Mechanics of Cytogels I: Oscillations in Physarum, Cell Motility 4: 469–503

Pachong SM and Müller-Nedebock KK, 2017. Active force maintains the stability of a contractile ring, Eur. Phys. J. E 40: 91

Palagi S, . . . , Fischer P (14 coauthors), 2016. Structured light enables biomimetic swimming and versatile locomotion of photoresponsive soft microrobots, Nat. Mater. 15: 647–653

Paluch EK, Nelson CM, Biais N, Fabry B, Moeller J, Pruitt BL, Wollnik C, Kudryasheva G, Rehfeldt F, and Federle W, 2015. Mechanotransduction: use the force(s), BMC Biology 13: 47

Papavassiliou D and Alexander GP, 2017. Exact solutions for hydrodynamic interactions of two squirming spheres, J. Fluid Mech. 813: 618–646

Park J-A, . . . , Fredberg JJ (27 coauthors), 2015. Unjamming and cell shape in the asthmatic airway epithelium, Nat. Mater. 14: 1040–1048

Parsons JT, Horwitz AR and Schwartz MA, 2010. Cell adhesion: integrating cytoskeletal dynamics and cellular tension, Nat. Rev. Molec. Cell Biol. 11: 633–643

Partridge JD, Ariel G, Schvartz O, Harshey RM, and Be'er A, 2018. The 3D architecture of a bacterial swarm has implications for antibiotic tolerance, Sci. Rep. 8: 15823

Patteson AE, Gopinath A, and Arratia PE, 2018. The propagation of active–passive interfaces in bacterial swarms, Nat. Commun. 9: 5373

Pearl EJ, Li J, and Green JBA, 2017. Cellular systems for epithelial invagination, Phil. Trans. R. Soc. B 372: 20150526

Peng C, Turiv T, Guo Y, Wei Q-H, and Lavrentovich OD, 2016. Command of active matter by topological defects and patterns, Science 354: 882–885

Peruani F, Deutsch A, and Bär M, 2006. Nonequilibrium clustering of self-propelled rods, Phys. Rev. E, 74: 030904(R)

Peruani F, Deutsch A, and Bär M, 2008. A mean-field theory for self-propelled particles interacting by velocity alignment mechanisms, Eur. Phys. J. Spec. Top. 157: 111–122

Peruani F, Schimansky-Geier L, and Bär M, 2010. Cluster dynamics and cluster size distributions in systems of self-propelled particles, Eur. Phys. J. Spec. Top. 191: 173–185

Peshkov A, Bertin E, Ginelli F, and Chaté H, 2014. Boltzmann-Ginzburg-Landau approach for continuous descriptions of generic Vicsek-like models, Eur. Phys. J. Spec. Top. 223: 1315–1344

Pieuchot L, . . . , Anselme K (16 coauthors), 2018. Curvotaxis directs cell migration through cell-scale curvature landscapes, Nat. Commun. 9: 3995

Pikovsky A, Rosenblum M, and Kurths J, 2001. *Synchronization. A universal concept in nonlinear sciences*, Cambridge University Press

Pilz da Cunha M, Debije MG, and Schenning APHJ, 2020. Bioinspired light-driven soft robots based on liquid crystal polymers, Chem. Soc. Rev. 49; 6568–6578

Pismen LM, 1999. *Vortices in nonlinear fields*, Oxford University Press

Pismen LM, 2006. *Patterns and Interfaces in Dissipative Dynamics*, Springer, Berlin

Pismen LM, 2013. Dynamics of defects in an active nematic layer, Phys. Rev. E 88: 050502(R)

Pismen LM and Simakov DSA, 2011. Genesis of two-dimensional patterns in cross-gradient fields, Phys. Rev. E 84: 061917

Pismen LM, 2014. Physicists probing active media: What is the measure of success? Eur. Phys. J. Spec. Top. 223: 1243–1246

Plygawko AT, Kan S, and Campbell K, 2020. Epithelial-mesenchymal plasticity: emerging parallels between tissue morphogenesis and cancer metastasis, Phil. Trans. R. Soc. B 375: 20200087

Pohl O and Stark H, 2014. Dynamic Clustering and Chemotactic Collapse of Self-Phoretic Active Particles, Phys. Rev. Lett. 112: 238303

Popescu MN, Uspal WE, Eskandari Z, and Dietrich S, 2018. Effective squirmer models for self-phoretic chemically active spherical colloids, Eur. Phys. J. E 41: 145

Popescu MN, 2020. Chemically Active Particles: From One to Few on the Way to Many, Langmuir 36: 6861–6870

Poujade M, Grasland-Mongrain E, Hertzog A, Jouanneau J, Chavrier P, Ladoux B, Buguin A, and Silberzan P, 2007. Collective migration of an epithelial monolayer in response to a model wound, Proc. Natl. Acad. Sci. USA 104: 15988–15993

Prost J, Jülicher F and Joanny J-F, 2015. Active gel physics, Nat. Phys. 11: 111–117

Purcell EM, 1977. Life at low Reynolds number, Am. J. Phys. 45: 3–11

Qiu T, Lee TC, Mark AG, Morozov KI, Munster R, Mierka O, Turek S, Leshansky AM, and Fischer P, 2014. Swimming by reciprocal motion at low Reynolds number, Nat. Commun. 5: 5119

Quincke G, 1896. Uber Rotationen im Constanten Electrischen Felde, Ann. Phys. Chem. 59: 417

Quinlan ME, 2016. Cytoplasmic Streaming in the *Drosophila* Oocyte, Annu. Rev. Cell Dev. Biol. 32: 173–195

Radszuweit M, Engel H, and Bär M, 2014. An Active Poroelastic Model for Mechanochemical Patterns in Protoplasmic Droplets of *Physarum polycephalum*, PLoS One 9: e99220

Rafiq NBM, . . . , Bershadsky AD (11 coauthors), 2019. A mechano-signalling network linking microtubules, myosin IIA filaments and integrin-based adhesions, Nat. Mater. 18: 638–649

Ramaswamy S and Simha RA, 2006. The mechanics of active matter: Broken-symmetry hydrodynamics of motile particles and granular layers, Solid State Comm. 139: 617–622

Reffay M, Parrini MC, Cochet-Escartin O, Ladoux B, Buguin A, Coscoy S, Amblard F, Camonis J, and Silberzan P, 2014. Interplay of RhoA and mechanical forces in collective cell migration driven by leader cells, Nat. Cell Biol. 16: 217–223

Riley EE, Das D, and Lauga E, 2018. Swimming of peritrichous bacteria is enabled by an elasto-hydrodynamic instability, Sci. Rep. 8: 10728

Rodríguez-Franco P, . . . , Trepat X (10 coathors), 2017. Long-lived force patterns and deformation waves at repulsive epithelial boundaries, Nat. Mater. 16: 1029–1037

Rogóż M, Zeng H, Xuan C, Wiersma DS, and Wasylczyk P, 2016. Light-Driven Soft Robot Mimics Caterpillar Locomotion in Natural Scale, Adv. Optical Mater. 4: 1689–1694

Romanczuk P, Bär M, Ebeling W, Lindner B, and Schimansky-Geier L, 2012. Active Brownian Particles From Individual to Collective Stochastic Dynamics, Eur. Phys. J. Spec. Top. 202: 1–162

Rørth P, 2011. Whence directionality: guidance mechanisms in solitary and collective cell migration. Dev. Cell 20: 9–18

Rothschild Lord, 1949. Measurement of sperm activity before artificial insemination, Nature 163: 358–359

Ruiz-Herrero T, Fai TG, and Mahadevan L, 2019. Dynamics of Growth and Form in Prebiotic Vesicles, Phys. Rev. Lett. 123: 038102

Runions A, Tsiantis M, and Prusinkiewicz P, 2017. A common developmental program can produce diverse leaf shapes, New Phytologist 216: 401–418

Sachse R, Westermeier A, Max Mylo M, Nadasdi J, Bischoff M, Speck T, and Poppinga S, 2020. Snapping mechanics of the Venus flytrap (*Dionaea muscipula*), Proc. Natl. Acad. Sci. USA 117: 16035–16042

Saez A, Ghibaudo M, Buguin A, Silberzan P, and Ladoux B, 2007. Rigidity-driven growth and migration of epithelial cells on microstructured anisotropic substrates, Proc. Natl. Acad. Sci. USA 104: 8281–8286

Saggiorato G, Alvarez L, Jikeli JF, Kaupp UB, Gompper G and Elgeti J, 2017. Human sperm steer with second harmonics of the flagellar beat, Nat. Commun. 8: 1415

Saha A, Nishikawa M, Behrndt M, Heisenberg C-P, Jülicher F, and Grill SF, 2016. Determining Physical Properties of the Cell Cortex, Biophys. J. 110: 1421–1429

Saha S, Golestanian R, and Ramaswamy S, 2014. Clusters, asters, and collective oscillations in chemotactic colloids, Phys. Rev. E 89: 062316

Saha S, Ramaswamy S, and Golestanian R, 2019. Pairing, waltzing and scattering of chemotactic active colloids, New J. Phys. 21: 063006

Saint-Hilaire ÉG, 1822. Considérations générales sur la vertèbre, Mém, du Mus. Hist. Nat. 9, 89–119

Salbreux G, Joanny J-F, Prost J and Pullarkat P, 2007. Shape oscillations of non-adhering fibroblast cells, Phys. Biol. 4: 268–284

Sander K, 1997. Of gradients and genes: Developmental concepts of Theodor Boveri and his students, *Landmarks in Developmental Biology*, pp. 56–58, Springer-Verlag, Berlin

Saraswathibhatla A and Notbohm J, Tractions and Stress Fibers Control Cell Shape and Rearrangements in Collective Cell Migration, Phys. Rev. X 10: 011016

Sartori P, Geyer VF, Scholich A, Jülicher F, and Howard J, 2016. Dynamic curvature regulation accounts for the symmetric and asymmetric beats of *Chlamydomonas* flagella, eLife 5: e13258

Saw TB, . . . , Ladoux B (11 coauthors), 2017. Topological defects in epithelia govern cell death and extrusion, Nature 554: 212–216

Schaller V, Weber C, Semmrich C, Frey E, and Bausch AR, 2010. Polar patterns of driven filaments, Nature 467: 73–77

Schimansky-Geier L, Mieth M, Rosé H, and Malchow H, 1995. Structure formation by active Brownian particles, Phys. Lett. A 207: 140–146

Schoute JC, 1913. Beiträge zur Blattstellungslehre. Rec. Trav. Bot. Neerl. 10: 153–325

Schwarz US and Safran SA, 2013. Physics of adherent cells, Rev. Mod. Phys. 85: 1327–1381

Schweitzer F, 2003. *Browning Agents and Active Particles*, Springer-Verlag, Berlin

Severson AF, von Dassow G, and Bowerman B, 2016. Oocyte Meiotic Spindle Assembly and Function, Curr. Top. Develop. Biol. 116: 65–98

Shang Y, Wang J, Ikeda T, and Jiang L, 2019. Bio-inspired liquid crystal actuator materials, J. Mater. Chem. C 7: 3413

Shankar S, Ramaswamy S, Marchetti MC, and Bowick MJ, 2018. Defect Unbinding in Active Nematics, Phys. Rev. Lett. 121: 108002

Shankar S and Marchetti MC, 2019. Hydrodynamics of Active Defects: From Order to Chaos to Defect Ordering, Phys. Rev. X 9: 041047

Shao D, Levine H, and Rappel W-J, 2012. Coupling actin flow, adhesion, and morphology in a computational cell motility model, Proc. Natl. Acad. Sci. USA 109: 6851–6856

Sharon E, Roman B, and Swinney HL, 2007. Geometrically driven wrinkling observed in free plastic sheets and leaves, Phys. Rev. E 75: 046211

Sharon E and Sahaf M, 2018. The Mechanics of Leaf Growth on Large Scales, in *Plant Biomechanics*, eds. Geitmann A and Gril J, pp. 109–126, Springer, Cham

Shi X and Ma Y, 2013. Topological structure dynamics revealing collective evolution in active nematics, Nat. Commun. 4: 3013

Shi X, Chaté H, and Ma Y, 2014. Instabilities and chaos in a kinetic equation for active nematics, New J. Phys, 16: 035003

Shin H, Bowick MJ, and Xing X, 2008. Topological Defects in Spherical Nematics, Phys. Rev. Lett. 101: 037802

Shipman PD and Newell AC, 2005. Polygonal planforms and phyllotaxis on plants, J. Theor. Biol. 236: 154–197

Shyer AE, Rodrigues AR, Schroeder GG, Kassianidou E, Kumar S, and Harland RM, 2017. Emergent cellular self-organization and mechanosensation initiate follicle pattern in the avian skin, Science 357: 811–815

Simakov DSA, Cheung LS, Pismen LM, and Shvartsman SY, 2012. EGFR-dependent network interactions that pattern *Drosophila* eggshell appendages, Development 139: 2814–2820

Simha RA and Ramaswamy S, 2002. Hydrodynamic fluctuations and instabilities in ordered suspensions of self-propelled particles, Phys. Rev. Lett. 89: 058101

Singh DP, Domínguez A, Choudhury U, Kottapalli SN, Popescu MN, Dietrich S, and Fischer P, 2020. Interface-mediated spontaneous symmetry breaking and mutual communication between drops containing chemically active particles, Nat. Commun. 11: 2210

Skoblikow NE and Zimin AA, 2018. Mineral Grains, Dimples, and Hot Volcanic Organic Streams: Dynamic Geological Backstage of Macromolecular Evolution, J. Molec. Evol. 86: 172–183

Snezhko A and Aranson IS, 2011. Magnetic manipulation of self-assembled colloidal asters, Nat. Mater. 10: 698–703

Sokolov A, Aranson IS, Kessler JO and Goldstein RE, 2007. Concentration dependence of the collective dynamics of swimming bacteria, Phys. Rev. Lett. 98: 158102

Sokolov A, Apodacac MM, Grzybowski BA, and Aranson IS, 2010. Swimming bacteria power microscopic gears, Proc. Natl. Acad. Sci. USA 107: 969–974

Solon J, Kaya-Copur A, Colombelli J and Brunner D, 2009. Pulsed forces timed by a ratchet-like mechanism drive directed tissue movement during dorsal closure, Cell 137: 1331–1342

Solon AP, Caussin JB, Bartolo D, Chat/'e H, and Tailleur J, 2015. Pattern formation in flocking models: A hydrodynamic description, Phys. Rev. E 92: 062111

Solon AP, Chat/'e H, and Tailleur J, 2015. From Phase to Micro-Phase Separation in Flocking Models:The Essential Role of Non-Equilibrium Fluctuations, Phys. Rev. Lett. 114: 068101

Solon AP, Fily Y, Baskaran A, Cates ME, Kafri Y, Kadar M, and Tailleur J, 2015. Pressure is not a state function for generic active fluids, Nat. Phys. 11: 673–678

Son K, Guasto JS and Stocker R, 2013. Bacteria can exploit a flagellar buckling instability to change direction, Nat. Phys. 9: 494–498

Sowa Y, Rowe AD, Leake MC, Yakushi T, Homma M, Ishijima A, Berry RM, 2005. Direct observation of steps in rotation of the bacterial flagellar motor, Nature 437: 916–919

Spedding G, 2011. The cost of flight in flocks, Nature 474: 458–459

Spemann H and Mangold H, 1924. Induction of embryonic primordia by implantation of organizers from a different species. Roux's Arch. Entw. Mech. 100: 599–638

Stanton MM, Simmchen J, Ma X, Miguel-López A, and Sánchez S, 2016. Biohybrid janus motors driven by *Escherichia coli*, Adv. Mater. Interf. 3: 1500505

Stavans J, 1993.The evolution of cellular structures, Rep. Prop. Phys. 56: 733–789

Stewart PS and Franklin MJ, 2008. Physiological heterogeneity in biofilms. Nat. Rev. Microbiol. 6: 199–210

Streichan SJ, Lefebvre MF, Noll N, Wieschaus EF, and Shraiman BI, 2018. Global Morphogenetic Flow Is Accurately Predicted by the Spatial Distribution of Myosin Motors, eLife 7: e27454

Strömbom D, Siljestam M, Park J, and Sumpter DJT, 2015. The shape and dynamics of local attraction, Eur. Phys. J. Spec. Top. 224: 3311–3323

Stürmer J, Seyrich M, and Stark H, 2019. Chemotaxis in a binary mixture of active and passive particles J. Chem. Phys.Ê150: 214901

Sumino Y and Yoshikawa K, 2014. Amoeba-like motion of an oil droplet, Eur. Phys. J. Spec. Top. 223: 1345–1352

Sun L, Chen Z, Bian F, and Zhao Y, 2020. Bioinspired Soft Robotic Caterpillar with Cardiomyocyte Drivers, Adv. Funct. Mater. 30: 1907820

Suzuki R and Bausch AR, 2017. The emergence and transient behaviour of collective motion in active filament systems, Nat. Commun. 8: 41

Taber LA, 2009. Towards a unified theory for morphomechanics, Phil. Trans. R. Soc. A 367: 3555–3583

Taber LA, Shi Y, Yang L, and Bayly PV, 2011. Poroelastic Model For Cell Crawling Including Mechanical Coupling between Cytoskeletal Contraction and Actin Polymerization, J. Mech. Mater. Struct. 6: 569–589

Tambe DT, . . . , Trepat X (12 coauthors) 2011. Collective cell guidance by cooperative intercellular forces, Nat. Mater. 10: 469–475

Taylor GI, 1951. Analysis of the swimming of microscopic organisms, Proc. Roy. Soc. Lond. A 209: 447–461

Tee YH, Shemesh T, Thiagarajan V, Hariadi RF, Anderson KL, Page C, Volkmann N, Hanein D, Sivaramakrishnan S, Kozlov MM, and Bershadsky AD, 2015. Cellular chirality arising from the self-organization of the actin cytoskeleton, Nat. Cell Biol. 17: 445–456

Tero A, Takagi S, Saigusa T, Ito K, Bebber DP, Fricker MD, Yumiki K, Kobayashi R, and Nakagaki T, 2010. Rules for Biologically Inspired Adaptive Network Design, Science 327: 439–442

Thampi SP, Golestanian R, and Yeomans JM, 2014. Vorticity, defects and correlations in active turbulence, Phil. Trans. R. Soc. A 372: 20130366

Theveneau E and Mayor R, 2013. Collective cell migration of epithelial and mesenchymal cells, Cell. Mol. Life Sci. 70: 3481–3492

Theveneau E, Steventon B, Scarpa E, Garcia S, Trepat X, Streit A, and Mayor R, 2013. Chase-and-run between adjacent cell populations promotes directional collective migration, Nat. Cell Biol. 15: 763–772

Thiery JP, 2002. EpithelialÐmesenchymal transitions in tumour progression, Nat. Rev. Cancer 2: 442–454

Thutupalli S, Seemann R, and Herminghaus S, 2011. Swarming behavior of simple model squirmers, New J. Phys. 13: 073021

Thutupalli S, 2014. *Towards Autonomous Soft Matter Systems*, Springer, Cham

Tjhung E, Cates ME, and Marenduzzo D, 2011. Nonequilibrium steady states in polar active fluids, Soft Matter 7: 7453–7464

Tjhung E, Tiribocchi A, Marenduzzo D, and Cates ME, 2015. A minimal physical model captures the shapes of crawling cells. Nat. Commun. 6: 5420

Tjhung E, Nardini C, and Cates ME, 2018. Cluster Phases and Bubbly Phase Separation in Active Fluids: Reversal of the Ostwald Process, Phys. Rev. X 8: 031080

Toner J and Tu Y, 1995. Long-Range Order in a Two-Dimensional Dynamical XY Model: How Birds Fly Together, Phys. Rev. Lett. 75: 4326–4329

Toner J and Tu Y, 1998. Flocks, herds, and schools: A quantitative theory of flocking. Phys. Rev. E 58: 4828–4858

Trepat X, Deng L, An SS, Navajas D, Tschumperlin DJ, Gerthoffer WT, Butler JP, and Fredberg JJ, 2007. Universal physical responses to stretch in the living cell. Nature 447, 592–595

Trong PK, Guck J, and Goldstein RE, 2012. Coupling of Active Motion and Advection Shapes Intracellular Cargo Transport, Phys. Rev. Lett. 109: 028104

Trong PK, Doerflinger H, Dunkel J, St Johnston D, and Goldstein RE, 2015. Cortical microtubule nucleation can organise the cytoskeleton of *Drosophila* oocytes to define the anteroposterior axis, eLife 4: e06088

Tsang ACH, Lam AY, and Riedel-Kruse IH, 2018. Polygonal motion and adaptable phototaxis via flagellar beat switching in the microswimmer *Euglena gracilis*, Nat. Phys. 14: 1212–1222

Turing AM, 1952. The chemical basis of morphogenesis, Phil. Trans. R. Soc. B 237: 37–73

Urrutia R, McNiven M, Albanesi J, Murphy D, and Kachar B, 1991. Purified kinesin promotes vesicle motility and induces active sliding between microtubules in vitro. Proc. Natl. Acad. Sci. USA 88: 6701–6705

Usherwood JR, Stavrou M, Lowe JC, Roskilly K, and Wilson AM, 2011. Flying in a flock comes at a cost in pigeons, Nature 474: 494–497

Uspal WE, Popescu MN, Dietrich S, and Tasinkevych M, 2015. Self-propulsion of a catalytically active particle near a planar wall: from reflection to sliding and hovering, Soft Matter 11: 6613–6632

van Oosten CL, Bastiaansen CWM, and DJ, 2009. Printed artificial cilia from liquid-crystal network actuators modularly driven by light, Nat. Mater. 8: 677–682

Vantomme G, Gelebart AH, Broer DJ, and Meijer EW, 2017. A four-blade light-driven plastic mill based on hydrazone liquid- crystal networks, Tetrahedron, 73: 4963–4967

Varma A, Montenegro-Johnson TB, and Michelin S, 2018. Clustering-induced self-propulsion of isotropic autophoretic particles, Soft Matter 14: 7155–7173

Vicsek T, Czirok A, Ben-Jacob E, Cohen I, Shochet O, 1995. Novel type of phase transition in a system of self-driven particles, Phys. Rev. Lett. 75: 1226–1229

Viktorinová I, Pismen LM, Aigouy B, and Dahmann C, 2011. Modelling planar polarity of epithelia: the role of signal relay in collective cell polarization, J. R. Soc. Interface 8: 1059–1063

Vilfan A, 2012. Generic flow profiles induced by a beating cilium, Eur. Phys. J. E 35: 72

Vishwakarma M, Di Russo J, Probst D, Schwarz US, Das T, and Spatz U, 2018. Mechanical interactions among followers determine the emergence of leaders in migrating epithelial cell collectives, Nat. Commun. 9: 3469

Vizsnyiczai G, Frangipane G, Maggi C, Saglimbeni F, Bianchi S, and Di Leonardo R, 2017. Light controlled 3D micromotors powered by bacteria, Nat. Commun. 8: 15974

Voas MG and Rebay I, 2004. Signal integration during development: Insights from the *Drosophila* eye, Developmental Dynamics 229: 162–175

Vogel R and Stark H, 2010. Force-extension curves of bacterial flagella, Eur. Phys. J. E 33: 259–271

Vogel R and Stark H, 2012. Motor-driven bacterial flagella and buckling instabilities, Eur. Phys. J. E 35: 15

Vogel SK, Petrasek Z, Heinemann F, and Schwille P, 2013. Myosin motors fragment and compact membrane-bound actin filaments, eLife 2: e00116

Vollmer J, Casares F, and Iber D, 2017. Growth and size control during development, Open Biol. 7: 170190

von Nägeli C, 1884. Mechanisch-physiologische Theorie der Abstammungslehre, Verlag Oldenbourg, München

Wada H and Matsumoto D, 2018. Twisting Growth in Plant Roots, in *Plant Biomechanics*, eds. Geitmann A and Gril J, pp. 127–140, Springer, Cham

Walker M, Rizzuto P, Godin M, and Pelling AE, 2020. Structural and mechanical remodeling of the cytoskeleton maintains tensional homeostasis in 3D microtissues under acute dynamic stretch, Sci. Rep. 10: 7696

Wang Q, Feng JJ, and Pismen LM, 2012. A Cell-Level Biomechanical Model of *Drosophila* Dorsal Closure, Biophys. J. 103: 2265–2274

Wang Y-C, Khan Z, Kaschube M, and Wieschaus EF, 2013. Differential positioning of adherens junctions is associated with initiation of epithelial folding, Nature 484: 390–393

Wani OM, Zeng H, and Priimagi A, 2017. A light-driven artificial flytrap, Nat. Commun. 8: 15546

Ware TH, McConney ME, Wie JJ, Tondiglia VP, and White TJ, 2015. Voxelated liquid crystal elastomers, Science 347: 982–984

Warner M and Terentjev EM, 2003. *Liquid Crystal Elastomers*, Clarendon Press, Oxford

Wei J and Yu Y, 2012. Photodeformable polymer gels and crosslinked liquid-crystalline polymers, Soft Matter 8: 8050–8059

Weihs D, 1973. Hydromechanics of Fish Schooling, Nature 241: 290–291

Wehner M, Truby RL, Fitzgerald DJ, Mosadegh B, Whitesides GM, Lewis JA, and Wood RJ, 2016. An integrated design and fabrication strategy for entirely soft, autonomous robots, Nature 536: 451–455

Whitfield CA and Hawkins RJ, 2016. Instabilities, motion and deformation of active fluid droplets, New J. Phys. 18: 123016

Williams BJ, Anand SV, Rajagopalan, and Saif MTA, 2014. A self-propelled biohybrid swimmer at low Reynolds number, Nat. Commun. 5: 3081

Whitfield CA, Adhyapak TC, Tiribocchi A, Alexander GP, Marenduzzo D, and Ramaswamy S, 2017. Hydrodynamic instabilities in active cholesteric liquid crystals, Eur. Phys. J. E 40: 50

Wolff L, Fernández P, and Kroy K, 2012. Resolving the Stiffening–Softening Paradox in Cell Mechanics, PLoS One 7: e40063

Wollrab V, Thiagarajan R, Wald A, Kruse K, and Riveline D, 2016. Still and rotating myosin clusters determine cytokinetic ring constriction, Nat. Commun. 7: 11860

Wolpert L, 1969. Positional information and the spatial pattern of cellular differentiation. J. Theor. Biol., 25: 1–47

Wolpert L, 2002. Principles of Development, Oxford University Press

Wu P-H, . . . , Wirtz D (22 coauthors), 2018. A comparison of methods to assess cell mechanical properties, Nat. Meth. 15: 491–498

Yan J, Fei C, Mao S, Moreau A, Wingreen NS, Košmrlj A, Stone HA, and Bassler BL, 2019. Mechanical instability and interfacial energy drive biofilm morphogenesis, eLife 8: e4392

Yaman YI, Demir E, Vetter R, and Kocabas, A, 2019. Emergence of active nematics in chaining bacterial biofilms, Nat. Commun. 10: 2285

Yeomans JM, Pushkin DO, and Shum H, 2014. An introduction to the hydrodynamics of swimming microorganisms, Eur. Phys. J. Spec. Top. 223: 1771–1785

Zakharov AP and Pismen LM, 2015. Reshaping nemato-elastic sheets, Eur. Phys. J. E 38: 75

Zakharov AP, Leshansky AM, and Pismen LM, 2016. Flexible helical yarn swimmers, Eur. Phys. J. E 39: 87

Zakharov AP and Pismen LM, 2017a. Phase separation and folding in swelled nematoelastic films, Phys. Rev. E 96: 012709

Zakharov AP and Pismen LM, 2017b. Textures and shapes in nematic elastomers under the action of dopant concentration gradients, Soft Matter 13: 2886–2892

Zakharov AP and Pismen LM, 2019. Programmable filaments and textiles, Phys. Rev. Mater. 3: 055603

Zhang J, Schadschneider A, and Seyfried A, 2014. Empirical Fundamental Diagrams for Bidirectional Pedestrian Streams in a Corridor, in *Pedestrian and Evacuation Dynamics 2012*, Weidmann U et al. (eds.), Springer Switzerland

Zhou S, Sokolov A, Lavrentovich OD, and Aranson IS, 2014. Living liquid crystals. Proc. Natl. Acad. Sci. USA 111: 1265–1270

Zhou S, 2017. *Lyotropic Chromonic Liquid Crystals*, Springer, Cham

Zhou S, Tovkach O, Golovaty D, Sokolov A, Aranson IS and Lavrentovich OD, 2017. Dynamic states of swimming bacteria in a nematic liquid crystal cell with homeotropic alignment, New J. Phys. 19: 055006

Ziebert F and Aranson IS, 2013. Effects of adhesion dynamics and substrate compliance on the shape and motility of crawling cells, PLoS One 8: e64511

Ziebert F and Aranson IS, 2014. Modular approach for modeling cell motility, Eur. Phys. J. Spec. Top. 223: 1265–1277

Ziebert F, Löber J, and Aranson IS, 2016. Macroscopic Model of Substrate-Based Cell Motility, in *Physical Models of Cell Motility*, ed. Aranson IS, pp. 1–67, Springer, Cham

Zöttl A and Stark H, 2016. Emergent behavior in active colloids, J. Phys. Condens. Matter 28: 253001

Zöttl A and Stark H, 2018. Simulating squirmers with multiparticle collision dynamics, Eur. Phys. J. E 41: 61

Zwicker D, Seyboldt R, Weber CA, Hyman AA, and Jülicher F, 2017. Nat. Phys. 13: 408–413

Illustration Credits

Illustrations, except for those where Springer holds the copyright or a permission has been granted by copyright holders, are published under the Creative Commons (CC) license. Illustrations from journals encumbering permissions or charging high fees could not be included, irrespectively of their quality.

Fig. 1.1 From Ginelli, 2016
Fig. 1.2a,c From Spedding, 2011
Fig. 1.2b Adapted from Weihs, 1973
Fig. 1.3 From Ginelli, 2016, CC
Fig. 1.4(a) Adapted from Katyal et al, 2020
Fig. 1.4(b) From Chaté et al, 2008
Fig. 1.5 From Ihle and Chou, 2014
Fig. 1.6 From Komareji and Bouffanais, 2013, CC, DOI: 10.1371/ journal.pone.0082578
Fig. 1.7 From Solon, Fily et al, 2015
Fig. 1.8 By ChrisJLygouras - Own work, CC BY-SA 4.0,
https://commons.wikimedia.org/w/index.php?curid=89860916
Fig. 1.9 From Chaté et al, 2008
Fig. 1.10 From Strömbom et al, 2015
Fig. 1.11 From Großmann et al, 2015
Fig. 1.12 From Zhang et al, 2014
Fig. 1.13 From Helbing et al, 2000
Fig. 1.14: From Tjhung et al, 2018, CC, DOI: 10.1103/PhysRevX.8.031080
Fig. 1.15 From Geyer et al, 2019, CC, DOI: 10.1103/PhysRevX.9.031043
Fig. 1.16a From Bricard et al, 2013
Fig. 1.16b,c From Geyer et al, 2019, CC, DOI: 10.1103/PhysRevX.9.031043
Fig. 1.17 From Bricard et al, 2015
Fig. 1.18 From Kaiser et al, 2011, reproduced by permission, courtesy Prof. Igor Aronson
Fig. 2.1 By Kebes – Own work, CC BY-SA 3.0,
https://commons.wikimedia.org/w/index.php?curid=4170779,
https://commons.wikimedia.org/w/index.php?curid=4170810,
https://commons.wikimedia.org/w/index.php?curid=4170873
Fig. 2.2 Adapted from Kléman and Lavrentovich, 2003
Fig. 2.3 From Romanczuk et al, 2012
Fig. 2.4a,b From Peruani et al, 2010

L. Pismen, *Active Matter Within and Around Us*, The Frontiers Collection,
https://doi.org/10.1007/978-3-030-68421-1

Fig. 2.5 From Abkenar et al, 2013, reproduced by permission, ©American Physical Society
Fig. 2.6 From Shi et al, 2014, CC, DOI:10.1088/1367-2630/16/3/035003
Fig. 2.7 From Blair et al, 2003, reproduced by permission, ©American Physical Society
Fig. 2.8 From Das et al, 2017, CC, DOI: 10.1038/s41598-017-07301-w
Fig. 2.9a-c From Zakharov and Pismen , 2015
Fig. 2.9d-e From Doostmohammadi et al, 2018, CC, DOI: 10.1038/s41467-018-05666-8
Fig. 2.10 From Fukuda, 1998
Fig. 2.11 From Shi and Ma, 2013
Fig. 2.12, Fig. 2.13 From Thampi et al, 2014, reproduced by permission, courtesy Prof. Julia Yeomans
Fig. 2.14 From Doostmohammadi et al, 2018, CC, DOI: 10.1038/s41467-018-05666-8
Fig. 2.15(l) From Pismen, 2013, ©American Physical Society
Fig. 2.16 From Shankar and Marchetti, 2019, CC, DOI: 10.1103/PhysRevX.9.041047
Fig. 2.17 From Khoromskaia and Alexander, 2017, CC, DOI: 10.1088/1367-2630/aa89aa
Fig. 2.18 From Bonelli et al, 2016
Fig. 2.19 From Bonelli et al, 2019, CC, 10.1038/s41598-019-39190-6
Fig. 2.20 From Whitfield, 2017, CC, DOI 10.1140/epje/i2017-11536-2
Fig. 2.21 Adapted from Fürthauer et al, 2012, CC, DOI 10.1140/epje/i2012-12089-6
Fig. 3.1 Adapted from Zöttl and Stark, 2018, CC, DOI 10.1140/epje/i2018-11670-3
Fig. 3.2(l) Adapted from By Ved1123 - Own work, CC BY-SA 4.0, https://commons.wikimedia.org/w/index.php?curid=67552518
Fig. 3.2(r) Adapted from Qiu et al, 2014
Fig. 3.3 From Kroy et al, 2016
Fig. 3.4a From Popescu et al, 2018, CC, DOI 10.1140/epje/i2018-11753-1
Fig. 3.4b,c From Campbell et al, 2019, CC, DOI 10.1038/s41467-019-11842-1
Fig. 3.5 From Michelin and Lauga, 2015
Fig. 3.6 From Ibrahim and Liverpool, 2016, CC, DOI: 10.1140/epjst/e2016-60148-1
Fig. 3.7 From Das et al, 2015
Fig. 3.8 From Dietrich et al, 2017, CC, DOI: 10.1088/1367-2630/aa7126
Fig. 3.9 From Saha et al, 2019, CC, DOI: 10.1088/1367-2630/ab20fd
Fig. 3.10 From Nasouri and Golestanian, 2020, CC, DOI: 10.1103/PhysRevLett.124.168003
Fig. 3.11 From Varma et al, 2018, CC, DOI: 10.1039/c8sm00690c
Fig. 3.12b From Moerman et al, 2017, reproduced by permission, ©American Physical Society
Fig. 3.12c From Golovin et al, 1995, reproduced by permission, ©American Chemical Society
Fig. 3.13 From Thutupalli, 2014
Fig. 3.14a,b From Moerman et al, 2017, reproduced by permission, ©American Physical Society
Fig. 3.14c, Fig. 3.15 From Meredith et al, 2020
Fig. 3.16 From Nagai et al, 2007, reproduced by permission, ©Elsevier
Fig. 3.17 From Sumino and Yoshikawa, 2014
Fig. 3.18 From Marenduzzo, 2016, CC, DOI: 10.1140/epjst/e2016-60084-6
Fig. 3.19 From Whitfield and Hawkins, 2016, CC, DOI: 10.1088/1367-2630/18/12/123016
Fig. 3.20, Fig. 3.21 From Zwicker et al, 2017
Fig. 3.22 From Ruiz-Herrero et al, 2019
Fig. 3.23 From Leaver et al, 2009
Fig. 3.24 From Thutupalli, 2014
Fig. 3.25 From Pohl et al, 2014, reproduced by permission, ©American Physical Society
Fig. 3.26 From Stürmer et al, 2019, reproduced by permission, courtesy Prof. Holger Stark
Fig. 3.27 From Romanczuk et al, 2012
Fig. 3.28 From Ginot et al, 2018, CC, DOI: 10.1038/s41467-017-02625-7
Fig. 3.29 From Bäuerle e5 al, 2020, CC, DOI: 10.1038/s41467-020-16161-4
Fig. 3.30 From Singh et al, 2020, CC, DOI: 10.1038/s41467-020-15713-y
Fig. 4.1 By LadyofHats, Public Domain
Fig. 4.2 From Berg, 2004. Inset: from Vogel and Stark, 2010
Fig. 4.3 Adapted from Hu et al, 2015, CC, DOI: 10.1039/c5sm01678a

Fig. 4.4 From Son et al, 2013
Fig. 4.5 By Adenosine - Own work, CC BY 3.0, https://commons.wikimedia.org/w/index.php?curid=2759817
Fig. 4.6 Adapted from Sartori et al, 2016, CC, DOI: 10.7554/eLife.13258
Fig. 4.7a From Kaupp and Alvarez, 2016, CC, DOI: 10.1140/epjst/e2016-60097-1
Fig. 4.7b-d From Vilfan, 2012
Fig. 4.8 Adapted from Gong et al, 2019, CC, DOI: 10.1098/rstb.2019.0149
Fig. 4.9 From Saggiorato et al, 2017, CC, DOI: 10.1038/s41467-017-01462-y
Fig. 4.10 From Kaupp and Alvarez, 2016, CC, DOI: 10.1140/epjst/e2016-60097-1
Fig. 4.11 From Tsang et al, 2018
Fig. 4.12 From Friedrich, 2016, CC, DOI: 10.1140/epjst/e2016-60056-4
Fig. 4.13 From Brumley et al, 2014, CC, DOI: 10.1098/rsif.2014.1358
Fig. 4.14a From Cicuta, 2020, CC, DOI: 10.1042/BST20190571
Fig. 4.14b By Shyamal - Own work, Public Domain
Fig. 4.14c,d From Brumley et al, 2015, CC, DOI: 10.7554/eLife.02750
Fig. 4.15 From Harvey et al, 2013, CC, DOI: 10.1088/1367-2630/15/3/035029
Fig. 4.16 From Fu et al, 2018, CC, DOI: 10.1038/s41467-018-04539-4
Fig. 4.17 From Jemielita et al, 2018, CC, DOI: 10.7554/eLife.42057.015
Fig. 4.18 From Cisneros et al, 2007
Fig. 4.19 From Patteson et al, 2018, CC, DOI: s41467-018-07781-y
Fig. 4.20 From Nishiguchi et al, 2018, CC, DOI: 10.1038/s41467-018-06842-6
Fig. 4.21 Adapted from Kaiser et al, 2015
Fig. 4.22, Fig. 4.23 Adapted from Zhou, 2017
Fig. 4.24 From Genkin et al, 2017, CC, DOI: 10.1103/PhysRevX.7.011029
Fig. 4.25 From Zhou et al, 2017, CC, DOI: 10.1088/1367-2630/aa695b
Fig. 4.26 From Be'er et al, 2020, CC, DOI: 10.1038/s42005-020-0327-1
Fig. 4.27 From Partridge et al, 2018, CC, DOI:10.1038/s41598-018-34192-2
Fig. 4.28 From Dell'Arciprete et al, 2018, CC, DOI: 10.1038/s41467-018-06370-3
Fig. 4.29a,c From Ben Jacob et al, 1997
Fig. 4.29b,d From Ben Jacob et al, 1995, reproduced by permission, ©American Physical Society
Fig. 4.30 From Atis et al, 2019, CC, DOI: 10.1103/PhysRevX.9.021058
Fig. 4.31 From Hartmann et al, 2018, by permission, courtesy Profs. Jörn Dunkel and Knut Drischer
Fig. 4.32 From Yaman et al, 2019, CC, DOI: 10.1038/s41467-019-10311-z
Fig. 4.33 From Yan et al, 2019, CC, DOI: 10.7554/eLife.43920
Fig. 4.34(l) From Stewart and Franklin, 2008
Fig. 4.34(c,r) From Nadell et al, 2010, CC, DOI: 10.1371/journal.pcbi.1000716
Fig. 4.35 From Liu et al, 2015
Fig. 5.1a By http://rsb.info.nih.gov/ij/images/, Public Domain
Fig. 5.1b CC-BY-SA-3.0 https://upload.wikimedia.org/wikipedia/commons/6/6f/ActinaFilamento.png
Fig. 5.1c By Nikolya188811 - Own work, CC BY-SA 4.0, https://commons.wikimedia.org/w/index.php?curid=76098276
Fig. 5.1d From Karsenti et al, 2006
Fig. 5.2a From Koenderink and Paluch, 2018, reproduced by permission, ©Elsevier
Fig. 5.2b-d From Murrell et al, 2015
Fig. 5.3a From Pachong and Müller-Nedebock, 2017
Fig. 5.3b From Wollrab et al, 2016, CC, DOI: 10.1038/ncomms11860
Fig. 5.4a By Thomas Splettstoesser (www.scistyle.com) - Own work (rendered with Maxon Cinema 4D), CC BY-SA 4.0, https://commons.wikimedia.org/w/index.php?curid=41014850
Fig. 5.4b From Goley and Welch, 2006
Fig. 5.4c By Rpgch, CC BY-SA 4.0, from Wikimedia Commons
Fig. 5.5c From Fletcher and Mullins, 2010
Fig. 5.6 From Bausch and Kroy, 2006

Fig. 5.7 From Dogterom and Koenderink, 2019
Fig. 5.8 From Nédélec et al, 1997
Fig. 5.9a,b From Schaller et al, 2010
Fig. 5.9c From Suzuki and Bausch, 2017, CC, DOI: 10.1038/s41467-017-00035-3
Fig. 5.10 From Schaller et al, 2010
Fig. 5.11 From Vogel et al, CC, DOI: 10.7554/eLife.00116
Fig. 5.12 From Ennomani et al, 2016, reproduced by permission, ©Elsevier
Fig. 5.13 From Ideses et al, 2008, CC DOI: 10.1371/journal.pone.0003297
Fig. 5.14 From Ideses et al, 2018, CC DOI: 10.1038/s41467-018-04829-x
Fig. 5.15a From Echarri et al, 2019, CC DOI: 10.1038/s41467-019-13782-2
Fig. 5.15b,c From Murrel et al, 2015
Fig. 5.16l From Geiger et al, 2009
Fig. 5.16r From Rafiq et al, 2019
Fig. 5.17a,b From Geiger et al, 2009
Fig. 5.17c From Pieuchot et al, 2018, CC, DOI: 10.1038/s41467-018-06494-6
Fig. 5.18 From Tee et al, 2015
Fig. 5.19a Parsons et al, 2010
Fig. 5.19b,c From Case and Waterman, 2015
Fig. 5.20 From Burnette et al, 2011
Fig. 5.21 From Barnhart et al, 2011, CC, DOI: 201110.1371/journal.pbio.1001059
Fig. 5.22 From Agostinelli et al, 2019, CC, DOI:10.3934/mine.2020011
Fig. 5.23 From Lämmermann et al, 2008
Fig. 5.24 From Bergert et al, 2015
Fig. 5.25 From Mueller et al, 2014, CC, DOI: 10.1371/journal.pbio.1001765
Fig. 5.26a-c From Trong et al, 2012, reproduced by permission, ©American Physical Society
Fig. 5.26d, Fig. 5.27a From Illukkumbura et al, 2020, reproduced by permission, ©Elsevier
Fig. 5.27b-d From Trong et al, 2015, CC, DOI: 10.7554/eLife.06088
Fig. 5.28 From Goudarzi et el, 2019, CC, DOI: 10.1371/journal.pone.0212699
Fig. 5.29a By HelenGinger - Own work, CC BY-SA 3.0,
https://commons.wikimedia.org/w/index.php?curid=25489646
Fig. 5.29b From Haupt and Hauser, 2020, CC, DOI: 10.1088/1367-2630/ab7edf
Fig. 6.1 From Bausch and Kroy, 2006
Fig. 6.2 From Kruse et al, 2005
Fig. 6.3 Adapted from Marenduzzo, 2016, CC, DOI: 10.1140/epjst/e2016-60084-6
Fig. 6.4a,b From Ziebert et al, 2016
Fig. 6.4c,d From Ziebert and Aranson, 2013, CC, DOI: 10.1371/journal.pone.0064511
Fig. 6.5 From Giomi and DeSimone, 2014, reproduced by permission, ©American Physical Society
Fig. 6.6 From Marth et al, 2015, reproduced by permission, courtesy Prof. Axel Voigt
Fig. 6.7 From Tjhung et al, 2015
Fig. 6.8 From Herant and Dembo, 2010, reproduced by permission, ©Elsevier
Fig. 6.9 From Charras et al, 2005
Fig. 6.10 From Lewis et al, 2014, reproduced by permission, courtesy Prof. Robert Guy
Fig. 6.11 From Kruse, 2016
Fig. 6.12 From Moeendarbary et al, 2013
Fig. 6.13 From Radszuweit et al, 2014, CC, DOI: 10.1371/journal.pone.0099220
Fig. 6.14 From Köpf and Pismen, 2013, reproduced by permission, ©Elsevier
Fig. 6.15, Fig. 6.16 From Notbohm et al, 2016, reproduced by permission, ©Elsevier
Fig. 6.17a From Lee and Wolgemuth, 2011, CC, DOI: 10.1371/journal.pcbi.1002007
Fig. 6.17b-d From Köpf and Pismen, 2013b
reproduced by permission of The Royal Society of Chemistry
Fig. 6.18 From Reffay et al, 2014
Fig. 6.19 From Morozov and Pismen, 2011, ©American Physical Society
Fig. 6.20 From Chaudhuri et al, 2007
Fig. 6.21 From Trepat et al, 2007

Fig. 6.22 From Wolff et al, 2012, CC, DOI: 10.1371/journal.pone.0040063
Fig. 6.23 From Walker et al, 2020, CC, DOI: 10.1038/s41598-020-64725-7
Fig. 6.24a From Sharon et al, 2007, reproduced by permission, ©American Physical Society
Fig. 6.24b From Kuksenok and Balazs, 2013, reproduced by permission, ©Wiley Periodicals
Fig. 6.25a,b From Ohm et al, 2012
Fig. 6.25c From Camacho-Lopez et al, 2004
Fig. 6.26 From Griniasty et al, 2019, reproduced by permission, ©American Physical Society
Fig. 6.27 From Köpf and Pismen, 2013c
Fig. 6.28a From Zakharov and Pismen, 2017a, ©American Physical Society
Fig. 6.28b,c, Fig. 6.29 From Zakharov and Pismen, 2017b
reproduced by permission of The Royal Society of Chemistry
Fig. 6.30a From van Oosten et al, 2016
Fig. 6.30b From Huang et al, 2015, CC, DOI: 10.1038/srep17414
Fig. 6.30c From Ambrosio et al, 2012, CC, DOI: 10.1038/ncomms1996
Fig. 6.31a From Zakharov and Pismen, 2016, ©American Physical Society
Fig. 6.31b From Rogóż et al, 2016, reproduced by permission, ©Wiley Periodicals
Fig. 6.32a From Palagi et al, 2016
Fig. 6.32b From Zakharov et al, 2016
Fig. 6.33a From Vantomme et al, 2017, CC, DOI: 10.1016/j.tet.2017.06.041
Fig. 6.33b From Wani et al, 2017, CC, DOI: 10.1038/ncomms15546
Fig. 7.1b,c From Miklius and Hilgenfeldt, 2011
Fig. 7.2 By Balu Ertl - Own work, CC BY-SA 4.0,
https://commons.wikimedia.org/w/index.php?curid=38534275
Fig. 7.3a-c, Fig. 7.4 From Barton et al, 2017, CC, DOI: 10.1371/journal.pcbi.1005569
Fig. 7.3d,e, Fig. 7.6 From Salm and Pismen, 2012, reproduced by permission, ©IOP Publishing
Fig. 7.7 From Tambe et al, 2011
Fig. 7.8 From Vishwakarma et al, 2018, CC, DOI: 10.1038/s41467-018-05927-6
Fig. 7.9 From Martin and Lewis, 1992
Fig. 7.10 From Brugués et al, 2014
Fig. 7.11 From Kim et al, 2013
Fig. 7.12 From Rodríguez-Franco et al, 2017
Fig. 7.13a From Liu and Nagel, 1998
Fig. 7.14 From Bi et al, 2016, CC, DOI: 10.1103/PhysRevX.6.021011
Fig. 7.15, Fig. 7.16 From Malinverno et al, 2017
Fig. 7.17 From Lång et al, 2018, CC, DOI: 10.1038/s41467-018-05578-7
Fig. 7.18 From Ladoux and Mège, 2017
Fig. 7.19 From Theveneau and Mayor, 2013
Fig. 7.20, Fig. 7.13b From Löber et al, 2014, CC, DOI: 10.1038/srep09172
Fig. 7.21 From Theveneau et al, 2013
Fig. 7.22 Adapted from Thiery, 2002
Fig. 7.23 From Plygawko et al, 2020, CC, DOI: 10.1098/rstb.2020.0087
Fig. 7.24 From Chen et al, 2018, reproduced by permission, ©American Physical Society
Fig. 7.25 From Duclos et al, 2017
Fig. 7.26 From Saw et al, 2017
Fig. 7.27 From Viktorinová et al, 2011, ©The Royal Society
Fig. 7.28 From Chandrasekaran and Bose, 2019,
reproduced by permission, ©American Physical Society
Fig. 7.29a From Mason et al, 2013
Fig. 7.29b,c From Lecuit and Lenne, 2007
Fig. 7.29d,e From Wang et al, 2013
Fig. 7.30 From Pearl et al, 2017, CC, DOI: 10.1098/rstb.2015.0526
Fig. 7.31 From Deforet et al, 2013
Fig. 7.32 From Okuda et al 2018
Fig. 7.33 From Misra et al, 2017, reproduced by permission, courtesy Prof. Stanislav Shvartsman

Fig. 7.34a From Ioannou et al, 2020, CC, DOI: 10.3389/fbioe.2020.00405
Fig. 7.34b From Gómez-Gálvez et al, 2018, CC, DOI: 10.1038/s41467-018-05376-1
Fig. 7.35a From Sharon and Sahaf, 2018
Fig. 7.35b From Xu et al, 2020, reproduced by permission, ©American Physical Society
Fig. 7.36a By LadyofHats, Public Domain
Fig. 7.36b From Beauzamy et al, 2014, CC, DOI: 10.1093/aob/mcu187
Fig. 7.37 From Harrington et al, 2011
Fig. 7.38 From Hofhuis et al, 2016, CC, DOI: 10.1016/j.cell.2016.05.002
Fig. 7.39 From Forterre et al, 2005
Fig. 8.2 From Murray, 1980
Fig. 8.3 From Meinhard, 1995
Fig. 8.5 From Severson et al, 2016, reproduced by permission, ©Elsevier
Fig. 8.6a-c From Alsous et al, 2017, reproduced by permission, ©Elsevier
Fig. 8.6d From Müller, 1997
Fig. 8.7 From Bor et al, 2015, reproduced by permission, ©Wiley Periodicals
Fig. 8.8b From Kourakis and Smith, 2005, reproduced by permission, ©Elsevier
Fig. 8.8a, Fig. 8.9b-d, Fig. 8.10 From Pismen and Simakov, 2011, ©American Physical Society
Fig. 8.11a,b From Dequéant and Pourquié, 2008
Fig. 8.12 From Fried and Iber, 2014, CC, DOI: 10.1038/ncomms6077
Fig. 8.13a From Farge, 2011, reproduced by permission, ©Elsevier
Fig. 8.13b Adapted from Hoffman et al, 2011
Fig. 8.14 From Chanet and Martin, 2014, reproduced by permission, ©Elsevier
Fig. 8.15 From Brinkmann et al, 2018, CC, DOI: 10.1371/journal.pcbi.1006259
Fig. 8.16 From Braun and Keren, 2018, reproduced by permission, ©Wiley Periodicals
Fig. 8.17 From Lecuit and Lenne, 2007
Fig. 8.18 From Martin et al, 2009
Fig. 8.19 From Streichan et al, 2018, CC, DOI: 10.7554/eLife.27454
Fig. 8.20 From Noll et al, 2020, CC, DOI: 10.1103/PhysRevX.10.011072
Fig. 8.21a From Ducuing and Vincent, 2016
Fig. 8.21b-d From Durney et al, 2018, reproduced by permission, ©the Biophysical Society
Fig. 8.21e From Wang et al, 2012, reproduced by permission, ©the Biophysical Society
Fig. 8.22a CC BY-SA 3.0, https://commons.wikimedia.org/w/index.php?curid=121444
Fig. 8.22b From Harrison et al, 2009, reproduced by permission, ©Elsevier
Fig. 8.23a Adapted from Linardić and Braybrook, 2017, CC, DOI: 10.1038/s41598-017-13767-5
Fig. 8.23b,c From Shipman and Newell, 2005, reproduced by permission, ©Elsevier
Fig. 8.24, Fig. 8.25 From Runions et al, 2017, CC, DOI: 10.1111/nph.14449
Fig. 8.26 From Leyser and Domagalska, 2011
Fig. 8.27 From Wada and Matsumoto, 2018
Fig. 8.28 From Gladman et al, 2016
Fig. 8.29a,b From Wehner et al, 2016
Fig. 8.29c From Lu et al, 2020, CC, DOI: 10.1002/advs.202000069
Fig. 8.30a,b From Behkam and Sitti, 2007, reproduced by permission, ©Wiley-VCH Verlag
Fig. 8.30c From Vizsnyiczai et al, 2017, CC, DOI: 10.1038/ncomms15974
Fig. 8.30d From Sun et al, 2020, reproduced by permission, ©Wiley-VCH Verlag
Fig. 8.30e From Williams et al, 2014
Fig. 8.31 From Noor et al, 2019, CC, DOI: 10.1002/advs.201900344

Printed in the USA
CPSIA information can be obtained
at www.ICGtesting.com
CBHW062327021024
15300CB00007B/307

9 783030 684228